NUMERICAL ANALYSIS
OF SYMMETRIC MATRICES

Prentice-Hall
Series in Automatic Computation

George Forsythe, editor

ANDREE, *Computer Programming Techniques, Analysis, and Mathematics*
ARBIB, *Theories of Abstract Automata*
ANSELONE, *Collectively Compact Operator Approximation Theory and Applications to Integral Equations*
BATES AND DOUGLAS, *Programming Language/One*, 2nd ed.
BLUMENTHAL, *Management Information Systems*
BOBROW AND SCHWARTZ, *Computers and the Policy-Making Community*
BOWLES, editor, *Computers in Humanistic Research*
CESCHINO AND KUNTZMAN, *Numerical Solution of Initial Value Problems*
CRESS, et al., *FORTRAN IV with WATFOR and WATFIV*
DANIEL, *The Approximate Minimization of Functionals*
DOSMONDE, *A Conversational Graphic Data Processing System*
DESMONDE, *Computers and Their Uses*, 2nd ed.
DESMONDE, *Real-Time Data Processing Systems*
EVANS, et al., *Simulation Using Digital Computers*
FIKE, *Computer Evaluation of Mathematical Functions*
FIKE, *PL/I for Scientific Programmers*
FORSYTHE AND MOLER, *Computer Solution of Linear Algebraic Systems*
GAUTHIER AND PONTO, *Designing Systems Programs*
GEAR, *Numerical Initial Value Problems in Ordinary Differential Equations*
GOLDEN, *FORTRAN IV Programming and Computing*
GOLDEN AND LEICHUS, *IBM/360 Programming and Computing*
GORDON, *System Simulation*
GREENSPAN, *Lectures on the Numerical Solution of Linear, Singular and Nonlinear Differential Equations*
GRUENBERGER, editor, *Computers and Communications*
GRUENBERGER, editor, *Critical Factors in Data Management*
GRUENBERGER, editor, *Expanding Use of Computers in the 70's*
GRUENBERGER, editor, *Fourth Generation Computers*
HARTMANIS AND STEARNS, *Algebraic Structure Theory of Sequential Machines*
HULL, *Introduction to Computing*
JACOBY, et al., *Iterative Methods for Nonlinear Optimization Problems*
JOHNSON, *System Structure in Data, Programs and Computers*
KANTER, *The Computer and the Executive*
KIVIAT, et al., *The SIMSCRIPT II Programming Language*
LORIN, *Parallelism in Hardware and Software: Real and Apparent Concurrency*
LOUDEN AND LEDIN, *Programming the IBM 1130*, 2nd ed.
MARTIN, *Design of Real-Time Computer Systems*
MARTIN, *Future Developments in Telecommunications*
MARTIN, *Man-Computer Dialogue*

MARTIN, *Programming Real-Time Computing Systems*
MARTIN, *Systems Analysis for Data Transmission*
MARTIN, *Telecommunications and the Computer*
MARTIN, *Teleprocessing Network Organization*
MARTIN AND NORMAN, *The Computerized Society*
MATHISON AND WALKER, *Computers and Telecommunications: Issues in Public Policy*
MCKEEMAN, et al., *A Compiler Generator*
MINSKY, *Computation: Finite and Infinite Machines*
MOORE, *Interval Analysis*
PLANE AND MCMILLAN, *Discrete Optimization:*
Integer Programming and Network Analysis for Management Decisions
PRITSKER AND KIVIAT, *Simulation with GASP II:*
a FORTRAN-Based Simulation Language
PYLYSHYN, editor, *Perspectives on the Computer Revolution*
RICH, *Internal Sorting Methods: Illustrated with PL/I/Programs*
RUSTIN, editor, *Computer Networks*
RUSTIN, editor, *Debugging Techniques in Large Systems*
RUSTIN, editor, *Formal Semantics of Programming Languages*
SACKMAN AND CITRENBAUM, editors, *Online Planning: Towards Creative Problem-Solving*
SALTON, editor, *The SMART Retrieval System:*
Experiments in Automatic Document Processing
SAMMET, *Programming Languages: History and Fundamentals*
SCHULTZ, *Digital Processing: A System Orientation*
SCHWARZ, et al., *Numerical Analysis of Symmetric Matrices*
SHERMAN, *Techniques in Computer Programming*
SIMON AND SIKLOSSY, editors, *Representation and Meaning:*
Experiments with Information Processing Systems
SYNDER, *Chebyshev Methods in Numerical Approximation*
STERLING AND POLLACK, *Introduction to Statistical Data Processing*
STOUTMEYER, *PL/I Programming for Engineering and Science*
STROUD, *Approximate Calculation of Multiple Integrals*
STROUD AND SECREST, *Gaussian Quadrature Formulas*
TAVLSS, editor, *The Computer Impact*
TRAUB, *Iterative Methods for the Solution of Polynomial Equations*
VAN TASSEL, *Computer Security Management*
VARGA, *Matrix Iterative Analysis*
VAZSONYI, *Problem Solving by Digital Computers with PL/I Programming*
WAITE, *Implementing Software for Non-Numeric Application*
WILKINSON, *Rounding Errors in Algebraic Processes*
ZIEGLER, *Time-Sharing Data Processing Systems*

NUMERICAL ANALYSIS OF SYMMETRIC MATRICES

by

DR. H. R. SCHWARZ

Lecturer, Institute for Applied Mathematics
Eidgenossische Technische Hochschule
Zurich, Switzerland

With the Participation of

Prof. H. Rutishauser

And

Prof. E. Stiefel

Translation by

Paul Hertelendy

PRENTICE-HALL, INC.

ENGLEWOOD CLIFFS, N.J.

Library of Congress Cataloging in Publication Data

Schwarz, Hans R.
Numerical analysis of symmetric matrices.

(Prentice-Hall series in automatic computation)
Bibliography: p.
1. Matrices.
2. Numerical analysis.
I. Title.

QA263.S3313 512.9'43 72-8630
ISBN 0-13-626556-1

Arranged with the permission of
B. G. Teubner Verlag, Stuttgart.
The only authorized English translation
of the original German edition which
appeared in the series *Introductions to Applied Mathematics and Mechanics;*
edited by Professor G. Gortler.

10 9 8 7 6 5 4 3 2 1

Printed in the United States of America

PRENTICE-HALL INTERNATIONAL, INC., *London*
PRENTICE-HALL OF AUSTRALIA, PTY., LTD., *Sydney*
PRENTICE-HALL OF CANADA, LTD., *Toronto*
PRENTICE-HALL OF INDIA PRIVATE LIMITED, *New Delhi*
PRENTICE-HALL OF JAPAN, INC., *Tokyo*

CONTENTS

PREFACE

This book has grown out of lectures given at the Eidgenössische Technische Hochschule (Federal Technical Institute) in Zurich by Professor H. Rutishauser and the undersigned. It is intended for mathematicians and physicists, as well as engineers and scientists, who are interested in the numerical mathematics of linear algebra.

The comprehension of the book presupposes familiarity with elements of linear algebra and the fundamentals of matrix operations as presented in freshman and sophomore college courses. It is also assumed that the reader is familiar with elementary concepts of numerical mathematics, such as are taken up in the book of Professor E. Stiefel, "Einführung in die numerische Mathematik" (Introduction to Numerical Mathematics), published by B. G. Teubner, Stuttgart. In two instances the data-fitting calculus of variations is introduced: in one case to sketch the classes of problems to be solved by methods described using a more introductory approach; and in the other to set down a proper preparation for the problem. Knowledge of the computer language ALGOL (translator's note: or FORTRAN) is useful but not mandatory, as the corresponding sections can be omitted.

Basically, the book treats only problems of linear algebra whose solution involves a problem with a symmetric matrix. This confinement to symmetric matrices is justified, first, by the fact that many problems of mathematical physics (especially all problems where damping is absent) lead to symmetric systems of equations or eigenvalue problems in the course of discretization. Furthermore, the numerical treatment of problems in symmetric matrices has at its disposition special and highly effective algorithms whose development is one of the main purposes of this book. The choice of content was made on the basis that, for symmetric problems, methods that are well based and tested are available; in contrast to more general methods, they are safer and simpler. They belong in the basic arsenal of every numerical analyst and computation group. Thus only those methods that have been explored numerically are included.

The main goal has been to present the fundamental ideas and theoretical bases of the techniques described in the context of computer applications. At the same time, intrinsic numerical difficulties of the methods (which are often all too neglected) are pointed out where appropriate. In particular, the problem of the condition of a matrix is taken up in detail. However, the problem of roundoff error, along with the corresponding error analysis and numerical stability study, has not been examined. This area is studied fully in Ref. 75. In certain cases algorithms are developed right up to the formulation of ALGOL (or FORTRAN) programs. The emphasis is entirely on a clear and simple explanation of the computational procedure. No attempt was made to construct detailed and ultrasophisticated programs which would merely obscure the understanding of the algorithms. For this reason, they are not ideal in every sense, but by and large general enough that even some special cases can be treated correctly. In particular, the programs and procedures serve the student as a departure point for more detailed numerical exercises in the scope of a term project, in order to reveal the limits of applicability and to introduce improvements.

At this point I would like to thank Professors Stiefel and Rutishauser for their valuable inspiration and discussions for the selection and presentation of the material. My thanks are also due to Mrs. Ph. Kent, who was kind enough to sift through and summarize the available literature on the method of alternating directions for me; and to my wife, who provided the fair copy of the manuscript. My thanks are also due to Mr. Dipl.-Math. J. Gärtner for his careful and conscientious assistance in proofreading. Finally, I would like to thank the publishers B. G. Teubner for accepting the book in the series "Introductions to Applied Mathematics and Mechanics"; for their demonstration of confidence in me; and for their amiable cooperation.

Zurich H. R. SCHWARZ

Translator's Note

This book represents an unabridged translation of "Numerik symmetrischer Matrizen," originally published in German by B. G. Teubner, Stuttgart, Germany, in 1968. Except for the labels and captions, both the tables and figures are copied directly out of the original. Dr. Schwarz has sent several corrections and additions which have been incorporated into the text, the most notable of which involve an extension of Sec. 3–5 and some alterations in the proof of Theorem 4–21. He has also furnished problems, which are interspersed at appropriate places in each chapter of the English language edition.

The translator would like to thank his wife for her generous assistance in the editing, and to note his appreciation to the late Prof. George Forsythe of Stanford University for his role in initiating the translation project, clarifying some crucial terminology, and in editing the ms. jointly with Prof. Cleve Moler.

The original book contained 16 computer programs in the ALGOL language incorporated into the text. The translator has converted each of these into its FORTRAN IV language counterpart for this edition. The FORTRAN IV equivalents are found in the appendix for the convenience of readers who are unfamiliar with ALGOL. The programs therein have been checked in a trial computation. An attempt was made, as far as possible, to retain the same variable names as in the ALGOL counterparts.

PAUL HERTELENDY

1 EUCLIDEAN VECTOR SPACE. NORMS. QUADRATIC FORMS. SYMMETRIC DEFINITE SYSTEMS OF EQUATIONS.

1-1 THE LINEAR VECTOR SPACE. MATRICES

This section is intended for readers familiar with the theory of linear algebra. Its purpose lies not in a systematic exposition of linear algebra but rather in the establishment of some fundamental results and notation to be used.

1-1-1 The n-Dimensional Vector Space

Scalars from the field of real or complex numbers will be designated by lower-case Latin or Greek letters. The set of all n-dimensional vectors $\boldsymbol{x}$

$$\boldsymbol{x} = \begin{bmatrix} x_1 \\ x_2 \\ \cdot \\ \cdot \\ \cdot \\ x_n \end{bmatrix} \tag{1–1}$$

with n *components* $x_1, x_2, \ldots, x_n$ belonging to the field of numbers constitutes the n-dimensional vector space V_n. For the vectors $\boldsymbol{x}$ of the space, vector addition is defined such that the commutative and associative laws are satisfied; multiplication of a vector by a scalar is also defined. A *linear combination* of vectors $\boldsymbol{x}_1, \boldsymbol{x}_2, \ldots, \boldsymbol{x}_m$ with a set of scalars $c_1, c_2, \ldots, c_m$ is defined as

$$\boldsymbol{y} = c_1\boldsymbol{x}_1 + c_2\boldsymbol{x}_2 + \cdots + c_m\boldsymbol{x}_m. \tag{1–2}$$

The vectors $\boldsymbol{x}_1, \boldsymbol{x}_2, \ldots, \boldsymbol{x}_m$ are termed *linearly dependent* if there exist certain quantities $c_1, c_2, \ldots, c_m$, not all of which are zero, such that the linear combination (1–2) yields the *zero vector*, all of whose components vanish:

$$c_1\boldsymbol{x}_1 + c_2\boldsymbol{x}_2 + \cdots + c_m\boldsymbol{x}_m = 0. \tag{1–3}$$

In the event that this vector equation cannot be fulfilled unless $c_1 = c_2 = \cdots = c_m = 0$, then the vectors $\boldsymbol{x}_1, \boldsymbol{x}_2, \ldots, \boldsymbol{x}_m$ are termed *linearly independent*. A set of more than n vectors in an n-dimensional vector space is always linearly dependent. Conversely, one can always construct n linearly independent vectors $\boldsymbol{x}_1, \boldsymbol{x}_2, \ldots, \boldsymbol{x}_n$ in V_n. A system of n linearly independent vectors forms a *basis* in V_n. Every arbitrary vector $\boldsymbol{x}$ can be represented by a linear combination of the basis vectors. The coefficients of the uniquely determined linear combination are called the *coordinates* of $\boldsymbol{x}$ with respect to the basis. The system of *n unit vectors* $\boldsymbol{e}_k$

$$\boldsymbol{e}_1 = \begin{bmatrix} 1 \\ 0 \\ 0 \\ \cdot \\ \cdot \\ \cdot \\ 0 \end{bmatrix}, \quad \boldsymbol{e}_2 = \begin{bmatrix} 0 \\ 1 \\ 0 \\ \cdot \\ \cdot \\ \cdot \\ 0 \end{bmatrix}, \quad \ldots, \quad \boldsymbol{e}_n = \begin{bmatrix} 0 \\ 0 \\ 0 \\ \cdot \\ \cdot \\ \cdot \\ 1 \end{bmatrix} \tag{1–4}$$

clearly forms a basis. Actually, every vector $\boldsymbol{x}$ with components $x_1, x_2, \ldots, x_n$ can be represented by

$$\boldsymbol{x} = x_1\boldsymbol{e}_1 + x_2\boldsymbol{e}_2 + \cdots + x_n\boldsymbol{e}_n. \tag{1–5}$$

In this particular case, the components of the vector are also its coordinates.

Henceforth we confine ourselves to vector spaces over the field of real numbers. Here for every pair of vectors $\boldsymbol{x}$ and $\boldsymbol{y}$, there exists an *inner product* $(\boldsymbol{x}, \boldsymbol{y})$. The real-valued function $(\boldsymbol{x}, \boldsymbol{y})$ exhibits the following four properties:

$$(\boldsymbol{x}, \boldsymbol{y}) = (\boldsymbol{y}, \boldsymbol{x}) \tag{1–6}$$

$$(\lambda\boldsymbol{x}, \boldsymbol{y}) = \lambda(\boldsymbol{x}, \boldsymbol{y}) \tag{1–7}$$

$$(\boldsymbol{x}_1 + \boldsymbol{x}_2, \boldsymbol{y}) = (\boldsymbol{x}_1, \boldsymbol{y}) + (\boldsymbol{x}_2, \boldsymbol{y}) \tag{1–8}$$

$$(\boldsymbol{x}, \boldsymbol{x}) \geq 0 \quad \text{and} \quad (\boldsymbol{x}, \boldsymbol{x}) = 0 \quad \text{only when} \quad \boldsymbol{x} = 0. \tag{1–9}$$

Corresponding to this inner product, we can define the *magnitude*, the *length*, or the *norm* $\|\boldsymbol{x}\|$ of a vector $\boldsymbol{x}$ as the positive square root of the real, nonnegative value of the inner product of the vector with itself:

$$\|\boldsymbol{x}\| = \sqrt{(\boldsymbol{x}, \boldsymbol{x})}. \tag{1–10}$$

With this metric the vector space V_n becomes a *normed metric* space, where concepts of length, distance, angle, and, especially, orthogonality of vectors are defined. A vector with a norm of unity is called *normalized.* The distance d between two vectors $\boldsymbol{x}$ and $\boldsymbol{y}$ is simply the magnitude of the difference of the two vectors

$$d = \|\boldsymbol{x} - \boldsymbol{y}\| = \sqrt{(\boldsymbol{x} - \boldsymbol{y}, \boldsymbol{x} - \boldsymbol{y})}. \tag{1–11}$$

The inner product of two vectors $\boldsymbol{x}$ and $\boldsymbol{y}$ fulfills the *Schwarz inequality*

$$|(\boldsymbol{x}, \boldsymbol{y})| \leq \|\boldsymbol{x}\| \|\boldsymbol{y}\|. \tag{1–12}$$

Thus a real angle ϕ always exists in the interval $0 \leq \phi \leq \pi$, such that

$$(\boldsymbol{x}, \boldsymbol{y}) = \|\boldsymbol{x}\| \|\boldsymbol{y}\| \cos \phi. \tag{1–13}$$

The angle between the vectors $\boldsymbol{x}$ and $\boldsymbol{y}$ is defined to be this ϕ. If the inner product of $\boldsymbol{x}$ and $\boldsymbol{y}$ vanishes, the vectors are said to be *orthogonal.* A set of $m \leq n$ nonzero vectors that are mutually orthogonal is called an *orthogonal system.* Its vectors are always linearly independent. A system of n orthogonal nonzero vectors always exists in the vector space V_n. We then refer to a *complete* orthogonal system, since an orthogonal system can consist at most of n vectors because of linear independence. If the mutually orthogonal vectors are normalized vectors, they form an *orthonormal system.* In the n-dimensional vector space, there always exists a basis of n orthonormal vectors. The coordinates of a vector $\boldsymbol{x}$, referred to as an *arbitrary* orthonormal basis, are the same as the inner products of the vector $\boldsymbol{x}$ with the basis vectors. For two vectors $\boldsymbol{x}$ and $\boldsymbol{y}$ with components x_k and y_k, the *Euclidean inner product* is given by

$$(\boldsymbol{x}, \boldsymbol{y}) = \sum_{k=1}^{n} x_k y_k. \tag{1–14}$$

A vector space V_n with such an inner product is called a *Euclidean vector space.* The specifying of an Euclidean inner product means that the n unit vectors constitute an orthonormal basis.

1-1-2 Linear Transformations. Matrices

By a *transformation of the vector space* V_n into itself is meant a unique correspondence of a vector $\boldsymbol{y}$ to every arbitrary vector $\boldsymbol{x}$. The transformation of the space is formally given by

$$\boldsymbol{y} = \tilde{A}\boldsymbol{x} \tag{1–15}$$

where $\tilde{A}$ indicates the *operator* that carries out the transformation.

Let us now specifically examine the class of *linear transformations.* An operator $\tilde{A}$ is linear if the transformation (1–15) possesses the following properties for every arbitrary scalar c belonging to the field of real numbers and for arbitrary vectors $\boldsymbol{x}$ and $\boldsymbol{y}$:

$$\tilde{A}(c\boldsymbol{x}) = c(\tilde{A}(\boldsymbol{x})) = c\tilde{\boldsymbol{A}}\boldsymbol{x} \tag{1–16}$$

$$\tilde{A}(\boldsymbol{x} + \boldsymbol{y}) = \tilde{\boldsymbol{A}}\boldsymbol{x} + \tilde{\boldsymbol{A}}\boldsymbol{y} \tag{1–17}$$

Given a particular basis $\boldsymbol{b}_1, \boldsymbol{b}_2, \ldots, \boldsymbol{b}_n$, every linear operator $\tilde{\boldsymbol{A}}$ in V_n is uniquely represented by a *square matrix* $\boldsymbol{A}$ derived from an examination of its transformed vectors $\tilde{\boldsymbol{A}}\boldsymbol{b}_k$:

$$\boldsymbol{A} = \begin{bmatrix} a_{11} & a_{12} & \cdots & a_{1n} \\ a_{21} & a_{22} & \cdots & a_{2n} \\ \cdot & \cdot & \cdots & \cdot \\ a_{n1} & a_{n2} & \cdots & a_{nn} \end{bmatrix} = (a_{ik}). \tag{1–18}$$

In its kth column, it contains the *coordinates* of the transformed vector $\tilde{\boldsymbol{A}}\boldsymbol{b}_k$ referred to the basis $\boldsymbol{b}_1, \boldsymbol{b}_2, \ldots, \boldsymbol{b}_n$ in accordance with the formula

$$\tilde{\boldsymbol{A}}\boldsymbol{b}_n = \sum_{i=1}^{n} a_{ik}\boldsymbol{b}_i \qquad k = 1, 2, \ldots, n. \tag{1–19}$$

In case the vectors $\boldsymbol{x}$ and $\boldsymbol{y}$ are represented by their coordinates referred to the basis, that is,

$$\boldsymbol{x} = \sum_{k=1}^{n} x_k\boldsymbol{b}_k \quad \text{and} \quad \boldsymbol{y} = \sum_{i=1}^{n} y_i\boldsymbol{b}_i, \tag{1–20}$$

there follows from $\boldsymbol{y} = \tilde{\boldsymbol{A}}\boldsymbol{x}$, as a consequence of the linearity of the operator $\tilde{\boldsymbol{A}}$ and the linear independence of the basis vectors, the relation

$$y_i = \sum_{k=1}^{n} a_{ik}x_k \qquad i = 1, 2, \ldots, n, \tag{1–21}$$

which in matrix notation is given by

$$\boldsymbol{y} = \boldsymbol{A}\boldsymbol{x}. \tag{1–22}$$

Matrix $\boldsymbol{A}$, which is a *representation* of the linear operator $\tilde{\boldsymbol{A}}$, can itself be thought of as an operator in V_n, so that we can speak of a matrix as an operator. Through this relationship between linear transformations and matrices, the rules of matrix manipulations are conventionally derived (see references 25, 38, and 80).

If the linear transformation $\boldsymbol{y} = \boldsymbol{A}\boldsymbol{x}$ can be uniquely inverted, in other words if it is *regular* or *nonsingular*, then there exists the *inverse transformation* $\boldsymbol{x} = \boldsymbol{A}^{-1}\boldsymbol{y}$, where $\boldsymbol{A}^{-1}$ is the *inverse matrix* of $\boldsymbol{A}$.

The representation of a linear operator by a matrix is altered in going from one basis $\boldsymbol{b}_1, \boldsymbol{b}_2, \ldots, \boldsymbol{b}_n$ to another $\boldsymbol{b}'_1, \boldsymbol{b}'_2, \ldots, \boldsymbol{b}'_n$. The coordinates of the second basis referred to the first are given by

$$\boldsymbol{b}'_k = c_{1k}\boldsymbol{b}_1 + c_{2k}\boldsymbol{b}_2 + \cdots + c_{nk}\boldsymbol{b}_n \qquad k = 1, 2, \ldots, n. \tag{1–23}$$

The coordinates x_i of an arbitrary vector referred to the first basis are related to the coordinates x'_i referred to the second basis by

$$x_i = \sum_{k=1}^{n} c_{ik}x'_k, \quad \text{or} \quad \boldsymbol{x} = \boldsymbol{C}\boldsymbol{x}' \qquad i = 1, 2, \ldots, n. \tag{1–24}$$

The transformation matrix $\boldsymbol{C} = (c_{ik})$ for the coordinates of a vector is nonsingular, since the transition from one basis to another is uniquely invertible. Let $\boldsymbol{x}$ and $\boldsymbol{y}$ be the coordinate vectors in the first basis, $\boldsymbol{x}'$ and $\boldsymbol{y}'$ the coordinate vectors of the corresponding vectors in the second basis, and let $\boldsymbol{A}$ and $\boldsymbol{B}$ be the representation of the same linear operator in the two bases. Then it follows that

$$\boldsymbol{y} = \boldsymbol{A}\boldsymbol{x}, \qquad \boldsymbol{y}' = \boldsymbol{B}\boldsymbol{x}', \qquad \boldsymbol{x} = \boldsymbol{C}\boldsymbol{x}', \qquad \boldsymbol{y} = \boldsymbol{C}\boldsymbol{y}', \tag{1-25}$$

$$\boldsymbol{C}\boldsymbol{y}' = \boldsymbol{A}(\boldsymbol{C}\boldsymbol{x}') = \boldsymbol{A}\boldsymbol{C}\boldsymbol{x}' \quad \text{or} \quad \boldsymbol{y}' = \boldsymbol{C}^{-1}\boldsymbol{A}\boldsymbol{C}\boldsymbol{x}'. \tag{1–26}$$

From the uniqueness of the representation with fixed basis, it follows from (1–25) and (1–26) that

$$\boldsymbol{B} = \boldsymbol{C}^{-1}\boldsymbol{A}\boldsymbol{C}. \tag{1–27}$$

Two matrices $\boldsymbol{A}$ and $\boldsymbol{B}$ related through a nonsingular matrix $\boldsymbol{C}$ by (1–27) are called *similar*. The transition from matrix $\boldsymbol{A}$ to matrix $\boldsymbol{B}$ through (1–27) is called a *similarity transformation*. Similar matrices represent the same linear operator but refer to different basis systems. This property can be reformulated by saying that every linear operator in an n-dimensional vector space corresponds to a certain class of similar matrices. Certain properties of a linear operator are independent of a particular representation. These properties are termed *invariant* under similarity transformations. Thus, similarity transformations play a significant role in the theory and practice of matrix calculations, since it is often desirable to represent a linear operator in the most convenient matrix form through a suitable choice of basis.

An *eigenvector* $\boldsymbol{x}$ of a linear transformation, such as matrix $\boldsymbol{A}$, is a nonzero vector satisfying the equation

$$\boldsymbol{A}\boldsymbol{x} = \lambda\boldsymbol{x} \tag{1–28}$$

where λ is a scalar. The value of λ is called an *eigenvalue* of $\boldsymbol{A}$. It is important to note that the eigenvalues of a matrix remain invariant under a similarity transformation. On the other hand, the eigenvectors are transformed in accordance with the new basis. The invariance of the eigenvalues of similar matrices is the key to many numerical techniques of calculation of eigenvalues (see Chapter 4).

The *zero matrix* is a matrix all of whose elements are zero. The *identity matrix* $\boldsymbol{I}$ consists of elements that are unity along the diagonal and zero elsewhere:

$$\boldsymbol{I} = \begin{bmatrix} 1 & 0 & 0 & \cdots & 0 \\ 0 & 1 & 0 & \cdots & 0 \\ 0 & 0 & 1 & \cdots & 0 \\ \cdot & \cdot & \cdot & \cdots & \cdot \\ 0 & 0 & 0 & \cdots & 1 \end{bmatrix}.$$

The columns of the identity matrix $\boldsymbol{I}$ are formed by the n unit vectors.

The *transpose* of a matrix, $\boldsymbol{A}^{\mathrm{T}}$, results from the interchanging of rows with corresponding columns in matrix $\boldsymbol{A}$. The resulting linear transformation is generally called the *adjoint*. For a Euclidean inner product with arbitrary vectors $\boldsymbol{x}$ and $\boldsymbol{y}$,

$$(\boldsymbol{A}\boldsymbol{x}, \boldsymbol{y}) = (\boldsymbol{x}, \boldsymbol{A}^{\mathrm{T}}\boldsymbol{y}).$$

The relocation of a linear operator from the first vector to the second in the inner product requires switching to the adjoint operator. A *symmetric matrix* $\boldsymbol{A}$ is identical with its transpose so that it can be transferred from the first vector to the second in a Euclidean product. This is the formal requirement for the linear operator to be *self-adjoint*. From this property it follows directly that the eigenvalues of a self-adjoint operator are real and that the eigenvectors of differing eigenvalues are mutually orthogonal.

In Sec. 4-3 we will show that the self-adjoint operator actually possesses a complete system of orthonormal eigenvectors. Self-adjoint operators obviously have special properties and are thus accorded special treatment in theory and practice.

To every symmetric matrix $\boldsymbol{A}$ belongs a corresponding *quadratic form*

$$Q(\boldsymbol{x}) = (\boldsymbol{A}\boldsymbol{x}, \boldsymbol{x}) = \sum_{i=1}^{n} \sum_{k=1}^{n} a_{ik} x_i x_k \tag{1–29}$$

for any arbitrary vector $\boldsymbol{x}$ with components $x_1, x_2, \ldots, x_n$. If it is true for arbitrary vectors $\boldsymbol{x}$ that

$$Q(\boldsymbol{x}) = \sum_{i=1}^{n} \sum_{k=1}^{n} a_{ik} x_i x_k \geq 0, \quad \text{and} = 0 \quad \text{only when} \quad \boldsymbol{x} = 0, \tag{1–30}$$

then the quadratic form is called *positive definite.* The corresponding symmetric matrix is also termed positive definite.

This link between linear self-adjoint operators in a Euclidean vector space and quadratic forms is important. The problems whose mathematical formulations can be reduced to the study of quadratic forms also possess the property of self-adjointness. The following will make constant use of this, presenting numerical procedures for the treatment of self-adjoint problems that are simpler when compared with more general methods of linear algebra.

PROBLEMS

1-1. Are the following three vectors of a four-dimensional Euclidean vector space linearly independent?

$$\boldsymbol{x}_1 = \begin{bmatrix} 2 \\ -3 \\ 0 \\ 4 \end{bmatrix} \qquad \boldsymbol{x}_2 = \begin{bmatrix} 0 \\ 4 \\ 3 \\ -1 \end{bmatrix} \qquad \boldsymbol{x}_3 = \begin{bmatrix} 0 \\ 0 \\ -2 \\ 6 \end{bmatrix}$$

1-2. What are the angles ϕ between the vectors in the previous problem?

1-3. Let $\boldsymbol{b}_1, \boldsymbol{b}_2, \ldots, \boldsymbol{b}_n$ be orthonormal vectors in V_n. Show that the coordinates of an arbitrary vector $\boldsymbol{x}$ of V_n referred to the basis $\boldsymbol{b}_i$ are given by

$$c_i = (\boldsymbol{b}_i, \boldsymbol{x})$$

1-4. What is the matrix of the linear transformation corresponding to a rotation of V_3 about the y-axis by an angle ϕ?

1-5. Prove the invariance of the eigenvalues of a matrix under a similarity transformation. How are the eigenvectors transformed?

1-2 NORMS. CONDITION OF A MATRIX

In order to examine the convergence of vector and matrix sequences, a general concept of distance is necessary. Section 1-1-1 introduced the length or norm of a vector $\boldsymbol{x}$ in a normed vector space as the square root of the inner product of $\boldsymbol{x}$ with itself. The concept of a norm, however, is broader and not restricted to a normed vector space.

DEFINITION 1–1

A norm $N(\boldsymbol{x}) = \|\boldsymbol{x}\|$ of a vector $\boldsymbol{x}$ is a real function of the vector $\boldsymbol{x}$ having the properties (1–31) *through* (1–33):

$$\|\boldsymbol{x}\| \geq 0 \quad \text{with} \quad \|\boldsymbol{x}\| = 0 \quad \text{only for} \quad \boldsymbol{x} = 0, \tag{1–31}$$

$$\|cx\| = |c|\,\|x\| \quad \text{for any arbitrary scalar } c, \tag{1–32}$$

$$\|x + y\| \leq \|x\| + \|y\| \quad \text{(triangle inequality).} \tag{1–33}$$

With this definition of a norm, the concept of length of a vector is broadened. Clearly, the conventional concept of vector length fulfills these three requirements of a norm.

Example 1-1. Some possible vector norms are

$$\|x\|_1 = \max_i |x_i|, \tag{1–34}$$

$$\|x\|_2 = \sum_{i=1}^{n} |x_i|, \tag{1–35}$$

$$\|x\|_3 = \left(\sum_{i=1}^{n} |x_i|^2\right)^{1/2}. \tag{1–36}$$

These are particular cases of *Hölder norms* $\sqrt[p]{\sum_{i=1}^{n} |x_i|^p}$ for $p = \infty$, $p = 1$, and $p = 2$, respectively. Equation (1–36) is the well-known *Euclidean* vector norm.

A sequence of vectors $x^{(k)}$ converges to a vector x if every component of the sequence converges. For an infinite vector sequence $x^{(k)}$ to converge to x, it is necessary and sufficient that

$$\lim_{k\to\infty} \|x^{(k)} - x\| = 0 \tag{1–37}$$

for an arbitrary norm. This result is immediately evident for (1–34) in Example 1-1. For the two other examples, it follows directly from the inequalities

$$\|x\|_1 \leq \|x\|_2 \leq n\|x\|_1 \quad \text{and} \quad \|x\|_1 \leq \|x\|_3 \leq \sqrt{n}\,\|x\|_1.$$

We can easily show that Eq. (1–37) is necessary and sufficient for the convergence of a vector sequence, given an arbitrary vector norm.

DEFINITION 1–2

The norm $N(A) = \|A\|$ of a matrix A is a real function of matrix elements satisfying the properties (1–38) *through* (1–41):

$$\|A\| \geq 0, \quad \text{with} \quad \|A\| = 0 \quad \text{only for} \quad A = 0, \tag{1–38}$$

$$\|cA\| = |c| \cdot \|A\| \quad \text{for any arbitrary scalar } c, \tag{1–39}$$

$$\|A + B\| \leq \|A\| + \|B\| \quad \text{(triangle inequality),} \tag{1–40}$$

$$\|A \cdot B\| \leq \|A\| \cdot \|B\|. \tag{1–41}$$

In comparison with the definition of a vector norm, the first three properties correspond formally, but a fourth is added. The first three required properties are those of distance. In this framework, postulate (1–41) is an intruder. The admissible functions are thereby restricted to be the so-called multiplicative matrix norms. Even though it is not obvious that such matrix norms can even exist in light of (1–41), there are actually numerous possibilities.

Example 1-2. Some possible matrix norms are

$$\|A\|_1 = n \cdot \max_{i,k} |a_{ik}|, \tag{1–42}$$

$$\|A\|_2 = \max_k \sum_{i=1}^{n} |a_{ik}|, \tag{1–43}$$

$$\|A\|_3 = \left(\sum_{i,k}^{n} |a_{ik}|^2\right)^{1/2}. \tag{1–44}$$

Equation (1–44) is variously referred to as the *Euclidean*, *Schur*, or *Frobenius* norm. Clearly, the three norms given fulfill the first three postulates of a norm. For (1–44), requirement (1–40) is fulfilled if matrix A is regarded as a vector in a n^2-dimensional space. Property (1–41) for matrix norm (1–42) is proven below; proofs for the two other cases are left to the reader.

$$\begin{aligned}\|A \cdot B\|_1 &= n \cdot \max_{i,k} \left| \sum_{j=1}^{n} a_{ij} b \right| \leq n \cdot \max_{i,k} \sum_{j=1}^{n} |a_{ij}| \cdot |b_{jk}| \\ &\leq n \cdot \max_{i,k} \sum_{j=1}^{n} (\max_{l,m} |a_{lm}|) \cdot (\max_{r,s} |b_{rs}|) \\ &= (n \cdot \max_{l,m} |a_{lm}|) \cdot (n \cdot \max_{r,s} |b_{rs}|) = \|A\|_1 \cdot \|B\|_1.\end{aligned}$$

A sequence of matrices $A^{(k)}$ converges to a matrix A if each individual element of the matrices $A^{(k)}$ converges. A necessary and sufficient condition for the convergence of a matrix sequence $A^{(k)}$ to a matrix A is

$$\lim_{k \to \infty} \|A^{(k)} - A\| = 0, \tag{1–45}$$

for an arbitrary matrix norm. Such is clearly the case for norm (1–42). For the two other examples, convergence can be shown by invoking the inequalities

$$\frac{1}{n} \cdot \|A\|_1 \leq \|A\|_2 \leq \|A\|_1 \quad \text{and} \quad \frac{1}{n} \cdot \|A\|_1 \leq \|A\|_3 \leq \|A\|_1.$$

It can readily be shown that (1–45) is necessary and sufficient for the convergence of a matrix sequence with an arbitrary matrix norm.

In most examinations of convergence, we encounter both matrices and vectors. Therefore, the norms used for vectors and matrices must be related in logical fashion to one another.

Definition 1–3

A matrix norm $\|A\|$ *is called compatible with a vector norm* $\|x\|$ *if for every arbitrary vector* x, *the relation* (1–46) *is fulfilled:*

$$\|Ax\| \leq \|A\| \cdot \|x\|. \tag{1–46}$$

The examples of matrix norms given earlier are compatible with the corresponding vector norms. The compatibility criterion (1–46) is shown below for the matrix norm (1–43) and the vector norm (1–35):

$$\begin{aligned}\|Ax\|_2 &= \sum_{i=1}^{n}\left|\sum_{k=1}^{n} a_{ik}x_k\right| \leq \sum_{i=1}^{n}\sum_{k=1}^{n}|a_{ik}|\cdot|x_k| = \sum_{k=1}^{n}|x_k|\cdot\left\{\sum_{i=1}^{n}|a_{ik}|\right\} \\ &\leq \left\{\max_k \sum_{i=1}^{n}|a_{ik}|\right\}\cdot\left\{\sum_{k=1}^{n}|x_k|\right\} = \|A\|_2\cdot\|x\|_2.\end{aligned}$$

For every given matrix norm $\|A\|$, a compatible vector norm can be constructed. To each vector x, we can relate a matrix X whose first column contains the elements of x and whose remaining elements vanish. We define the norm of the vector x as

$$\|x\| = \|X\|. \tag{1–47}$$

The function of vector components thus defined is actually a vector norm in consequence of the properties (1–38) through (1–40) of a matrix norm. The compatibility of the vector norm with the given matrix norm is a direct result of (1–41) and the definition of a vector norm:

$$\|Ax\| = \|AX\| \leq \|A\|\cdot\|X\| = \|A\|\cdot\|x\|.$$

Observe that AX represents a matrix of the same form as X, containing the components of the vector Ax in its first column.

Although the vector norm (1–34) is compatible with the corresponding matrix norm (1–42), it cannot be determined by the process just described. We may note that for a given matrix norm there are, in general, several compatible vector norms.

Application of compatible norms leads to the following theorem.

Theorem 1-1. *For every matrix norm,* $\|A\|$ *is an upper bound for the magnitude of the eigenvalues of* A.

Proof: Let A be an arbitrary matrix, λ an eigenvalue, and x a corresponding eigenvector so that $Ax = \lambda x$. Furthermore, let $\|A\|$ be an arbitrary matrix norm and $\|x\|$ a correspondingly compatible vector norm. Then

$$\left.\begin{aligned} \|Ax\| &= \|\lambda x\| = |\lambda|\cdot\|x\| \\ \|Ax\| &\leq \|A\|\cdot\|x\| \end{aligned}\right\} \quad |\lambda|\cdot\|x\| \leq \|A\|\cdot\|x\|.$$

The vector x, being an eigenvector, is nonvanishing, i.e., $\|x\| > 0$, and from this follows the assertion of the theorem, $|\lambda| \leq \|A\|$, for every eigenvalue λ of A.

Conversely, a matrix norm can always be constructed compatible with any arbitrary vector norm. The corresponding matrix norm is not, in general, unique. One particular compatible matrix norm can be defined as follows.

DEFINITION 1–4

The value of the real function

$$N(A) = \max_{x\neq 0} \frac{\|Ax\|}{\|x\|} = \max_{\|x\|=1} \|Ax\| \tag{1–48}$$

is the matrix norm subordinate to the vector norm $\|x\|$. Reference 15 shows that function (1–48) is a matrix norm.

Example 1-3. Find the matrix norm subordinate to the Euclidean vector norm. The Euclidean vector norm is $\|x\| = (x, x)^{1/2}$. For a real matrix A, the squared quantity $N(A)^2$ is, in light of (1–48),

$$N(A)^2 = \max_{\|x\|=1} \|Ax\|^2 = \max_{\|x\|=1} (Ax, Ax) = \max_{\|x\|=1} (x, A^{\mathrm{T}}Ax).$$

The matrix $A^{\mathrm{T}}A$ is symmetric (thus self-adjoint) and positive semidefinite, since $(Ax, Ax) \geq 0$ for every vector x. Accordingly, the matrix $A^{\mathrm{T}}A$ has n real, nonnegative eigenvalues $\mu_1 \geq \mu_2 \geq \mu_3 \geq \cdots \geq \mu_n \geq 0$ and a corresponding orthonormal system of eigenvectors $x_1, x_2, x_3, \ldots, x_n$. An arbitrary normed vector can be represented by a linear combination of the eigenvectors

$$x = c_1x_1 + c_2x_2 + c_3x_3 + \cdots + c_nx_n.$$

Orthonormality of eigenvectors yields

$$\|x\|^2 = (x, x) = c_1^2 + c_2^2 + c_3^2 + \cdots + c_n^2 = 1.$$

We can thus estimate the value of $(\boldsymbol{x}, A^{\mathrm{T}}A\boldsymbol{x})$:

$$(\boldsymbol{x}, A^{\mathrm{T}}A\boldsymbol{x}) = \left(\sum_{i=1}^{n} c_i\boldsymbol{x}_i, A^{\mathrm{T}}A \sum_{i=1}^{n} c_i\boldsymbol{x}_i\right) = \left(\sum_{i=1}^{n} c_i\boldsymbol{x}_i, \sum_{i=1}^{n} c_i\mu_i\boldsymbol{x}_i\right)$$

$$= \sum_{i=1}^{n} c_i^2\mu_i \leq \mu_1 \sum_{i=1}^{n} c_i^2 = \mu_1.$$

Specifically, $(\boldsymbol{x}_1, A^{\mathrm{T}}A\boldsymbol{x}_1) = \mu_1$ so that $N(A)^2 = \max_{\|\boldsymbol{x}\|=1} (\boldsymbol{x}, A^{\mathrm{T}}A\boldsymbol{x}) = \mu_1 \geq 0$. The matrix norm subordinate to the Euclidean vector norm is thus $\|A\| = \sqrt{\mu_1}$, where μ_1 signifies the largest eigenvalue of $A^{\mathrm{T}}A$. This norm is called the *spectral norm* of A.

Examine next a symmetric, positive definite matrix A. Its eigenvalues are $\lambda_1 \geq \lambda_2 \geq \lambda_3 \geq \cdots \geq \lambda_n > 0$. Since $A^{\mathrm{T}} = A$, the eigenvalues of $A^{\mathrm{T}}A = A^2$ are $\lambda_1^2 \geq \lambda_2^2 \geq \lambda_3^2 \geq \cdots \geq \lambda_n^2 > 0$. Consequently, the spectral norm of a symmetric, positive definite matrix is simply

$$\|A\| = \lambda_1, \tag{1–49}$$

where λ_1 indicates the largest eigenvalue of A. By construction it is also the smallest matrix norm compatible with the Euclidean vector norm. Because of Theorem 1-1, this is the smallest of all possible matrix norms.

Application: condition of a matrix. The solution of a linear system of equations $A\boldsymbol{x} + \boldsymbol{b} = 0$ cannot be evaluated more accurately than is permitted by the inaccuracy in the calculation of $A\boldsymbol{x} + \boldsymbol{b}$ for a given vector $\boldsymbol{x}$ in the vicinity of the exact solution vector $\boldsymbol{x}_l$. Let $\delta\boldsymbol{x}$ be an arbitrary vector whose Euclidean norm is one unit in the last significant figure of the largest component (in absolute value) of $\boldsymbol{x}_l$. The magnitude of the error in the calculation of

$$A\boldsymbol{x} + \boldsymbol{b} = A(\boldsymbol{x}_l + \delta\boldsymbol{x}) + \boldsymbol{b} = A\boldsymbol{x}_l + \boldsymbol{b} + A\delta\boldsymbol{x} = A\delta\boldsymbol{x}$$

is given by the norm $\|A\delta\boldsymbol{x}\|$. This value gives an absolute measure of error. The relative error is given by $\|A\delta\boldsymbol{x}\|/\|\delta\boldsymbol{x}\|$. The quotient yields the inaccuracy of the largest component (in absolute value) of $\boldsymbol{x}_l$ in units of the last significant figure. The maximum of this expression is, by Eq. (1–48), just the spectral norm of A subordinate to the Euclidean vector norm. Thus, in summation, the determination of $A\boldsymbol{x}_l + \boldsymbol{b}$ yields, in the most unfavorable case, a vector whose norm equals $\epsilon = \|A\|$ units of the last significant figure of the largest component of $\boldsymbol{x}_l$. Conversely, every vector $\boldsymbol{x}'$ can be considered a solution to the system of equations as long as $\|A\boldsymbol{x}' + \boldsymbol{b}\| \leq \epsilon$. In consequence of $\boldsymbol{b} = -A\boldsymbol{x}_l$, this applies to all vectors $\boldsymbol{x}'$ such that $\|A(\boldsymbol{x}' - \boldsymbol{x}_l)\| \leq \epsilon$. The goal is now to estimate the magnitude of $\|\boldsymbol{x}' - \boldsymbol{x}_l\|$ in order to draw conclusions concerning the error contained in the solution. The inequality just given applies for every vector $\boldsymbol{d} = A(\boldsymbol{x}' - \boldsymbol{x}_l)$ such that $\|\boldsymbol{d}\| \leq \epsilon$.

Thus for $\|x' - x_l\|$, it follows that

$$\|x' - x_l\| = \|A^{-1}d\| \leq \|A^{-1}\| \cdot \|d\| \leq \|A^{-1}\| \epsilon = \|A^{-1}\| \cdot \|A\|.$$

The quantity

$$\kappa = \|A^{-1}\| \cdot \|A\| \tag{1–50}$$

is called the *condition number* of matrix A. Examination shows that every vector x' differing from the exact solution by κ units in the last significant figure of the largest component can be considered a solution. Applied to the numerical solution of a system of linear equations, this qualitative statement implies the following: In calculating with m significant figures and known condition number κ, it is possible to have an error in the numerical solution of magnitude κ units of the last significant figure of the largest component. This magnitude of error may be expected in all the components.

Example 1-4. For an identity matrix **I**, it is true that

$$\|I\| = \max_{\|x\|=1} \|Ix\| = 1, \qquad \|I^{-1}\| = \|I\| = 1, \qquad \text{and} \quad \kappa = 1.$$

This is the smallest possible condition number. It is also indicative of the fact that in solving such a system of equations, no numerical error will occur.

Example 1-5. The matrix

$$A = \begin{bmatrix} 5 & 7 \\ 7 & 10 \end{bmatrix}$$

is symmetric and positive definite. The eigenvalues of the inverse matrix A^{-1} are reciprocals of those of A. Based on the spectral norm, the condition number for a symmetric, positive definite matrix A is

$$\kappa = \lambda_{\max}/\lambda_{\min}, \tag{1–51}$$

where $\lambda_{\max}$ and $\lambda_{\min}$ represent the largest and smallest eigenvalues of A. For this example, $\lambda_{\max} = 14.933$, $\lambda_{\min} = 0.06697$, and the condition number is a moderate $\kappa \cong 223$.

Example 1-6. For the symmetric positive definite matrix

$$A = \begin{bmatrix} 1 & 10 \\ 10 & 101 \end{bmatrix}$$

calculations yield $\lambda_{\max} = 101.99$, and $\lambda_{\min} \cong 0.009805$. The condition number is poor: $\kappa \cong 10{,}402$.

Example 1-7. The symmetric positive definite matrix

$$A = \begin{bmatrix} 5 & 7 & 3 \\ 7 & 11 & 2 \\ 3 & 2 & 6 \end{bmatrix}$$

has eigenvalues $\lambda_{\max} \cong 16.662$ and $\lambda_{\min} = 0.0112$, and condition number $\kappa \cong 1487$.

PROBLEMS

1-6. Show that the matrix norms (1-43) and (1-44) possess the property (1-41).

1-7. What are the matrix norms compatible with the vector norms (1-34) and (1-35)?

1-8. Find the condition number of the following matrices:

(a)
$$A_1 = \begin{bmatrix} 1 & 0 & 0 & 0 \\ 0 & 25 & 0 & 0 \\ 0 & 0 & 64 & 0 \\ 0 & 0 & 0 & 100 \end{bmatrix}$$

(b)
$$A_2 = \begin{bmatrix} 2 & -1 & 0 & 0 \\ -1 & 3 & -1 & 0 \\ 0 & -1 & 3 & -1 \\ 0 & 0 & -1 & 2 \end{bmatrix}$$

(c)
$$A_3 = \begin{bmatrix} 1 & 1 & 1 & 1 \\ 1 & 2 & 3 & 4 \\ 1 & 3 & 6 & 10 \\ 1 & 4 & 10 & 20 \end{bmatrix}$$

In case (a), is the condition number meaningful? How large are the numerical errors in the solution of systems of equations having a diagonal matrix?

1-9. *Scaling* of a matrix A means the transition to $B = D \cdot A \cdot D$, where D is a diagonal matrix. The ith row and ith column of A are multiplied, respectively, by the ith diagonal element d_i. Verify that if A is symmetric, B is, too, but A and B are not in general similar! How is the condition number of A_1 in the previous problem affected by application of the scaling matrix

$$D = \begin{bmatrix} 1 & 0 & 0 & 0 \\ 0 & 0.2 & 0 & 0 \\ 0 & 0 & 0.125 & 0 \\ 0 & 0 & 0 & 0.1 \end{bmatrix}$$

1-10. Which readily evident scaling significantly reduces the condition number of the matrix

$$A = \begin{bmatrix} 1 & 10 \\ 10 & 101 \end{bmatrix}$$

Can a 2×2 matrix be scaled in a simple fashion such that its condition number be a minimum?

1-3 NECESSARY AND SUFFICIENT CONDITIONS FOR THE DEFINITENESS OF A QUADRATIC FORM

This section examines necessary and sufficient conditions for the definiteness of a given quadratic form with real coefficients in real variables. The criteria mentioned are based on a variety of considerations. Less time-consuming criteria often fail to answer the question of definiteness in general terms, yet suffice either to suggest definiteness or to identify immediately the indefiniteness.

1-3-1 Direct Criteria. Necessary Conditions

The positive definiteness of a quadratic form is often assured through its physical significance, such as is the kinetic energy of a system of mass particles or the deformation energy in a discretized problem. In these cases, numerical criteria would be superfluous.

1–3–1–1. PARTICULAR EXAMPLES

Example 1-8. The quadratic form of the identity matrix $\boldsymbol{I}$ with n rows,

$$Q(\boldsymbol{x}) = \sum_{i=1}^{n} x_i^2$$

is certainly greater than zero, being the sum of the squares of the real variables x_i, provided that not all x_i vanish. It vanishes if and only if $x_i = 0$ for $i = 1, 2, \ldots, n$.

Example 1-9. The infinite *Pascal matrix* contains the binomial coefficients in diagonals, running from lower left to upper right. The fact that every principal minor of the infinite Pascal matrix is positive definite will be verified in a principal minor of four rows. The four-row principal minor and the corresponding quadratic form are

$$A = \begin{bmatrix} 1 & 1 & 1 & 1 \\ 1 & 2 & 3 & 4 \\ 1 & 3 & 6 & 10 \\ 1 & 4 & 10 & 20 \end{bmatrix} \qquad \begin{aligned} Q(\boldsymbol{x}) = x_1^2 + 2x_1x_2 &+ 2x_1x_3 + 2x_1x_4 \\ + 2x_2^2 &+ 6x_2x_3 + 8x_2x_4 \\ &+ 6x_3^2 + 20x_3x_4 \\ &\qquad + 20x_4^2 \end{aligned}$$

The quadratic form may be represented as a sum of squares of linear forms in the variables x_k:

$$Q(\boldsymbol{x}) = (x_1 + x_2 + x_3 + x_4)^2 + (x_2 + 2x_3 + 3x_4)^2 + (x_3 + 3x_4)^2 + x_4^2.$$

This grouping shows that $Q(\boldsymbol{x}) \geq 0$. Conversely, for $Q(\boldsymbol{x}) = 0$, each individual linear form must vanish, and thus, because of the particular form, it follows necessarily that $x_k = 0$ $(k = 1, 2, 3, 4)$. The matrix A is thus positive definite.

Example 1-10. The matrix

$$A = \begin{bmatrix} 1 & -1 & 0 \\ -1 & 1 & 0 \\ 0 & 0 & 1 \end{bmatrix} \qquad \begin{aligned} Q(\boldsymbol{x}) &= x_1^2 - 2x_1x_2 + x_2^2 + x_3^2 \\ &= (x_1 - x_2)^2 + x_3^2 \end{aligned} \tag{1–52}$$

is not positive definite, despite the indicated grouping into a sum of perfect squares. In fact, whenever the values of the variables x_1 and x_2 are identical and $x_3 = 0$, $Q(\boldsymbol{x}) = 0$. The quadratic form can vanish without all variables x_k vanishing. Since, through the grouping (1–52), it clearly cannot take on negative values, it is called *positive semidefinite*.

1–3–1–2. NECESSARY CONDITIONS

Theorem 1-2. *A positive definite matrix must have nonvanishing positive diagonal elements.*

Proof: The quadratic form of a positive definite matrix takes on a positive value for any choice of variables $x_i \not\equiv 0$. Specifically, it must follow for $x_k = 1$, $x_i = 0$ $(i \neq k)$. The quadratic form takes the value $Q(\boldsymbol{x}) = a_{kk} > 0$, and the desired criterion follows.

A matrix with one or more vanishing or negative diagonal elements cannot be positive definite.

Theorem 1-3. *The elements of a positive definite matrix $A = (a_{ik})$ must fulfill the relation*

$$a_{ik}^2 < a_{ii}a_{kk} \quad \text{for all} \quad i \neq k.$$

Proof: For two arbitrary but differing indices $i \neq k$, choose x_i arbitrarily, $x_k = 1$ and $x_j = 0$ for $j \neq i, k$. The value of the quadratic form $Q(\boldsymbol{x}) = a_{ii}x_i^2 + 2a_{ik}x_i + a_{kk}$ must be positive for all x_i. The quadratic equation $a_{ii}x_i^2 + 2a_{ik}x_i + a_{kk} = 0$ cannot have real solutions of x_i; in other words, the discriminant $a_{ik}^2 - a_{ii}a_{kk}$ must necessarily be negative.

The matrix (1–52) does not fulfill the necessary condition of Theorem 1-3 and thus cannot be positive definite.

Theorem 1-4. *The largest element in absolute value of a positive definite matrix A must lie in the diagonal.*

Proof: The converse statement, that the largest element in absolute value of a positive definite matrix lies outside the diagonal, is contradictory to Theorem 1-3.

Naturally, these three criteria are in no way sufficient. For instance, the matrix

$$A = \begin{bmatrix} 3 & 2 & -2 \\ 2 & 3 & 2 \\ -2 & 2 & 3 \end{bmatrix} \quad \text{with} \quad \begin{aligned} Q(\boldsymbol{x}) = 3x_1^2 + 4x_1x_2 &- 4x_1x_3 \\ + 3x_2^2 &+ 4x_2x_3 \\ &+ 3x_3^2 \end{aligned}$$

fulfills all three criteria without being positive definite. For $\boldsymbol{x}_i = (1, -1, 1)^{\mathrm{T}}$, $Q(\boldsymbol{x}) = -3$.

1-3-2 Criterion of Dominant Positive Diagonal Elements

In contrast to 1–3–1–2, a sufficient, but not necessary, condition for positive definiteness can be given as follows.

Definition 1–5

A matrix $\boldsymbol{A}$ *is called strongly diagonal dominant if in every row the absolute value of the diagonal element is greater than the sum of the absolute values of the nondiagonal elements, i.e.,*

$$|a_{ii}| > \sum_{\substack{k=1 \\ k \neq i}}^{n} |a_{ik}| \quad \text{for} \quad i = 1, 2, \ldots, n. \tag{1–53}$$

A matrix is called *weakly diagonal dominant* if one or more but not all of the inequalities in (1–53) are equalities.

In the following definition, alterations of a matrix are permitted in which rows are permuted, and the corresponding columns undergo the identical permutation.

Definition 1–6

A matrix $\boldsymbol{A}$ *is termed reducible if by a simultaneous row and column permutation it can be brought into a form*

$$\begin{bmatrix} \boldsymbol{U} & \boldsymbol{V} \\ \boldsymbol{O} & \boldsymbol{W} \end{bmatrix}$$

where $\boldsymbol{U}$ *and* $\boldsymbol{W}$ *signify square matrices,* $\boldsymbol{V}$ *a rectangular matrix, and* $\boldsymbol{O}$ *a rectangular zero matrix. Otherwise, it is termed irreducible.*

Since the simultaneous permutation of rows and columns leaves the matrix symmetry invariant, reducibility in a symmetric matrix means that it breaks down into submatrices such that all of the nonzero elements are contained in square submatrices along the diagonal.

This readily visualized definition of an irreducible matrix is equivalent with

Definition 1–7

Let $W = \{1, 2, \ldots, n\}$ be the set of whole numbers from 1 to n, and let S and T be two arbitrary, nonempty, disjoint subsets of W such that

$$S \cup T = W, \qquad S \cap T = \emptyset, \qquad S \neq \emptyset, \qquad T \neq \emptyset. \tag{1–54}$$

*A matrix **A** is reducible if a separation of W into S and T [consistent with (1–54)] is possible such that $a_{ij} = 0$ for all $i \in S$ and $j \in T$. A matrix **A** is irreducible if for every separation of W into S and T [by (1–54)], there is always some element $a_{ij} \neq 0$ such that $i \in S$ and $j \in T$.*

Theorem 1-5. *An irreducible symmetric matrix that is weakly diagonal dominant and has positive diagonal elements is positive definite.*

Given: 1. $\boldsymbol{A} = (a_{ik})$ is symmetric and irreducible.

2. $a_{ii} \geq \sum_{\substack{k=1\\k\neq i}}^{n} |a_{ik}|$ with $i = 1, 2, \ldots, n$, but unequal for at least one value of i.

Assertion: $\boldsymbol{A}$ is positive definite.

Proof: From the second premise, it follows for an arbitrary real vector $\boldsymbol{x}$ that

$$\sum_{i=1}^{n} a_{ii}x_i^2 \geq \sum_{i=1}^{n}\left(\sum_{\substack{k=1\\k\neq i}}^{n} |a_{ik}|\right)x_i^2. \tag{1–55}$$

The quadratic form $Q(\boldsymbol{x}) = (\boldsymbol{A}\boldsymbol{x}, \boldsymbol{x})$ corresponding to matrix $\boldsymbol{A}$ will first be suitably broken down and then compared to two quantities, using (1–55):

$$\begin{aligned} Q(\boldsymbol{x}) &= \sum_{i=1}^{n} a_{ii}x_i^2 + \sum_{i=1}^{n}\sum_{\substack{k=1\\k\neq i}}^{n} a_{ik}x_i x_k \\ &\geq \sum_{i=1}^{n} a_{ii}x_i^2 - \sum_{i=1}^{n}\sum_{\substack{k=1\\k\neq i}}^{n} |a_{ik}||x_i||x_k| \\ &\geq \sum_{i=1}^{n}\left(\sum_{\substack{k=1\\k\neq i}}^{n} |a_{ik}|\right)|x_i|^2 - \sum_{i=1}^{n}\sum_{\substack{k=1\\k\neq i}}^{n} |a_{ik}||x_i||x_k|. \end{aligned}$$

We end up with the inequality

$$Q(\boldsymbol{x}) \geq \sum_{i=1}^{n}\sum_{\substack{k=1\\k\neq i}}^{n} |a_{ik}||x_i|(|x_i| - |x_k|). \tag{1–56}$$

Because of the symmetry of A, (1–56) yields

$$Q(x) \geq \sum_{i=1}^{n} \sum_{\substack{k=1 \\ k \neq i}}^{n} |a_{ik}| |x_k| (|x_k| - |x_i|). \tag{1–57}$$

Now, $Q(x)$ will certainly be greater than or equal to the arithmetic mean of the two quantities given in (1–56) and (1–57):

$$Q(x) \geq \frac{1}{2} \sum_{i=1}^{n} \sum_{\substack{k=1 \\ k \neq i}}^{n} |a_{ik}| (|x_i| - |x_k|)^2 \geq 0. \tag{1–58}$$

In light of (1–58), the value of $Q(x)$ for an arbitrary real vector x is nonnegative. So far, only the symmetry and the weak diagonal dominance have been exploited. Now examine the conditions whereby $Q(x)$ can vanish. Then the nonnegative lower bound of (1–58) must vanish. This can occur in three cases:

(a) $a_{ik} = 0$ for all $i \neq k$; $i, k = 1, 2, \ldots, n$. The given matrix A must then be a diagonal matrix and thus reducible. This must be excluded, as it fails to fulfill the given premises.

(b) All components of the vector have the same magnitude and take on, in the only interesting (nonzero) case, the values $|x_i| = |x_k|$ for all $i, k = 1, 2, \ldots, n$. By the second premise, the equality cannot apply for all index values i; thus (1–55) and consequently (1–56) and (1–57) become genuine inequalities. Therefore, by (1–58), if $x \neq 0$, $Q > 0$.

(c) Not all components of the vector have the same magnitude. Expression (1–58) vanishes only when $a_{ik} = 0$ for every pair of indices $i \neq k$ for which $|x_i| \neq |x_k|$. Let S be the set of index values i and j such that $|x_i| = |x_j| \neq 0$. Let T be the set of the first n remaining indices such that $|x_k| \neq |x_i|$ with $i \in S$ and $k \in T$. The two sets S and T fulfill the conditions (1–54), since each contains at least one element. Since $Q(x) = 0$, we must conclude in this case that $a_{ik} = 0$ for all $i \in S$ and $k \in T$. But this contradicts the irreducibility of matrix A.

A strongly diagonal dominant symmetric matrix with positive diagonal elements is positive definite. In fact, the irreducibility of the matrix is no longer necessary for this case since (1–55), and thus (1–56) and (1–57) as well, become strict inequalities so that, by (1–58), $Q > 0$ unconditionally for all $x \neq 0$.

The weak dominance of the positive diagonal elements of a symmetric, irreducible matrix is a sufficient condition for its being positive definite. Many applications, such as the discretization of a boundary value problem (see Chapter 5), involve weakly diagonal dominant symmetric matrices, for which Theorem 1-5 permits effortless identification of positive definiteness.

Example 1-11. The symmetric irreducible matrix

$$A = \begin{bmatrix} 2 & -1 & 0 & 0 \\ -1 & 2 & -1 & 0 \\ 0 & -1 & 2 & -1 \\ 0 & 0 & -1 & 2 \end{bmatrix}$$

fulfills the premises of Theorem 1-5 and is thus positive definite.

The case of positive, weakly dominant diagonal elements is sufficient for positive definiteness, but it is certainly not necessary. Section 1–3–3 will show that the matrix

$$A = \begin{bmatrix} 3 & 2 & 2 \\ 2 & 3 & 2 \\ 2 & 2 & 3 \end{bmatrix}$$

is positive definite. It is certainly not diagonal dominant.

Ostrowski (Ref. 40) achieved a generalization of Theorem 1-5:

Theorem 1-6. *A symmetric matrix $A = (a_{ik})$ with positive diagonal elements whose elements fulfill the $\binom{n}{2}$ conditions*

$$a_{ii}a_{kk} > \left(\sum_{\substack{j=1 \\ j \neq i}}^{n} |a_{ij}|\right)\left(\sum_{\substack{j=1 \\ j \neq k}}^{n} |a_{kj}|\right), \qquad i \neq k, \qquad i, k = 1, 2, \ldots, n$$

is positive definite.

1-3-3 Systematic Reduction to a Sum of Squares

This section will develop a necessary and sufficient condition for the unmistakable numerical identification of the positive definiteness of a given matrix, using its quadratic form. In the quadratic form

$$Q(x) = \sum_{i=1}^{n} \sum_{k=1}^{n} a_{ik} x_i x_k, \tag{1–59}$$

for which the necessary conditions for positive definiteness are fulfilled, especially $a_{11} > 0$, all terms dependent on x_1, such as $a_{11}x_1^2$, $2a_{12}x_1x_2$, $2a_{13}x_1x_3, \ldots, 2a_{1n}x_1x_n$, may be separated out through the completion of squares so that (1–59) may be written as

$$\left.\begin{aligned} Q(x) &= \left(\sqrt{a_{11}}\, x_1 + \sum_{k=2}^{n} \frac{a_{1k}}{\sqrt{a_{11}}} x_k\right)^2 + \sum_{i=2}^{n} \sum_{k=2}^{n} a_{ik}^{(1)} x_i x_k \\ \text{with} \qquad a_{ik}^{(1)} &= a_{ik} - \frac{a_{1i}a_{1k}}{a_{11}}, \qquad i, k = 2, 3, \ldots, n. \end{aligned}\right\} \tag{1–60}$$

The symmetry of A is carried over to the elements $a_{ik}^{(1)}$ $(i, k = 2, 3, \ldots, n)$. The given quadratic form $Q(x)$, according to (1–60), is split into a sum of a perfect square of a linear form and a new quadratic form

$$Q^{(1)}(x) = \sum_{i=2}^{n} \sum_{k=2}^{n} a_{ik}^{(1)} x_i x_k \tag{1–61}$$

in the $(n - 1)$ variables $x_2, x_3, \ldots, x_n$.

Theorem 1-7. *The matrix* $A = (a_{ik})$ *with* $a_{11} > 0$ *is positive definite if and only if after the reduction* (1–60), *the matrix* $A^{(1)} = (a_{ik}^{(1)})$, *having* $(n - 1)$ *rows with* $i, k = 2, 3, \ldots, n$, *is positive definite.*

Proof: (a) That it is necessary: Let A be positive definite. Assume that $A^{(1)}$ is not positive definite. Then there exist values for $x_2, x_3, \ldots, x_n$ not all of which vanish such that $Q^{(1)}(x) = 0$. If

$$x_1 = -\sum_{k=2}^{n} \frac{a_{1k}}{a_{11}} x_k, \tag{1–62}$$

$Q(x)$ also vanishes, which leads to a contradiction. Thus the matrix $A^{(1)}$ must necessarily be positive definite.

(b) That it is sufficient: Let $A^{(1)}$ be positive definite. $Q(x) = 0$ requires that both terms of the sum in (1–60) vanish. $Q^{(1)}(x) = 0$ implies $x_2 = x_3 = \cdots = x_n = 0$ because of the supposition, and thus the first term of the sum is zero only for $x_1 = 0$. Thus $Q(x)$ must necessarily be positive definite.

Establishing the definiteness of a given matrix A is reduced through Theorem 1-7 to the same problem for $A^{(1)}$, whence the order is diminished by one. In the event that $a_{22}^{(1)} > 0$, $Q^{(1)}(x)$ can be further subdivided through the separation of all terms containing x_2:

$$\left.\begin{aligned} Q^{(1)}(x) &= \left(\sqrt{a_{22}^{(1)}}\, x_2 + \sum_{k=3}^{n} \frac{a_{2k}^{(1)}}{\sqrt{a_{22}^{(1)}}} x_k\right)^2 + Q^{(2)}(x) \\ Q^{(2)}(x) &= \sum_{i=3}^{n} \sum_{k=3}^{n} a_{ik}^{(2)} x_i x_k, \quad a_{ik}^{(2)} = a_{ik}^{(1)} - \frac{a_{2i}^{(1)} a_{2k}^{(1)}}{a_{22}^{(1)}} \end{aligned}\right\} \tag{1–63}$$

By continuing this method in a consistent fashion, two possibilities may arise:

(a) After carrying out the jth step $(j = 1, 2, \ldots, n - 1)$, we find $a_{j+1,j+1}^{(j)} \leq 0$. Then

$$Q^{(j)}(x) = \sum_{i=j+1}^{n} \sum_{k=j+1}^{n} a_{ik}^{(j)} x_i x_k$$

is not positive definite, and thus neither is $Q(x)$ by Theorem 1-7.

(b) We carry out $n - 1$ steps with positive $a_{jj}^{(j-1)}$ $(j = 1, 2, \ldots, n - 1)$, leading to $Q^{(n-1)}(\boldsymbol{x}) = a_{nn}^{(n-1)}x_n^2$ with $a_{nn}^{(n-1)} > 0$. Then $Q^{(n-1)}(\boldsymbol{x})$ is positive definite, and by Theorem 1-7 so is $Q(\boldsymbol{x})$ also.

For the sake of uniformity, the last quadratic form $Q^{(n-1)}(\boldsymbol{x})$ may be written as a perfect square $(\sqrt{a_{nn}^{(n-1)}}\, x_n)^2$. This method yields the reduction of a positive definite form to a sum of squares of linear forms. The decomposition cannot be completed by using only real numbers if some radicand is less than or equal to zero. Then the quadratic form is not positive definite. By way of summation, the following theorem may be formulated.

Theorem 1-8. *A quadratic form in n variables is positive definite if and only if all square roots appearing in the process of reduction to a sum of squares are real and positive, that is, if and only if all n radicands are greater than zero.*

Example 1-12. The positive definiteness of the matrix

$$A = \begin{bmatrix} 3 & 2 & 2 \\ 2 & 3 & 2 \\ 2 & 2 & 3 \end{bmatrix} \quad \text{with} \quad \begin{aligned} Q(\boldsymbol{x}) = 3x_1^2 &+ 4x_1x_2 + 4x_1x_3 \\ &+ 3x_2^2 \quad + 4x_2x_3 \\ &\qquad\qquad + 3x_3^2 \end{aligned}$$

is evident because it can be decomposed into a sum of squares with positive radicands, as follows:

$$Q(\boldsymbol{x}) = \left(\sqrt{3}\,x_1 + \frac{2}{\sqrt{3}}x_2 + \frac{2}{\sqrt{3}}x_3\right)^2 + Q^{(1)}(\boldsymbol{x})$$

$$Q^{(1)}(\boldsymbol{x}) = \frac{5}{3}x_2^2 + \frac{4}{3}x_2x_3 + \frac{5}{3}x_3^2 \qquad A^{(1)} = \begin{bmatrix} \frac{5}{3} & \frac{2}{3} \\ \frac{2}{3} & \frac{5}{3} \end{bmatrix}$$

$$= \left(\sqrt{\frac{5}{3}}x_2 + \frac{2}{\sqrt{15}}x_3\right)^2 + Q^{(2)}(\boldsymbol{x})$$

$$Q^{(2)}(\boldsymbol{x}) = \frac{7}{5}x_3^2 = \left(\sqrt{\frac{7}{5}}x_3\right)^2 \qquad A^{(2)} = \left[\frac{7}{5}\right]$$

$$Q(\boldsymbol{x}) = \left(\sqrt{3}\,x_1 + \frac{2}{\sqrt{3}}x_2 + \frac{2}{\sqrt{3}}x_3\right)^2 + \left(\sqrt{\frac{5}{3}}x_2 + \frac{2}{\sqrt{15}}x_3\right)^2 + \left(\sqrt{\frac{7}{5}}x_3\right)^2$$

Example 1-13. Absence of definiteness in a matrix is reflected in the reduction process by a negative diagonal element in one of the reduced matrices:

$$A = \begin{bmatrix} 2 & 2 & -2 \\ 2 & 3 & 2 \\ -2 & 2 & 3 \end{bmatrix}, \qquad A^{(1)} = \begin{bmatrix} \frac{5}{3} & \frac{10}{3} \\ \frac{10}{3} & \frac{5}{3} \end{bmatrix}, \qquad A^{(2)} = [-5], \qquad a_{33}^{(2)} < 0.$$

The absence of definiteness in A is already discernible in $A^{(1)}$, since the largest element (in absolute value) does not lie on the diagonal.

PROBLEMS

1-11. Why are the following matrices clearly not positive definite?

(a)
$$A_1 = \begin{bmatrix} 11 & 7 & 1 \\ 7 & -2 & 3 \\ 1 & 3 & 10 \end{bmatrix},$$
(b)
$$A_2 = \begin{bmatrix} 1 & 3 & -4 & 6 \\ 3 & 4 & 6 & -5 \\ -4 & 6 & 5 & 4 \\ 6 & -5 & 4 & 3 \end{bmatrix},$$
(c)
$$A_3 = \begin{bmatrix} 1 & 2 & -3 & 0 \\ 2 & 5 & 2 & -4 \\ -3 & 2 & 6 & 3 \\ 0 & -4 & 3 & 5 \end{bmatrix}.$$

1-12. Using systematic reduction of the quadratic forms to a sum of squares, determine whether the following matrices are positive definite:

(a)
$$A_1 = \begin{bmatrix} 1 & 2 & 3 \\ 2 & 4 & 6 \\ 3 & 6 & 9 \end{bmatrix},$$
(b)
$$A_2 = \begin{bmatrix} 1 & 4 & 2 \\ 4 & 25 & 2 \\ 2 & 2 & 7 \end{bmatrix},$$
(c)
$$A_3 = \begin{bmatrix} 81 & -27 & 18 & -9 \\ -27 & 45 & 18 & 21 \\ 18 & 18 & 45 & 0 \\ -9 & 21 & 0 & 63 \end{bmatrix}.$$

1-13. Let A and B be two positive definite matrices. Prove that their sum is also a positive definite matrix.

1-14. Show that the $n \times n$ matrix A having elements $a_{ik} = 1/(i + k - 1)$ is positive definite. *Hint*: Note that $a_{ik} = \int_0^1 x^{i+k-2}\,dx$, and express the quadratic form for A as an integral.

1-4 SYMMETRIC TRIANGULAR DECOMPOSITION: CHOLESKY'S METHOD

The reduction of a positive definite quadratic form to a sum of squares of linear forms can be interpreted as a symmetric triangular decomposition of the corresponding positive definite matrix. By way of preparation, let us turn to some triangular matrices important in numerical analysis and list some of their properties.

1-4-1 Triangular Matrices

There are two classes of triangular matrices:

(a) *Lower Triangular Matrices.* All elements above the diagonal are zero. The name refers to the triangle of the matrix which, in general, contains nonzero elements.

$$L = \begin{bmatrix} l_{11} & 0 & 0 & \cdots & 0 \\ l_{21} & l_{22} & 0 & \cdots & 0 \\ l_{31} & l_{32} & l_{33} & \cdots & 0 \\ \cdot & \cdot & \cdot & \cdots & 0 \\ \cdot & \cdot & \cdot & \cdots & 0 \\ l_{n1} & l_{n2} & l_{n3} & \cdots & l_{nn} \end{bmatrix} \tag{1-64}$$

(b) *Upper Triangular Matrices.* Upper triangular matrices are the transposes of lower triangular matrices. All the elements below the diagonal vanish.

$$R = \begin{bmatrix} r_{11} & r_{12} & r_{13} & \cdots & r_{1n} \\ 0 & r_{22} & r_{23} & \cdots & r_{2n} \\ 0 & 0 & r_{33} & \cdots & r_{3n} \\ \cdot & \cdot & \cdot & \cdots & \cdot \\ \cdot & \cdot & \cdot & \cdots & \cdot \\ 0 & 0 & 0 & \cdots & r_{nn} \end{bmatrix} \tag{1-65}$$

Diagonal matrices are special cases belonging to both classes.

Theorem 1-9. *The determinant of a triangular matrix is the product of its diagonal elements.*

This statement is a direct consequence of the definition of a determinant. A direct consequence of Theorem 1-9 is:

Theorem 1-10. *A triangular matrix is nonsingular if and only if all its diagonal elements are nonzero.*

Theorem 1-11. *The set of nonsingular lower (or upper) triangular matrices of fixed order n forms a multiplicative group.*

Proof: The demonstration will be confined to nonsingular lower triangular matrices, as the proof for upper triangular matrices is analogous.

1. The product matrix $C = AB$ of two nonsingular lower triangular matrices $A = (a_{ik})$ and $B = (b_{ik})$ of order n is also a nonsingular lower triangular matrix. An arbitrary element c_{ik} of the product matrix C above the diagonal $(i < k)$ is

$$c_{ik} = \sum_{j=1}^{n} a_{ij} b_{jk} = \sum_{j=1}^{i} a_{ij} b_{jk} = 0,$$

since $a_{ij} = 0$ for $j > i$ and $b_{jk} = 0$ for $j < k$. For a diagonal element, $c_{kk} = a_{kk} b_{kk}$ $(k = 1, 2, \ldots, n)$; thus by Theorem 1-10, nonsingularity is assured.

2. The identity matrix $\boldsymbol{I}$ is a multiplicative element of unity. As a diagonal matrix with nonvanishing diagonal elements, it belongs to the set of nonsingular lower triangular matrices.

3. For every lower triangular matrix $\boldsymbol{A} = (a_{ik})$ there exists an inverse $\boldsymbol{B} = (b_{jk})$, which is itself a lower triangular matrix. The existence of an inverse is guaranteed by nonsingularity. We need now show only that $b_{jk} = 0$ for $j < k$ follows from

$$a_{ij} = 0 \quad \text{for} \quad i < j \quad \text{and} \quad \sum_{j=1}^{n} a_{ij} b_{jk} = \begin{cases} 1 & \text{for} \quad i = k \\ 0 & \text{for} \quad i \neq k \end{cases}. \tag{1–66}$$

This will be shown by complete induction on the row index of the inverse $\boldsymbol{B}$.

Induction premise: $b_{jk} = 0$ for $i < k$ and $i = 1, 2, \ldots, m$, where $m < n$.

Induction assertion: $b_{m+1,\,k} = 0$ for $k = m + 2,\ m + 3, \ldots, n$.

Induction proof: Statement (1–66) and the inductive premise yield, for $i = m + 1$ and $k > i$,

$$\sum_{j=1}^{n} a_{m+1,j} b_{jk} = \sum_{j=1}^{m+1} a_{m+1,j} b_{jk} = a_{m+1,\,m+1} b_{m+1,\,k} = 0.$$

The nonsingularity of $\boldsymbol{A}$ then leads to the inductive assertion.

Induction check: Statement (1–66) yields directly, for $i = 1$,

$$b_{11} = 1/a_{11} \quad \text{and} \quad b_{1k} = 0 \quad \text{for} \quad k = 2, 3, \ldots, n.$$

The proof of the group property is thus complete.

Theorem 1-12. *The inversion of a nonsingular lower (or upper) triangular matrix is an explicit process.*

Proof: Theorem 1-11 leads to the following equations for determining the nonvanishing elements b_{ik} of the inverse $\boldsymbol{B} = (b_{ik})$ of a nonsingular lower triangular matrix $\boldsymbol{A} = (a_{ik})$, if we progress row by row:

$$\left.\begin{aligned} i = 1&: \quad a_{11} b_{11} = 1, \\ i = 2&: \quad a_{21} d_{11} + a_{22} b_{21} = 0, \quad a_{22} b_{22} = 1, \\ i = 3&: \quad a_{31} b_{11} + a_{32} b_{21} + a_{33} b_{31} = 0, \\ &\qquad a_{32} b_{22} + a_{33} b_{32} = 0, \quad a_{33} b_{33} = 1, \\ \text{etc.}& \end{aligned}\right\} \tag{1–67}$$

Equations (1–67) can be successively solved for the unknowns $b_{11}, b_{21}, b_{22}, b_{31}, b_{32}$, etc. The required divisions by a_{kk} can be carried out for nonsingular lower triangular matrices. In general, the formulas will have the form

$$\left.\begin{aligned} b_{ik} &= -\left(\sum_{j=k}^{i-1} a_{ij} b_{jk}\right) \Big/ a_{ii}, \quad i > k \\ b_{ii} &= 1/a_{ii} \end{aligned}\right\} \quad i = 1, 2, \ldots, n. \tag{1–68}$$

The analogous arguments for upper triangular matrices will be left to the reader.

The computation of the inverse of a nonsingular triangular matrix is important for numerical applications. The complete process can be given as an ALGOL procedure (Refs. 2, 3, 4, 12, 37, 39, 50, 55).[1]

ALGOL Procedure for Inversion of a Nonsingular Lower Triangular Matrix. The procedure parameters are defined as:

- *n* Order of matrix ***A***
- *a* Elements of the nonsingular lower triangular matrix ***A***
- *b* Elements of the inverse of ***A***.

ALGOL Procedure No. 1

```
procedure linverse (n, a, b);
          value n;  integer n;  array a, b;
begin integer  i, j, k;  real s;
   for i: = 1  step 1 until n do
   begin comment calculating the ith row of B;
      for k: = 1 step 1 until i-1 do
      begin s: = 0;
         for j: = k step 1 until i—1 do
           s: = s + a[i, j] × b[j, k];
         b[i, k]: = −s/a[i, i]
      end k;
      b[i, i]: = 1/a[i, i]
   end i
end linverse
```

Note that when $i = 1$, the loop involving k is empty, and no computation is to be done. Only the elements of $\boldsymbol{A}$ in the lower triangle need be given, and only the corresponding elements of $\boldsymbol{B}$ are found.

Observe further that in the calculation sequence for b_{ik}, element a_{ik} is only used until the corresponding b_{ik} with the matching index value is calculated. If matrix $\boldsymbol{A}$ is not needed after inversion, one may identify b with a in the procedure and delete b from the procedure heading.

1-4-2 Cholesky's Method

The reduction of a positive definite quadratic form to a sum of squares as in 1–3–3 yields the representation

$$Q(\boldsymbol{x}) = \sum_{i=1}^{n} \sum_{k=1}^{n} a_{ik}^{(0)} x_i x_k = \sum_{i=1}^{n} \left(\sqrt{a_{ii}^{(i-1)}}\, x_i + \sum_{k=i+1}^{n} \frac{a_{ik}^{(i-1)}}{\sqrt{a_{ii}^{(i-1)}}} x_k \right)^2. \qquad (1\text{–}69)$$

[1] FORTRAN counterparts of all ALGOL subprograms appearing in this book are given in the Appendix.

In the decomposition (1–69) there appear n linear forms

$$y_i = \sum_{k=i}^{n} r_{ik} x_k, \qquad i = 1, 2, \ldots, n \tag{1–70}$$

whose coefficients r_{ik} are defined

$$r_{ii} = \sqrt{a_{ii}^{(i-1)}} \quad \text{and} \quad r_{ik} = \frac{a_{ik}^{(i-1)}}{\sqrt{a_{ii}^{(i-1)}}}, \qquad k > i. \tag{1–71}$$

By this formula, the values of r_{ik} are given only for $k \geq i$. If we set $r_{ik} = 0$ for $k < i$, the coefficients can be collected in an upper triangular matrix $\boldsymbol{R}$ that defines a linear transformation $\boldsymbol{y} = \boldsymbol{R}\boldsymbol{x}$. In combination with (1–69) and (1–70), we arrive in sequence at the equations

$$\begin{aligned} Q(\boldsymbol{x}) &= (\boldsymbol{A}\boldsymbol{x}, \boldsymbol{x}) = \sum_{i=1}^{n} \left(\sum_{k=i}^{n} r_{ik} x_k \right)^2 = \sum_{i=1}^{n} y_i^2 = (\boldsymbol{y}, \boldsymbol{y}) \\ &= (\boldsymbol{R}\boldsymbol{x}, \boldsymbol{R}\boldsymbol{x}) = (\boldsymbol{R}^{\mathrm{T}}\boldsymbol{R}\boldsymbol{x}, \boldsymbol{x}). \end{aligned} \tag{1–72}$$

Because of the uniqueness of the representation, Eq. (1–72) yields the relation

$$\boldsymbol{A} = \boldsymbol{R}^{\mathrm{T}}\boldsymbol{R}. \tag{1–73}$$

Conclusion: The reduction of a positive definite quadratic form to a sum of squares yields the decomposition of the corresponding positive definite matrix $\boldsymbol{A}$ into the product of two triangular matrices that are transposes of each other, according to Eq. (1–73).

This technique is attributable to the geodetic surveyor Cholesky (Ref. 5), after whom it is usually named.

ALGOL Procedure for Cholesky's Method. With coefficients r_{ik} as given in (1–71), a generalization of (1–63) yields the reduction formulas for the transition from $\boldsymbol{A}^{(p-1)}$ to $\boldsymbol{A}^{(p)}$.

$$a_{ik}^{(p)} = a_{ik}^{(p-1)} - r_{pi} r_{pk} \begin{cases} i, k = p+1, p+2, \ldots, n; \\ \quad p = 1, 2, \ldots, n-1. \end{cases} \tag{1–74}$$

If we observe that for fixed i and k only the last one of the sequence $a_{ik}^{(0)}$, $a_{ik}^{(1)}, \ldots, a_{ik}^{(p)}$ is of importance, the upper index may be omitted on the understanding that a_{ik} indicates the current computed value. Under this observation, the given matrix is constantly altered. Through symmetry considerations, $\boldsymbol{A}$ need be given only as an upper triangular matrix, and the reduction need be carried out only on or above the diagonal. These conclusions are incorporated into the procedure *cholesky* that follows.

The parameters are defined as:

n	Order of the matrix $\boldsymbol{A}$
a	Elements of matrix $\boldsymbol{A}$
r	Elements of the upper triangular matrix $\boldsymbol{R}$, $\boldsymbol{A} = \boldsymbol{R}^T\boldsymbol{R}$
indef	Exit in the event that $\boldsymbol{R}$ is not positive definite.

ALGOL Procedure No. 2

```
procedure cholesky (n, a, r, indef);
        value n;  integer n;  array a, r;  label indef;
begin integer i, k, p;
    for p: = 1 step 1 until n do
    begin if a[p, p] ≤ 0 then goto indef;
        comment solving for r[p, k];
        r[p, p]: = sqrt(a[p, p]);
        for k: = p + 1 step 1 until n do
            r[p, k]: = a[p, k]/r[p, p];
        comment reduction of the elements a[i, k];
        for i: = p + 1 step 1 until n do
            for k: = i step 1 until n do
                a[i, k]: = a[i, k] − r[p, i] × r[p, k]
    end p
end cholesky
```

Note: In procedure *cholesky* it is possible to identify the elements r_{ik} with a_{ik} such that by appropriate modification following completion of the program, the elements of $\boldsymbol{A}$ contain those of $\boldsymbol{R}$. The given matrix $\boldsymbol{A}$ is destroyed in any case.

1-4-3 Solution of Symmetric Definite Systems of Equations

A linear system of n equations in n unknowns x_k

$$\boldsymbol{A}\boldsymbol{x} + \boldsymbol{b} = 0 \quad \text{or} \quad \sum_{k=1}^{n} a_{ik}x_k + b_i = 0 \qquad i = 1, 2, \ldots, n \tag{1–75}$$

is called *symmetric definite* if matrix $\boldsymbol{A}$ in the system is both symmetric and positive definite. For the solution of such systems of equations, Cholesky's method is very suitable. With the decomposition $\boldsymbol{A} = \boldsymbol{R}^T\boldsymbol{R}$, (1–75) becomes

$$\boldsymbol{R}^T\boldsymbol{R}\boldsymbol{x} + \boldsymbol{b} = 0, \quad \text{or} \quad \boldsymbol{R}^T(\boldsymbol{R}\boldsymbol{x}) + \boldsymbol{b} = 0. \tag{1–76}$$

With the auxiliary vector $\boldsymbol{y} = \boldsymbol{R}\boldsymbol{x}$, the solution of (1–75) is equivalent to the problem of solving the two systems of equations (1–77) and (1–78) in succession:

$$R^{T}y + b = 0 \quad \text{for } y\text{, with } b \text{ given.} \tag{1–77}$$

$$Rx - y = 0 \quad \text{for } x\text{, with } y \text{ given.} \tag{1–78}$$

Because of the triangular form of matrix R, the two systems look like this in detail:

$$\left.\begin{array}{l} r_{11}y_1 \qquad\qquad\qquad\qquad\qquad\quad + b_1 = 0 \\ r_{12}y_1 + r_{22}y_2 \qquad\qquad\qquad\qquad + b_2 = 0 \\ r_{13}y_1 + r_{23}y_2 + r_{33}y_3 \qquad\qquad + b_3 = 0 \\ \cdot \qquad \cdot \qquad \cdot \qquad\qquad\qquad\qquad \cdot \quad \cdot \\ r_{1n}y_1 + r_{2n}y_2 + r_{3n}y_3 + \cdots + r_{nn}y_n + b_n = 0 \end{array}\right\} \tag{1–79}$$

$$\left.\begin{array}{r} r_{11}x_1 + r_{12}x_2 + r_{13}x_3 + \cdots + r_{1n}x_n - y_1 = 0 \\ r_{22}x_2 + r_{23}x_3 + \cdots + r_{2n}x_n - y_2 = 0 \\ r_{33}x_3 + \cdots + r_{3n}x_n - y_3 = 0 \\ \cdots \quad \cdot \quad \cdot \quad \cdot \\ r_{nn}x_n - y_n = 0 \end{array}\right\} \tag{1–80}$$

In consequence of the *cholesky* decomposition, the diagonal elements of R are greater than zero for a symmetric definite system of Eqs. (1–75). For given values of b_i, (1–79) yields the auxiliary unknowns y_k in *ascending* order by the explicit formulas

$$y_k = -\left(b_k + \sum_{i=1}^{k-1} r_{ik}y_i\right)\Big/ r_{kk}, \qquad k = 1, 2, \ldots, n. \tag{1–81}$$

For $k = 1$ in (1–81), the sum is taken to be zero. Analogously, for the computed values of y_k, (1–80) yields the unknowns x_i, but this time in *descending* order in explicit fashion, according to the formulas

$$x_i = \left(y_i - \sum_{k=i+1}^{n} r_{ik}x_k\right)\Big/ r_{ii}, \qquad i = n, n-1, \ldots, 1. \tag{1–82}$$

In (1–82), clearly, the sum is taken to be zero when $i = n$.

These two independent processes, namely to find the auxiliary vector y and from it the solution vector x, are termed *forward* and *backward substitution* because of the sequence in which the equations are used and the respective unknowns solved for.[1]

[1]Cholesky's method for solution of symmetric definite systems of equations is just a modification of the Gauss algorithm (Refs. 64, 75) in which the elimination is carried out maintaining symmetry. But a linear system of equations is soluble without the extraction of square roots, as is well known. The square roots that appear stem from the reduction of the quadratic form to a sum of squares.

ALGOL Procedure for Forward and Backward Substitution. The Cholesky decomposition of the positive definite matrix $\boldsymbol{A} = \boldsymbol{R}^{\mathrm{T}}\boldsymbol{R}$ is assumed known.

The parameters are defined as:

n Order of the system of equations; number of unknowns
r Elements of matrix $\boldsymbol{R}$
b Elements of the constant vector in $\boldsymbol{Ax} + \boldsymbol{b} = 0$
x Elements of the solution vector.

ALGOL Procedure No. 3

```
procedure choleskysol (n, r, b, x);
          value n;  integer n;  array r, b, x;
begin integer i, k;  real s;  array y[1 : n];
    comment Forward substitution;
    for k: = 1 step 1 until n do
    begin s: = b[k];
        for i: = 1 step 1 until k—1 do
            s: = s + r[i, k] × y[i];
        y[k]: = −s/r[k, k]
    end k;
    comment Backward substitution;
    for i: = n step −1 until 1 do
    begin s: = y[i];
        for k: = i + 1 step 1 until n do
            s: = s − r[i, k] × x[k];
        x[i]: = s/r[i, i]
    end i
end choleskysol
```

Note: With this procedure it is possible, on one hand, to identify the elements y_k with b_k should the given constant vector be alterable. On the other hand, the elements x_i may be identified with y_i. If use is made of both of these facts, the vectors $\boldsymbol{x}$ and $\boldsymbol{y}$ may be eliminated so that the solution vector appears in place of the constant vector $\boldsymbol{b}$.

The process of forward and backward substitution does not alter the triangular matrix $\boldsymbol{R}$. Thus with the given decomposition $\boldsymbol{A} = \boldsymbol{R}^{\mathrm{T}}\boldsymbol{R}$, another system of equations with the same coefficient matrix $\boldsymbol{A}$ but differing constant vector $\boldsymbol{b}$ may be solved merely through forward and backward substitution. For this reason, the steps for solution of a symmetric definite system of equations were summarized as two separate procedures. One application thereof is in the following.

Subsequent Iteration of a Solution. As a consequence of the inevitable rounding off error in any numerical calculation and because of a possible disadvantageous condition of matrix $\boldsymbol{A}$ of the system of equations, we usually arrive at an approximate solution $\boldsymbol{x}'$ in place of the exact solution vector $\boldsymbol{x}$

in $\boldsymbol{Ax} + \boldsymbol{b} = 0$. With $\boldsymbol{x}'$, the system of equations is not fulfilled exactly. On the contrary, it satisfies

$$\boldsymbol{Ax}' + \boldsymbol{b} = \boldsymbol{r} \tag{1–83}$$

where $\boldsymbol{r}$ is a *residual vector* that is generally nonzero. The substitution $\boldsymbol{x} = \boldsymbol{x}' + \boldsymbol{d}$ for the solution $\boldsymbol{x}$ with the correction vector $\boldsymbol{d}$ leads to

$$\boldsymbol{Ax} + \boldsymbol{b} = \boldsymbol{Ax}' + \boldsymbol{Ad} + \boldsymbol{b} = \boldsymbol{Ad} + \boldsymbol{r} = 0, \tag{1–84}$$

in other words, to a system of equations for $\boldsymbol{d}$ with the same matrix $\boldsymbol{A}$ but with the residual vector $\boldsymbol{r}$ as the new constant vector. By the observation made above, the correction $\boldsymbol{d}$ can be found without much effort. We thus arrive at a more precise solution $\boldsymbol{x}'' = \boldsymbol{x}' + \boldsymbol{d}$, which again need not be the exact solution. In principle, the subsequent correction may be repeated. The use of higher numerical precision is required in the computation of the residual vector $\boldsymbol{r}$.

Example 1-14. We illustrate Cholesky's method and subsequent iteration of a solution by simulated use of a computer with six significant figures. The deviation of the approximate solution $\boldsymbol{x}'$ from the exact solution $\boldsymbol{x}$ will be related to the condition of the system of equations.

$$\boldsymbol{A} = \begin{bmatrix} 5 & 7 & 3 \\ 7 & 11 & 2 \\ 3 & 2 & 6 \end{bmatrix}, \quad \boldsymbol{b} = \begin{bmatrix} 0 \\ -1 \\ 0 \end{bmatrix}, \quad \boldsymbol{x} = \begin{bmatrix} -36 \\ 21 \\ 11 \end{bmatrix}$$

$$\boldsymbol{R} = \begin{bmatrix} 2.23607 & 3.13049 & 1.34164 \\ & 1.09546 & -2.00828 \\ & & 0.408429 \end{bmatrix}, \quad \boldsymbol{y} = \begin{bmatrix} 0 \\ 0.912859 \\ 4.48861 \end{bmatrix},$$

$$\boldsymbol{x}' = \begin{bmatrix} -35.9671 \\ 20.9809 \\ 10.9899 \end{bmatrix}.$$

The residual vector $\boldsymbol{r}$ and the solution improved through subsequent iteration are

$$\boldsymbol{r} = 10^{-4} \cdot \begin{bmatrix} 5 \\ 0 \\ -1 \end{bmatrix}, \quad \boldsymbol{y}_r = 10^{-4} \cdot \begin{bmatrix} -2.23670 \\ 6.39001 \\ 41.2138 \end{bmatrix},$$

$$\boldsymbol{d} = 10^{-4} \cdot \begin{bmatrix} -328.700 \\ 190.826 \\ 100.908 \end{bmatrix} = \begin{bmatrix} -0.03287 \\ 0.01908 \\ 0.01009 \end{bmatrix}, \quad \boldsymbol{x}'' = \boldsymbol{x}' + \boldsymbol{d} = \begin{bmatrix} 36.0000 \\ 21.0000 \\ 11.0000 \end{bmatrix}.$$

The second iterated solution $\boldsymbol{x}''$ agrees with the exact solution to six significant figures. Note the magnitude of the residual vector $\boldsymbol{r}$ and the correction vector $\boldsymbol{d}$. The magnitude of the residual vector by itself says nothing about the size of the error in the approximate solution $\boldsymbol{x}'$. For this, the condition number of matrix $\boldsymbol{A}$ is important. For the symmetric, positive definite matrix $\boldsymbol{A}$ with $\lambda_{\max} \cong 16.6622$ and $\lambda_{\min} \cong 0.0112$, the condition is about $\kappa \cong 1487$. By Sec. 1-2, an inaccuracy of magnitude of 0.1487 may be expected in the solution $\boldsymbol{x}'$; in other words, the four last significant figures cannot be guaranteed to be accurate. In actual fact, the error turned out to be about four times smaller. The same discussion is applicable to the correction vector $\boldsymbol{d}$. It is therefore clear why in this example $\boldsymbol{x}''$ is the exact solution to six significant figures. With a worse condition number, correspondingly more iterations are needed to arrive at a desired precision of solution (see Ref. 81).

1-4-4 Inversion of a Positive Definite Matrix

With the Cholesky decomposition of a symmetric, positive definite matrix $\boldsymbol{A} = \boldsymbol{R}^{\mathrm{T}}\boldsymbol{R}$, the inverse is given by

$$\boldsymbol{B} = \boldsymbol{A}^{-1} = (\boldsymbol{R}^{\mathrm{T}}\boldsymbol{R})^{-1} = \boldsymbol{R}^{-1}(\boldsymbol{R}^{\mathrm{T}})^{-1} = \boldsymbol{R}^{-1}(\boldsymbol{R}^{-1})^{\mathrm{T}}.$$

The inversion of $\boldsymbol{A}$ takes place essentially in three steps:

1. Cholesky decomposition of $\boldsymbol{A} = \boldsymbol{R}^{\mathrm{T}}\boldsymbol{R}$ by Sec. 1–4–2.
2. Inversion of the upper triangular matrix $\boldsymbol{R}$. By Sec. 1–4–1, this is an explicit process, and it yields an upper triangular matrix $\boldsymbol{R}^{-1} = \boldsymbol{S} = (s_{ik})$ with $s_{ik} = 0$ for $i > k$.
3. Computation of $\boldsymbol{B} = \boldsymbol{S}\boldsymbol{S}^{\mathrm{T}} = (b_{ik})$.

Like $\boldsymbol{A}$, $\boldsymbol{B}$ is symmetric. It is sufficient to compute its elements on and above the diagonal. Because of the triangular form of $\boldsymbol{S}$, it follows

$$b_{ik} = \sum_{j=k}^{n} s_{ij}s_{kj} \qquad i \leq k, \tag{1–85}$$

and for the diagonal elements of the inverse, it follows

$$b_{kk} = \sum_{j=k}^{n} s_{kj}^2 \qquad k = 1, 2, \ldots, n. \tag{1–86}$$

The diagonal elements of $\boldsymbol{B} = \boldsymbol{A}^{-1}$ can thus be computed from $\boldsymbol{S}$ independent of the remaining elements of $\boldsymbol{B}$. If we are concerned only with the diagonal elements or, more generally, with some particular elements of the inverse, the inversion method described offers a notable advantage.

Application: In unconstrained fitting of data (see Chapter 3), the average errors of the adjusted quantities are often required. These are found through the diagonal elements of the inverse of the matrix of the normal equations (see ref. 80), while the remaining elements of the inverse are not of direct interest in this context. The normal equations are symmetric definite (see Sec. 3–2); thus their solution is possible by Cholesky's method. In that case, the triangular matrix $\boldsymbol{R}$ must be calculated, and thus the desired diagonal elements of the inverse can readily be found without having to carry out an entire inversion.

1-4-5 Symmetric Definite Band Matrices

A matrix whose elements outside a band along the main diagonal vanish is called a *band matrix*. It is characterized by the existence of an integer $m < n$ such that

$$a_{ik} = 0 \quad \text{for all } i \text{ and } k \quad \text{with } |i - k| > m \tag{1–87}$$

The number m is the *band width*.

Example 1-15. Some band matrices with various band widths are:

1. $m = 0$, $\boldsymbol{A}$ is a *diagonal matrix*.
2. $m = 1$, $\boldsymbol{A}$ is a *tridiagonal* or *Jacobi matrix*. Its nonzero elements are only on the main diagonal and the immediately adjacent parallel lines; for instance,

$$\boldsymbol{A} = \begin{bmatrix} 2 & 1 & 0 & 0 \\ 1 & 3 & 1 & 0 \\ 0 & 1 & 3 & 1 \\ 0 & 0 & 1 & 2 \end{bmatrix}. \tag{1–88}$$

Tridiagonal matrices will turn up later in various contexts.

3. $m = 2$. Band matrices of this sort turn up in discretized problems of the bending of beams or their vibrations. The band matrices here consist of five neighboring diagonal lines of elements.

Let us now consider *symmetric definite band matrices*. For instance, (1–88) is, by Theorem 1–5, a symmetric definite band matrix.

Theorem 1-13. *The Cholesky decomposition of a symmetric definite band matrix* $\boldsymbol{A} = \boldsymbol{R}^{\mathrm{T}}\boldsymbol{R}$ *leaves the band width unchanged. For the upper diagonal matrix* $\boldsymbol{R} = (r_{ik})$, *it follows*

$$r_{ik} = 0 \qquad k - i < m, \tag{1–89}$$

where m *is the band width of* $\boldsymbol{A}$.

Proof: The assertion of the theorem follows directly for the first line of $\boldsymbol{R}$ by (1–71). Clearly, if $a_{1k} = 0$ for $k - 1 > m$, then $r_{1k} = a_{1k}/\sqrt{a_{11}} = 0$. It remains only to show that the band width remains unaltered by the reduction from $\boldsymbol{A}$ to $\boldsymbol{A}^{(1)}$. By (1–74),

$$a_{ik}^{(1)} = a_{ik} - r_{1i}r_{1k} \qquad i, k = 2, 3, \ldots, n.$$

Because of symmetry, a proof for elements above the diagonal suffices. For an arbitrary element $a_{ik}^{(1)}$ with $i \geq 2$ and $k - i > m$, $a_{ik} = 0$. Furthermore, since $k > m + i \geq m + 2$, $r_{1k} = 0$ and thus $a_{ik}^{(1)} = 0$ follows. A step-by-step continuation of this argument yields (1–89).

The fact that the band form of a symmetric definite matrix $\boldsymbol{A}$ is retained in a Cholesky decomposition to the matrix $\boldsymbol{R}$ simplifies the calculations considerably. Note, however, that the band width is increased in multiplying two band matrices and that the band property completely disappears in the process of inversion.

ALGOL Procedure for Cholesky Decomposition of Symmetric-Definite Band Matrices. In order to accommodate a later application (Sec. 4–6–4), the procedure is organized so that the matrix $\boldsymbol{A}$ is saved. This is done by defining $r_{ij} = a_{ij}$ initially, and the decomposition on r_{ij} is carried out so that, at the end, the resultant r_{ij} represent the elements of $\boldsymbol{R}$ (see note in Sec. 1–4–2).

The symmetry and band property are exploited through a special indexing of matrix elements; the nonzero elements on or above the diagonal are subjected to an index substitution as in Eq. (1–90).

$$a_{ik} \xrightarrow{k=i+j} a_{i,k-i} = a_{ij}, \qquad k \geq i \tag{1–90}$$

The nonzero diagonal lines are thereby transferred to the columns of a rectangular array in which, for the new indices, the first index i runs from 1 to n, and the second index j runs from 0 to m. The diagonal elements are now indexed a_{i0} ($i = 1, 2, \ldots, n$), the elements of the diagonal line just to the right a_{i1} ($i = 1, 2, \ldots, n - 1$), and so on to the mth diagonal line with a_{im} ($i = 1, 2, \ldots, n - m$). This rectangular array is not entirely occupied by matrix elements.

In order to simplify some of the loop instructions, the assumption is made that the elements a_{ij} for $n < i + j \leq n + m$, which lie outside the matrix and do not actually exist, are all zero. With all these conventions, a matrix of order $n = 6$ with $m = 2$ looks like the rectangular array below at right.

$$A = \begin{bmatrix} 32 & 16 & 8 & & & \\ 16 & 16 & 8 & 4 & & \\ 8 & 8 & 8 & 4 & 2 & \\ & 4 & 4 & 4 & 2 & 1 \\ & & 2 & 2 & 2 & 1 \\ & & & 1 & 1 & 1 \end{bmatrix} \qquad \boxed{\begin{matrix} 32 & 16 & 8 \\ 16 & 8 & 4 \\ 8 & 4 & 2 \\ 4 & 2 & 1 \\ 2 & 1 & 0 \\ 1 & 0 & 0 \end{matrix}}$$

The procedure parameters are:

a	Elements of the band matrix A with indexing as in (1–90)
n	Order of the matrix
m	Band width, as given in (1–87)
r	Elements of the matrix R in $A = R^T R$, using the indexing of (1–90)
indef	Exit for case of A not positive definite.

ALGOL Procedure No. 4

```
procedure choleskyband (a, n, m, r, indef);
          value n, m;  integer n, m;  array a, r;  label indef;
begin integer i, j, p, min;
   comment Initialize storage;
   for i: = 1 step 1 until n do
      for j: = 0 step 1 until m do
         r[i, j]: = a[i, j];
   comment Actual Cholesky decomposition with elements r[i, j];
for p: = 1 step 1 until n do
   begin if r[p, 0] ≤ 0 then goto indef;
      r[p, 0]: = sqrt(r[p, 0]);
      for j: = 1 step 1 until m do
         r[p, j]: = r[p, j]/r[p, 0];
      min: = if n ≤ p + m then n else p + m;
      for i: = p + 1 step 1 until min do
         for j: = 0 step 1 until m + p − i do
            r[i, j]: = r[i, j] − r[p, i − p] × r[p, i − p + j]
   end p
end choleskyband
```

In order to understand the process of reduction better, the schematic representation of Fig. 1 with $m = 4$ is handy. The significant elements of the band on and above the diagonal are indicated by a cross, and the extra elements outside the matrix by a circle.

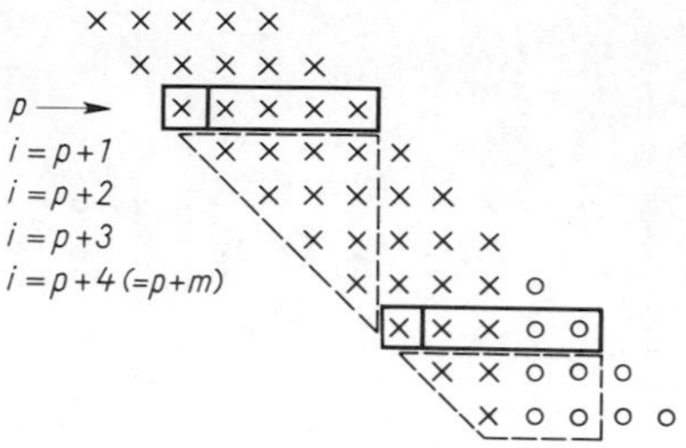

Fig. 1. For the reduction of a band matrix.

Because of the band character, the actual reduction affects only the elements within a triangular region that converts to a trapezoidal region if $p > n - m$. The reader may readily verify that the elements outside the matrix drawn into the computation retain their zero values.

The solution of a symmetric definite system of equations with a band matrix A using Sec. 1–4–3 is now trivial, and the formulation of the technique as an ALGOL procedure is now at hand.

PROBLEMS

1-15. What are the formulas for finding the inverse of the lower diagonal matrix given below?

$$L = \begin{bmatrix} a_1 & 0 & 0 & 0 \cdots 0 & 0 \\ b_1 & a_2 & 0 & 0 \cdots 0 & 0 \\ 0 & b_2 & a_3 & 0 \cdots 0 & 0 \\ 0 & 0 & b_3 & a_4 \cdots 0 & 0 \\ \cdot & \cdot & \cdot & \cdot \cdots \cdot & \\ 0 & 0 & 0 & 0 \quad a_{n-1} & 0 \\ 0 & 0 & 0 & 0 \quad b_{n-1} & a_n \end{bmatrix}$$

1-16. Develop a program for inverting a nonsingular upper triangular matrix. Begin by determining the explicit formulas for the successive computation of elements of the inverse.

1-17. Solve the system of type $Ax + b = 0$, with

$$\begin{aligned} 4x_1 - 2x_2 + 6x_3 - 10 &= 0 \\ -2x_1 + 17x_2 + x_3 - 39 &= 0 \\ 6x_1 + x_2 + 19x_3 - 53 &= 0 \end{aligned}$$

by the Cholesky method. Then find the diagonal elements of the inverse A^{-1}.

1-18. Solve the band equations given, using Cholesky's method.

$$\begin{aligned}
x_1 + 2x_2 - x_3 \qquad\qquad\qquad\qquad\qquad\qquad\qquad\qquad + 2 &= 0 \\
2x_1 + 8x_2 + 4x_3 - 2x_4 \qquad\qquad\qquad\qquad\qquad + 10 &= 0 \\
-x_1 + 4x_2 + 19x_3 + 9x_4 - 3x_5 \qquad\qquad\qquad - 35 &= 0 \\
-2x_2 + 9x_3 + 26x_4 + 5x_5 - 3x_6 - 47 &= 0 \\
-3x_3 + 5x_4 + 14x_5 + x_6 + 34 &= 0 \\
-3x_4 + x_5 + 6x_6 + 3 &= 0
\end{aligned}$$

2 RELAXATION METHODS

Many boundary value problems of mathematical physics may be directly formulated as variational problems (using *energy methods*), yielding through discretization a quadratic function F which must be minimized or maximized (see refs. 11, 13, 17). Through differentiation of F, this requirement yields a system of linear equations. The physical significance of the quadratic function as energy means that as a rule the homogeneous quadratic portion represents a positive definite quadratic form. For this reason, the resultant system of equations is symmetric definite. The solution of symmetric definite systems of equations thus constitutes a basic step in a large class of problems, and it behooves us to examine it thoroughly.

Section 1–4–3 described a direct method of solution, using the Cholesky technique. In contrast, the system of equations will now be solved iteratively by *relaxation*, in which an approximation will be progressively improved.

2-1 FUNDAMENTALS IN RELAXATION CALCULATIONS

2-1-1 The Symmetric Definite System of Equations as a Minimum Problem

We are given a symmetric definite system of equations to be solved for the vector $\boldsymbol{x}$:

$$\boldsymbol{A}\boldsymbol{x} + \boldsymbol{b} = 0, \qquad \sum_{k=1}^{n} a_{ik}x_k + b_i = 0 \qquad i = 1, 2, \ldots, n. \tag{2–1}$$

If we substitute any arbitrary *trial vector* $\boldsymbol{v}$ for $\boldsymbol{x}$ in (2–1), we arrive at a *residual vector* $\boldsymbol{r} = \boldsymbol{A}\boldsymbol{v} + \boldsymbol{b}$ with the components $r_1, r_2, \ldots, r_n$. The purpose of any relaxation method is to alter the trial vector $\boldsymbol{v}$ systematically in such a way that the residuals eventually disappear.

Together with the system of equations (2–1), consider the quadratic function

$$F(\boldsymbol{v}) = \frac{1}{2}\sum_{i=1}^{n}\sum_{k=1}^{n} a_{ik}v_i v_k + \sum_{i=1}^{n} b_i v_i = \frac{1}{2}(\boldsymbol{A}\boldsymbol{v}, \boldsymbol{v}) + (\boldsymbol{b}, \boldsymbol{v}). \tag{2–2}$$

By assumption, the quadratic form

$$Q(\boldsymbol{v}) = \sum_{i=1}^{n}\sum_{k=1}^{n} a_{ik}v_i v_k = (\boldsymbol{A}\boldsymbol{v}, \boldsymbol{v}) \tag{2–3}$$

is positive definite. Next take the partial derivative of the quadratic form $F(\boldsymbol{v})$ with respect to v_i:

$$\frac{\partial F}{\partial v_i} = \sum_{k=1}^{n} a_{ik}v_k + b_i. \tag{2–4}$$

This is the ith component of the residual vector $\boldsymbol{r}$ for the trial vector $\boldsymbol{v}$. Combining yields (2–5):

$$\boldsymbol{r} = \text{grad } F = \boldsymbol{A}\boldsymbol{v} + \boldsymbol{b} \tag{2–5}$$

Theorem 2-1. *The solution of a symmetric definite system of equations* (2–1) *is equivalent to the problem of finding the minimum of a quadratic function* $F(\boldsymbol{v})$ *of equation* (2–2).

Proof: The equations (2–1) are considered solved when the residual vector $\boldsymbol{r}$ vanishes. By (2–4) and (2–5), the values v_i are to be found such that all partial derivatives of $F(\boldsymbol{v})$ vanish. For these values, $F(\boldsymbol{v})$ takes on a stationary value. In consequence of the positive definiteness of the quadratic form $Q(\boldsymbol{v})$ required initially, the quadratic form $F(\boldsymbol{v})$ has precisely one stationary point, and it is a minimum. The coordinates of the *minimum point* of the function $F(\boldsymbol{v})$ form the solution vector of the corresponding symmetric definite system of equations. Conversely, the solution of a symmetric definite system of equations clearly makes the corresponding quadratic function a minimum.

In the event that the matrix $\boldsymbol{A}$ of a symmetric system of equations is nonsingular but indefinite, its solution makes the corresponding quadratic function (2–2) stationary but not minimal. Geometrically, the solution represents a saddle point. Therefore, the following observations using the minimum property are confined to symmetric definite systems. The relaxation methods do not converge if applied to indefinite cases.

Example 2-1. To the symmetric definite system of equations

$$x_1 + 10x_2 + 1 = 0$$
$$10x_1 + 101x_2 + 11 = 0$$

belongs the quadratic function

$$\begin{aligned} F(\boldsymbol{v}) &= \tfrac{1}{2}(v_1^2 + 20v_1v_2 + 101v_2^2) + v_1 + 11v_2 \\ &= \tfrac{1}{2}[(v_1 + 10v_2 + 1)^2 + (v_2 + 1)^2 - 2]. \end{aligned}$$

$F(\boldsymbol{v})$ evidently takes on a minimum at $x_1 = v_1 = 9$ and $x_2 = v_2 = -1$. By Theorem 2–1, this pair of values corresponds to the solution of the system of equations. The value of the minimum is $F = -1$. Although the quadratic form is positive definite, the minimum of the quadratic function may very well be negative. Let us examine two points in the neighborhood of the solution and find the corresponding values of F.

(a) $v_1 = 8.9$, $v_2 = -1.1$. The distance from the solution point is about 0.14. Also $F(\boldsymbol{v}) = -0.39$, a value which lies 0.61 above the minimum.

(b) $v_1 = 7$, $v_2 = -0.8$. The distance from the solution point is about 2. In addition, $F(\boldsymbol{v}) = -0.98$, or only 0.02 above the minimum!

The variable growth of the quadratic function $F(\boldsymbol{v})$ in different directions is linked with the unfavorable condition $\kappa = 10{,}400$ of the matrix A computed in Sec. 1–2. The contours of the function $F(\boldsymbol{v}) =$ constant are concentric ellipses with their center points at $(9, -1)$. The semiaxes of the ellipses are proportional to the reciprocals of the square roots of the eigenvalues $\lambda_{max} \cong 102$ and $\lambda_{min} \cong 0.01$. The minimum is very flat in the direction of the major axis. This is one reason why the solution of a system of equations with an adverse condition is numerically inaccurate.

2-1-2 The Basic Principle of Relaxation

Theorem 2–1 constitutes the foundation of the general relaxation principle for solution of a symmetric definite system of equations. Starting with a trial vector $\boldsymbol{v}$, we select some nonzero *direction vector* $\boldsymbol{p}$ and correct $\boldsymbol{v}$ in the direction $\boldsymbol{p}$ with the intention of approaching the minimum by decreasing the quadratic function $F(\boldsymbol{v})$. Examine the one-dimensional set of new trial vectors $\boldsymbol{v}'$ which are linearly related to a parameter t to be specified later:

$$\boldsymbol{v}' = \boldsymbol{v} + t\boldsymbol{p}. \tag{2–6}$$

Given fixed $\boldsymbol{v}$ and some selected value of $\boldsymbol{p}$, the quadratic function F is a quadratic function in t only.

$$F = F(\boldsymbol{v} + t\boldsymbol{p}) = \tfrac{1}{2}(A(\boldsymbol{v} + t\boldsymbol{p}), (\boldsymbol{v} + t\boldsymbol{p})) + (\boldsymbol{b}, \boldsymbol{v} + t\boldsymbol{p})$$
$$= \tfrac{1}{2}(A\boldsymbol{v}, \boldsymbol{v}) + t(A\boldsymbol{v}, \boldsymbol{p}) + \tfrac{1}{2}t^2(A\boldsymbol{p}, \boldsymbol{p}) + (\boldsymbol{b}, \boldsymbol{v}) + t(\boldsymbol{b}, \boldsymbol{p}).$$

Invoking Eq. (2–2) and (2–5) yields

$$F = \tfrac{1}{2}t^2(A\boldsymbol{p}, \boldsymbol{p}) + t(\boldsymbol{r}, \boldsymbol{p}) + F(\boldsymbol{v}). \tag{2–7}$$

The parameter t is selected such that F is a minimum within the set under examination. The necessary condition for this is

$$\frac{dF}{dt} = t(A\boldsymbol{p}, \boldsymbol{p}) + (\boldsymbol{r}, \boldsymbol{p}) = 0.$$

In other words,

$$t_{\min} = -\frac{(\boldsymbol{r}, \boldsymbol{p})}{(A\boldsymbol{p}, \boldsymbol{p})}, \qquad \text{with} \qquad \boldsymbol{r} = A\boldsymbol{v} + \boldsymbol{b}. \tag{2–8}$$

Assurance of actually finding the minimum of F by this choice of $t_{\min}$ for the *relaxation direction* $\boldsymbol{p}$ follows from the second derivative $d^2F/dt^2 = (A\boldsymbol{p}, \boldsymbol{p})$, which is positive for every nonzero direction vector because of the positive definiteness of A. The function F, quadratic in t as shown in Eq. 2–7, forms a parabola opening upward, so that the value of F for all values of t between 0 and $2t_{\min}$ is smaller than $F(\boldsymbol{v})$.

The point $\boldsymbol{v}'$, which one reaches in the relaxation direction $\boldsymbol{p}$ with $t = t_{\min}$, is called a *minimum point*. The decrease of the quadratic function in going from $\boldsymbol{v}$ to the minimum point, using Eqs. (2–7) and (2–8), is

$$\Delta F = F(\boldsymbol{v} + t_{\min}\boldsymbol{p}) - F(\boldsymbol{v}) = -\frac{1}{2}\frac{(\boldsymbol{r}, \boldsymbol{p})^2}{(A\boldsymbol{p}, \boldsymbol{p})} < 0 \quad \text{for} \quad (\boldsymbol{r}, \boldsymbol{p}) \neq 0. \tag{2–9}$$

This is the maximum possible reduction of F in the direction $\boldsymbol{p}$.

By Eq. (2–8), the relaxation direction vector $\boldsymbol{p}$ must not be chosen orthogonal to the residual vector $\boldsymbol{r}$. Otherwise, we remain at the trial point $\boldsymbol{v}$ with $t_{\min} = 0$.

Theorem 2-2. *At the minimum point $\boldsymbol{v}'$ with $t = t_{\min}$, the new residual vector $\boldsymbol{r}' = A\boldsymbol{v}' + \boldsymbol{b}$ is orthogonal to the relaxation direction $\boldsymbol{p}$.*

Proof: With $\boldsymbol{r}' = A\boldsymbol{v}' + \boldsymbol{b} = A(\boldsymbol{v} + t\boldsymbol{p}) + \boldsymbol{b} = \boldsymbol{r} + tA\boldsymbol{p}$ it follows that $(\boldsymbol{r}', \boldsymbol{p}) = (\boldsymbol{r}, \boldsymbol{p}) + t(A\boldsymbol{p}, \boldsymbol{p}) = 0$ for $t = t_{\min}$ by Eq. (2–8).

Geometrical Interpretation of the General Relaxation Principle. In the case $n = 2$ and a Cartesian coordinate system (v_1, v_2), the contours $F(\boldsymbol{v}) = \text{constant}$ form concentric ellipses whose common center point coincides with the minimum point of $F(\boldsymbol{v})$ and constitutes the solution point. At a trial point $\boldsymbol{v}$, the residual vector $\boldsymbol{r}$ is orthogonal to the contour through the point $\boldsymbol{v}$, as it is the gradient of F by Eq. (2–5). In one relaxation step we pass from $\boldsymbol{v}$ in direction $\boldsymbol{p}$ to $\boldsymbol{v}'$ where F is a minimum along the relaxation direction. Here $\boldsymbol{r}'$ is perpendicular to $\boldsymbol{p}$; that is, the point $\boldsymbol{v}'$ is the point of contact of the relaxation direction with the contour (see Fig. 2).

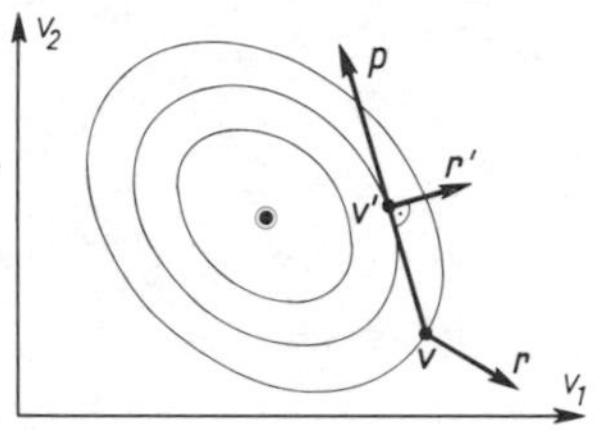

Figure 2. Interpretation of the relaxation principle.

The particular relaxation methods described below are variations of the general relaxation principle, as the various methods differ only in the choice of relaxation direction $\boldsymbol{p}$ for the individual relaxation steps and in the path followed (through the choice of t).

PROBLEMS

2-1. For the symmetric definite systems of equations

(a) $9x_1 - x_2 - 7 = 0$
$-x_1 + 9x_2 - 17 = 0$

(b) $31x_1 + 29x_2 - 33 = 0$
$29x_1 + 31x_2 - 27 = 0$

construct the appropriate quadratic functions whose minima are the solutions of the systems. Determine several contours of the quadratic functions. In what way do the two systems differ? Show the relationship between the geometric result and the condition number.

2-2 THE SUCCESSIVE DISPLACEMENT METHOD

2-2-1 Hand Relaxation

The method developed for hand calculation is mostly of historical significance. Let k be the index of the equation with the largest absolute residual $|r_k| = \max_{i=1,\ldots,n} |r_i|$. Then we correct merely the kth component v_k of the approximation $\boldsymbol{v}$, such that the residual of the kth equation consequently disappears. In the context of the general relaxation principle, this means the following. The kth unit vector $\boldsymbol{e}_k$ corresponding to the greatest residual in absolute value is chosen as relaxation direction $\boldsymbol{p}$, and we proceed as far as the minimum point. Using Eqs. (2–6) and (2–8), with $\boldsymbol{p} = \boldsymbol{e}_k$, we arrive successively at

$$t_{\min} = -\frac{(\boldsymbol{r}, \boldsymbol{p})}{(A\boldsymbol{p}, \boldsymbol{p})} = -\frac{(\boldsymbol{r}, \boldsymbol{e}_k)}{(A\boldsymbol{e}_k, \boldsymbol{e}_k)} = -\frac{r_k}{a_{kk}}, \tag{2–10}$$

$$\boldsymbol{v}' = \boldsymbol{v} - \frac{r_k}{a_{kk}}\boldsymbol{e}_k, \tag{2–11}$$

or, in terms of components,

$$v'_k = v_k - \frac{r_k}{a_{kk}}, \qquad v'_i = v_i \qquad i \neq k. \tag{2–12}$$

The value of $t_{\min}$ is also the correction Δv_k of the kth component of $\boldsymbol{v}$. By Theorem 2–2, the new residual vector $\boldsymbol{r}'$ is perpendicular to the relaxation direction, thus $(\boldsymbol{r}', \boldsymbol{e}_k) = 0$. Therefore, $r'_k = 0$; in other words, the kth equation is satisfied exactly. The new residual vector is

$$\boldsymbol{r}' = \boldsymbol{r} + t_{\min}A\boldsymbol{e}_k = \boldsymbol{r} - \frac{r_k}{a_{kk}}A\boldsymbol{e}_k,$$

or, in terms of components,

$$r'_i = r_i - \frac{r_k}{a_{kk}}a_{ik} = r_i + \Delta v_k a_{ik} \qquad i = 1, 2, \ldots, n, \tag{2–13}$$

and, by Eq. (2–9), the reduction in $F(\boldsymbol{v})$ by this relaxation step amounts to

$$\Delta F = -\frac{1}{2}\frac{(\boldsymbol{r}, \boldsymbol{e}_k)^2}{(A\boldsymbol{e}_k, \boldsymbol{e}_k)} = -\frac{1}{2}\frac{r_k^2}{a_{kk}}. \tag{2–14}$$

The reduction ΔF is less than zero provided that $r_k \neq 0$.

Convergence of hand relaxation for a symmetric definite system of equations is based on the fact that the quadratic function $F(\boldsymbol{v})$ decreases at each step in ΔF, as per Eq. (2–14). Furthermore, $F(\boldsymbol{v})$ has a finite minimum and is thus bounded from below. Therefore, the monotonically decreasing sequence of values of F, which are bounded from below, are convergent, and ΔF must necessarily converge toward zero. Since in Eq. (2–14) r_k identifies the residual of greatest magnitude, the residual vector $\boldsymbol{r}$ converges to the zero vector, and the trial vector $\boldsymbol{v}$ toward the solution vector.

Although convergence is guaranteed, experience in calculations reveals that the solution is often approached very slowly. Rapidity of convergence is assured if the matrix A of the system of equations is not just positive definite, but also strongly diagonal dominant.

The relaxation method is very well adapted to hand calculations, since the calculations are simple and systematic. In order to determine the new residuals, it is not necessary to substitute the new approximation into all equations. The equations (2–13) permit recursive computation. Thus the computation per step requires one division, n multiplications, and $n+1$ additions. In hand calculations, the residual with greatest magnitude is evident at first glance. For computers, however, this relaxation method is a poorly suited technique, since the necessary comparisons are more costly than the arithmetic operations that actually need to be carried out. Procedures which determine relaxation directions without many comparisons are more suitable for computers. In the past, hand relaxation was used along with various refinements for the manual solution of discretized elliptic boundary value problems (see Sec. 5–2–1 and Refs. 57, 59, 60, 61).

2-2-2 The Successive Displacement Method (Gauss-Seidel)

In contrast to hand relaxation calculations, the relaxation direction $\boldsymbol{p}$ now runs *cyclically* through the coordinate directions in the sequence $\boldsymbol{e}_1$, $\boldsymbol{e}_2, \ldots, \boldsymbol{e}_n$, regardless of the residuals. Thus we may speak of a *cyclic iteration process*. Evidently *Gauss* was the first to use this iteration method in successive steps. It was used later by *Seidel*, so that the method is called the *Gauss-Seidel iteration method*.* In the general kth relaxation step ($k = 1, 2, \ldots$), the relaxation direction is given by

$$\boldsymbol{p}^{(k)} = \boldsymbol{e}_j \quad \text{with} \quad j \equiv k \pmod{n}. \tag{2–15}$$

This direction is followed to the minimum point. Correspondingly, the formulas (2–10) to (2–14) remain valid, since the Gauss-Seidel procedure differs from hand calculations only in the selection of directions. In the kth step, the jth component v_j with $j \equiv k \pmod{n}$ is corrected such that the current residual r_j vanishes. In this variation it is inadvisable to compute all the new residuals recursively by Eq. (2–13), since the new component v'_j is found with the same effort by Eq. (2–16).

$$v'_j = v_j - \frac{r_j}{a_{jj}} = -\left\{\sum_{\substack{l=1\\ l\neq j}}^{n} a_{jl}v_l + b_j\right\}\Big/ a_{jj}, \qquad j = 1, 2, \ldots, n \text{ (cyclic)}. \tag{2-16}$$

The residual vector is not actually involved in the computations.

*Editor's note: Professor A. M. Ostrowski's study states that Gauss did not use a cyclic order of relaxation, and that Seidel specifically recommended against using it. G.E.F.

Table 1 GAUSS-SEIDEL ITERATION

Cycle	v_1	v_2	v_3	v_4
0	0	0	0	0
1	−0.5000	1.1667	2.7222	1.3611
2	0.0834	2.2685	3.5432	1.7716
3	0.6343	2.7258	3.8325	1.9163
4	0.8629	2.8985	3.9383	1.9692
5	0.9493	2.9625	3.9772	1.9886
6	0.9813	2.9862	3.9916	1.9958
7	0.9931	2.9949	3.9969	1.9985
⋮	↓	↓	↓	↓
∞	1.0000	3.0000	4.0000	2.0000

Example 2-2. For simplicity, let us examine the symmetric definite system of equations (2–17) with tridiagonal coefficient matrix A. Starting with the zero vector as the approximation, we arrive at the successively improved components as given in Table 1. They are found row by row, left to right, by Eq. (2–16).

$$A = \begin{bmatrix} 2 & -1 & 0 & 0 \\ -1 & 3 & -1 & 0 \\ 0 & -1 & 3 & -1 \\ 0 & 0 & -1 & 2 \end{bmatrix}, \quad \boldsymbol{b} = \begin{bmatrix} 1 \\ -4 \\ -7 \\ 0 \end{bmatrix}. \quad \text{Solution } \boldsymbol{x} = \begin{bmatrix} 1 \\ 3 \\ 4 \\ 2 \end{bmatrix} \tag{2–17}$$

After seven complete cycles involving 28 successive steps, the solution is accurate to two to three decimals.

Next, let us look at the *convergence of the Gauss-Seidel method.* The n individual relaxation steps with the directions $\boldsymbol{e}_1, \boldsymbol{e}_2, \ldots, \boldsymbol{e}_n$ will be considered as one cycle, and the trial vectors $\boldsymbol{v}^{(0)}, \boldsymbol{v}^{(1)}, \boldsymbol{v}^{(2)}, \ldots, \boldsymbol{v}^{(m)}, \ldots$, resulting after each cycle will be examined. Since we take up the equations in cyclic fashion and fulfill each equation in turn with the appropriate components, the transition from trial vector $\boldsymbol{v}^{(m)}$ to vector $\boldsymbol{v}^{(m+1)}$ is as depicted in (2–18):

$$\left.\begin{array}{l} a_{11}v_1^{(m+1)} + a_{12}v_2^{(m)} + a_{13}v_3^{(m)} + \cdots + a_{1n}v_n^{(m)} + b_1 = 0 \\ a_{21}v_1^{(m+1)} + a_{22}v_2^{(m+1)} + a_{23}v_3^{(m)} + \cdots + a_{2n}v_n^{(m)} + b_2 = 0 \\ a_{31}v_1^{(m+1)} + a_{32}v_2^{(m+1)} + a_{33}v_3^{(m+1)} + \cdots + a_{3n}v_n^{(m)} + b_3 = 0 \\ \quad \cdot \qquad\quad \cdot \qquad\quad \cdot \qquad \cdots \qquad \cdot \qquad \cdot \quad \cdot \\ a_{n1}v_1^{(m+1)} + a_{n2}v_2^{(m+1)} + a_{n3}v_3^{(m+1)} + \cdots + a_{nn}v_n^{(m+1)} + b_n = 0 \\ \qquad m = 0, 1, 2, \ldots \end{array}\right\} \tag{2–18}$$

To simplify the notation, matrix A will be decomposed into the sum of three matrices,

$$A = E + D + F, \tag{2–19}$$

with

$$E = \begin{bmatrix} 0 & 0 & 0 & \cdots & 0 \\ a_{21} & 0 & 0 & \cdots & 0 \\ a_{31} & a_{32} & 0 & \cdots & \cdot \\ \cdot & \cdot & \cdot & \cdots & \cdot \\ a_{n1} & a_{n2} & a_{n3} & \cdots & 0 \end{bmatrix}, \quad F = E^{\mathrm{T}}, \quad D = \begin{bmatrix} a_{11} & 0 & 0 & \cdots & 0 \\ 0 & a_{22} & 0 & \cdots & 0 \\ 0 & 0 & a_{33} & \cdots & \cdot \\ \cdot & \cdot & \cdot & \cdots & \cdot \\ 0 & 0 & 0 & \cdots & a_{nn} \end{bmatrix}.$$

The diagonal elements of D are greater than zero because of the prescribed positive definiteness of A. The iteration requirement (2–18) thus reads

$$(D + E)v^{(m+1)} + Fv^{(m)} + b = 0. \tag{2–20}$$

In consequence of the triangular form of the matrix $(D + E)$, this only appears to be an implicit vector equation. It can be written in explicit form equivalent to the component calculation rule (2–16) equally well as

$$v^{(m+1)} = -D^{-1}(Ev^{(m+1)} + Fv^{(m)} + b).$$

In addition, the fact that $(D + E)$ represents a lower triangular matrix yields the iteration requirement

$$v^{(m+1)} = -(D + E)^{-1}Fv^{(m)} - (D + E)^{-1}b. \tag{2–21}$$

In the sequel we turn to the *general iteration method* and take it to mean the scheme

$$v^{(m+1)} = Mv^{(m)} + c, \tag{2–22}$$

where M is a fixed matrix and c is a constant vector. The matrix M is called the *iteration matrix* of the method. In the *Gauss-Seidel* method, for example, the iteration matrix M and the constant vector c are given through

$$M = -(D + E)^{-1}F, \qquad c = -(D + E)^{-1}b. \tag{2–23}$$

In order to examine convergence of a general iteration method, the behavior of the difference between the exact solution, to be designated by x, and an approximation $v^{(m)}$ is crucial. Let us introduce the *error vector*

$$f^{(m)} = x - v^{(m)}$$

For the solution $\boldsymbol{x}$, Eq. (2–22) turns into the equation $\boldsymbol{x} = \boldsymbol{M}\boldsymbol{x} + \boldsymbol{c}$, which indicates that the iteration procedure maps the solution into itself. Subtraction of Eq. (2–22) transforms this equation directly into the recursion formula for the error vectors

$$\boldsymbol{f}^{(m+1)} = \boldsymbol{M}\boldsymbol{f}^{(m)} \qquad m = 0, 1, 2, \ldots, \tag{2–24}$$

from which follows, recursively, the rule

$$\boldsymbol{f}^{(m)} = \boldsymbol{M}^m \boldsymbol{f}^{(0)} \tag{2–25}$$

Thus the iteration matrix $\boldsymbol{M}$ by itself determines convergence, and, in the event of convergence, it establishes the number of iteration steps necessary to reach a given degree of accuracy in the approximation $\boldsymbol{v}^{(m)}$.

In the ensuing, two theorems about general iteration methods will be formulated and, in the same context, another concept will be introduced.

Theorem 2-3. *A sufficient condition for the convergence of an iteration method is that some norm of the iteration matrix* $\boldsymbol{M}$ *is smaller than unity.*

Proof: For any arbitrary matrix norm there exists a compatible vector norm. Then it follows by Eq. (2–25) and by the fundamental characteristics of norms

$$||\boldsymbol{f}^{(m)}|| = ||\boldsymbol{M}^m \boldsymbol{f}^{(0)}|| \leq ||\boldsymbol{M}^m||\cdot||\boldsymbol{f}^{(0)}|| \leq ||\boldsymbol{M}||^m \, ||\boldsymbol{f}^{(0)}||. \tag{2–26}$$

The value $||\boldsymbol{f}^{(0)}||$ is determined by the initial approximation $\boldsymbol{v}^{(0)}$. Under the premise that there is a matrix norm $||\boldsymbol{M}|| < 1$, it follows from Eq. (2–26) that $\lim_{m\to\infty} ||\boldsymbol{f}^{(m)}|| = 0$, and thus that $\boldsymbol{v}^{(m)}$ converges to the solution vector $\boldsymbol{x}$.

Theorem 2-4. *A necessary and sufficient condition for convergence of an iteration procedure is that the magnitudes of the eigenvalues of the iteration matrix* $\boldsymbol{M}$ *are all less than unity.*

Proof: Assume for the sake of simplicity that the matrix $\boldsymbol{M}$, which is neither symmetric nor positive definite, has n independent eigenvectors $\boldsymbol{x}_1$, $\boldsymbol{x}_2, \ldots, \boldsymbol{x}_n$ with eigenvalues $\lambda_1, \lambda_2, \ldots, \lambda_n$. The error vector $\boldsymbol{f}^{(0)}$ can thus be represented as a linear combination of the eigenvectors

$$\boldsymbol{f}^{(0)} = \sum_{i=1}^{n} c_i \boldsymbol{x}_i.$$

For the mth error vector, because of the recursion formula, it follows that

$$\boldsymbol{f}^{(m)} = \sum_{i=1}^{n} c_i \lambda_i^m \boldsymbol{x}_i \qquad m = 1, 2, \ldots. \tag{2–27}$$

(a) *Necessity*. If $\lim_{m\to\infty} \boldsymbol{f}^{(m)} = 0$ for any arbitrary initial vector $\boldsymbol{v}^{(0)}$ and thus any arbitrary error vector $\boldsymbol{f}^{(0)}$, then by Eq. (2–27) the magnitudes of the eigenvalues $|\lambda_i|$ $(i = 1, 2, \ldots n)$ must necessarily be less than unity.

(b) *Sufficiency*. Should $|\lambda_i| < 1$ $(i = 1, 2, \ldots, n)$, the convergence of $\boldsymbol{f}^{(m)}$ toward the zero vector follows from Eq. (2–27) for any arbitrary vector $\boldsymbol{f}^{(0)}$.

If the matrix $\boldsymbol{M}$ has multiple eigenvalues and less than n eigenvectors that are independent, as may happen with unsymmetric matrices, the theorem can still be proven with more elaborate mathematical tools (see Ref. 24).

The result (2–27) permits a qualitative statement about the asymptotic convergence behavior of $\boldsymbol{f}^{(m)}$. The *dominant* eigenvalue λ_1 of matrix $\boldsymbol{M}$, that is, the one with largest magnitude, generally governs the rate of convergence, since by Eq. (2–27) the values λ_i^m of all smaller eigenvalues approach zero faster with increasing m. Thus every vector norm $\|\boldsymbol{f}^{(m)}\|$ converges asymptotically to zero like a geometric series with the quotient $|\lambda_1|$.

Definition 2–1

The magnitude of the dominant eigenvalue of the matrix $\boldsymbol{M}$, $\rho(\boldsymbol{M}) = \max_i |\lambda_i|$, *is called the spectral radius of* $\boldsymbol{M}$. *For a general iteration method,* such as Eq. (2–22),

$$R(\boldsymbol{M}) = -\log_{10} \rho(\boldsymbol{M}) \tag{2–28}$$

is called the convergence rate of the iteration.

These concepts constitute a measure of the asymptotic quantity of iteration steps necessary to reduce the error $\boldsymbol{f}^{(m)}$ by some given factor. For sufficiently large m, and for k greater than zero, we arrive at the approximate result

$$\|\boldsymbol{f}^{(m+k)}\| \sim \rho(\boldsymbol{M})^k \, \|\boldsymbol{f}^{(m)}\|. \tag{2–29}$$

Of special interest is the value of k such that $\|\boldsymbol{f}^{(m+k)}\|$ equals one-tenth of $\|\boldsymbol{f}^{(m)}\|$. Substitution of this requirement into (2–29) yields, after an elementary computation,

$$k \sim -\frac{1}{\log_{10}\rho(\boldsymbol{M})} = \frac{1}{R(\boldsymbol{M})}. \tag{2–30}$$

The number of iteration k necessary to reduce the error $\boldsymbol{f}^{(m)}$ by a factor of 10 is approximately inversely proportional to the convergence rate $R(\boldsymbol{M})$. To gain an additional significant decimal place in $\boldsymbol{v}^{(m)}$ demands k iterations. Table 2 summarizes convergence rates $R(\boldsymbol{M})$ and rounded-off values of k

Table 2 SPECTRAL RADII AND CONVERGENCE RATES

$\rho(M)$	$R(M)$	k (rounded off)
0.30	0.5229	2
0.50	0.3010	4
0.70	0.1549	7
0.80	0.0969	11
0.90	0.0558	18
0.95	0.0223	45
0.99	0.00436	230

for several spectral radii $\rho(M)$. The smaller the spectral radius of M, the better the convergence. If, however, the spectral radius is close to one, many iterations are necessary.

Example 2-3. For matrix A of Eq. (2–17), the iteration matrix M for the Gauss-Seidel method is given as:

$$A = \begin{bmatrix} 2 & -1 & 0 & 0 \\ -1 & 3 & -1 & 0 \\ 0 & -1 & 3 & -1 \\ 0 & 0 & -1 & 2 \end{bmatrix}, \qquad (D + E) = \begin{bmatrix} 2 & 0 & 0 & 0 \\ -1 & 3 & 0 & 0 \\ 0 & -1 & 3 & 0 \\ 0 & 0 & -1 & 2 \end{bmatrix}$$

$$(D + E)^{-1} = \frac{1}{36}\begin{bmatrix} 18 & 0 & 0 & 0 \\ 6 & 12 & 0 & 0 \\ 2 & 4 & 12 & 0 \\ 1 & 2 & 6 & 18 \end{bmatrix}, \qquad F = \begin{bmatrix} 0 & -1 & 0 & 0 \\ 0 & 0 & -1 & 0 \\ 0 & 0 & 0 & -1 \\ 0 & 0 & 0 & 0 \end{bmatrix}$$

$$M = -(D + E)^{-1}F = \frac{1}{36}\begin{bmatrix} 0 & 18 & 0 & 0 \\ 0 & 6 & 12 & 0 \\ 0 & 2 & 4 & 12 \\ 0 & 1 & 2 & 6 \end{bmatrix}$$

The eigenvalues of M are the zeros of the characteristic polynomial

$$P_4(\lambda) = \lambda^2\left(\lambda^2 - \frac{4}{9}\lambda + \frac{1}{36}\right).$$

Consequently, $\lambda_1 = 0.369$, $\lambda_2 = 0.075$, $\lambda_3 = \lambda_4 = 0$ from which we derive the values $\rho(M) = 0.369$, $R(M) = 0.433$ and $k \cong 2.3$. Asymptotically speaking, more than two full cycles are needed in this example to gain one more decimal place. The convergence is good. The rule (2–29) has not yet manifested itself in the initial stage of the iteration, as it does after several cycles (see Table 1). After 12 cycles, $\boldsymbol{v}^{(12)}$ agrees with the solution to four decimals.

For the Gauss-Seidel method, the following is particularly applicable:

Theorem 2-5. *The successive displacement technique always converges for a symmetric definite system of equations.*

A proof of this theorem will not be given here, as it is a part of a more general result in Sec. 2–2–3. Some other techniques of proofs should, however, be pointed out. In Ref. 51 there is a direct convergence proof which utilizes the decrease in the quadratic function. It represents a generalization of the hand relaxation proof dealing with this topic in Sec. 2–2–1. Theorem 2–5 can also be traced back to Theorem 2–4, since we can conclude from the positive definiteness of matrix $\boldsymbol{A}$ that the eigenvalues of the iteration matrix $\boldsymbol{M}$ are smaller than unity in magnitude (see Refs. 15, 66). Frequently, the convergence of the Gauss-Seidel method is proven under the supposition that matrix $\boldsymbol{A}$ is strongly diagonal dominant. From it immediately appear elementary statements about rapidity of convergence (see Refs. 9, 15, 80).

2-2-3 The Method of Successive Overrelaxation

In the case of larger systems of equations, the successive displacement routine converges poorly in general, since the spectral radius of the iteration matrix $\boldsymbol{M}$ lies close to unity. Convergence can, however, be improved if we proceed along the relaxation direction $\boldsymbol{p}^{(k)} = \boldsymbol{e}_j$ ($j \equiv k \pmod{n}$) of the successive displacement procedure not just to the minimal point, but beyond it by a certain amount. It seems paradoxical at first to refrain from minimizing the quadratic function at each displacement with the goal of achieving a better convergence. By way of generalizing Eq. (2–16), we now choose a parameter $t = \omega t_{\min}$, where ω indicates a factor which does not change, so that now the iteration rule in the kth relaxation step reads

$$v_j^{(k)} = v_j^{(k-1)} - \omega \frac{r_j^{(k-1)}}{a_{jj}} \qquad j \equiv k (\text{mod } n); \quad k = 1, 2, \ldots. \tag{2–31}$$

Here $r_j^{(k-1)}$ indicates the residual of the jth equation after the $(k-1)$th iteration step, and $v_j^{(k)}$ the jth component of the trial vector $\boldsymbol{v}_j^{(k)}$ after the kth displacement. This remains the only component of the approximation vector to be altered. If we substitute for $r_j^{(k-1)}$ the corresponding expression in Eq. (2–31), it follows that

$$v_j^{(k)} = v_j^{(k-1)} - \omega \left\{ \sum_{l=1}^{n} a_{jl} v_l^{(k-1)} + b_j \right\} \Big/ a_{jj} \quad j \equiv k (\text{mod } n); \quad k = 1, 2, \ldots \tag{2–32}$$

In the case of $\omega = 1$, Eq. (2–32) reverts to Eq. (2–16) for successive displacements in accord with Gauss-Seidel.

If $\omega > 1$, the current equation in the kth relaxation step will not be fulfilled exactly. Rather, the jth component is overcorrected. We thus speak of *overrelaxation*, and the constant ω is called the *relaxation parameter*, or relaxation factor.

Example 2-4. The system (2–17) will be calculated for three different relaxation parameters. In Table 3 are logged the successively improved components of the trial vectors, which may be read from left to right in each row. Each row corresponds to a trial vector having undergone the indicated number of cycles, with each cycle consisting of $n = 4$ individual overrelaxation displacements. For comparison purposes, the case $\omega = 1$ (Gauss-Seidel) is repeated. Compared to the successive displacements procedure, a better convergence is clearly evident in the overrelaxation case. The convergence for $\omega = 1.1$ and $\omega = 1.2$ is roughly equally good, for reasons to be given in Sec. 2–2–4.

Table 3 Overrelaxation with Various Relaxation Parameters

	$\omega = 1.00$				$\omega = 1.05$			
Cycle	v_1	v_2	v_3	v_4	v_1	v_2	v_3	v_4
0	0	0	0	0	0	0	0	0
1	−0.5000	1.1667	2.7222	1.3611	−0.5250	1.2162	2.8756	1.5097
2	0.0834	2.2685	3.5432	1.7716	0.1398	2.3946	3.6728	1.8527
3	0.6343	2.7258	3.8325	1.9163	0.7252	2.8195	3.9016	1.9557
4	0.8629	2.8985	3.9383	1.9692	0.9190	2.9462	3.9706	1.9868
5	0.9493	2.9625	3.9772	1.9886	0.9758	2.9839	3.9912	1.9960
6	0.9813	2.9862	3.9916	1.9958	0.9928	2.9952	3.9974	1.9988
7	0.9831	2.9949	3.9969	1.9985	0.9978	2.9986	3.9992	1.9996
	↓	↓	↓	↓	↓	↓	↓	↓
∞	1.0000	3.0000	4.0000	2.0000	2.0000	3.0000	4.0000	2.0000
	$\omega = 1.10$				$\omega = 1.20$			
Cycle	v_1	v_2	v_3	v_4	v_1	v_2	v_3	v_4
0	0	0	0	0	0	0	0	0
1	−0.5500	1.2650	3.0305	1.6668	−0.6000	1.3600	3.3440	2.0064
2	0.2008	2.5249	3.8006	1.9237	0.3360	2.8000	4.0538	2.0310
3	0.8187	2.9079	3.9582	1.9846	1.0128	3.0666	4.0283	2.0108
4	0.9675	2.9819	3.9919	1.9971	1.0374	3.0130	4.0039	2.0002
5	0.9934	2.9964	3.9984	1.9994	1.0003	2.9991	3.9989	1.9993
6	0.9987	2.9993	3.9997	2.0000	0.9994	2.9995	3.9997	2.0000
7	0.9998	2.9999	4.0000	2.0000	0.9998	2.9999	4.0000	2.0000
	↓	↓	↓	↓	↓	↓	↓	↓
∞	1.0000	3.0000	4.0000	2.0000	1.0000	3.0000	4.0000	2.0000

Theorem 2-6. *In symmetric definite systems of equations, overrelaxation methods converge to the solution for any fixed value of* ω *in the range* $0 < \omega < 2$.

Proof: Let $j = k \pmod{n}$ and let $r_j^{(k-1)}$ be the jth component of the residual vector $\boldsymbol{r}^{(k-1)}$ after the $(k-1)$th step. Through the kth relaxation step, with $\boldsymbol{p}^{(k)} = \boldsymbol{e}_j$ and the choice of $t = \omega t_{\min} = -\omega r_j^{(k-1)}/a_{jj}$, the new trial vector $\boldsymbol{v}^{(k)}$ becomes

$$\boldsymbol{v}^{(k)} = \boldsymbol{v}^{(k-1)} - \omega \frac{r_j^{(k-1)}}{a_{jj}} \boldsymbol{e}_j \qquad j \equiv k \pmod{n}. \tag{2–33}$$

From the general result (2–7), we arrive at the equation

$$F(\boldsymbol{v}^{(k)}) - F(\boldsymbol{v}^{(k-1)}) = (\tfrac{1}{2}\omega^2 - \omega)\frac{r_j^{(k-1)2}}{a_{jj}} \tag{2–34}$$

for the change in the quadratic function $F(\boldsymbol{v})$ in the transition from $\boldsymbol{v}^{(k-1)}$ to $\boldsymbol{v}^{(k)}$. Because matrix A is positive definite, $a_{jj} > 0$. Furthermore, the square of $r_j^{(k-1)}$ is certainly nonnegative. The factor $(\omega^2/2 - \omega)$ is negative for all values of ω such that $0 < \omega < 2$. For the values of ω under consideration, the function $F(\boldsymbol{v}^{(k)})$ is monotonically nonincreasing with increasing k. It may be stationary for $r_j^{(k-1)} = 0$. For a complete cycle of n steps, it may be stationary if and only if all residuals vanish; that is, in the event that the trial vector $\boldsymbol{v}^{(k)}$ is identical to the solution. The function which is monotonically nonincreasing has a minimum, thus the sequence $F(\boldsymbol{v}^{(k)})$ must necessarily converge. From this must follow

$$\lim_{k\to\infty} [F(\boldsymbol{v}^{(k)}) - F(\boldsymbol{v}^{(k-1)})] = 0, \tag{2–35}$$

or

$$\lim_{k\to\infty} r_j^{(k-1)} = 0, \quad \text{if} \quad j \equiv k \pmod{n}. \tag{2–36}$$

By (2–36), there exists for every $\epsilon > 0$ a number $K_1(\epsilon)$ such that

$$|r_j^{(k-1)}| < \epsilon \text{ for } k > K_1(\epsilon), \quad \text{if} \quad j \equiv k \pmod{n}. \tag{2–37}$$

This result does not yield the convergence of the sequence of residual vector $\boldsymbol{r}^{(k-1)}$ to zero directly, as the inequality is valid for a fixed jth component only with sufficiently large k and after n displacements. From Eq. (2–33) and because of (2–36), it follows for a Euclidean norm that

$$\lim_{k\to\infty} \|\boldsymbol{v}^{(k)} - \boldsymbol{v}^{(k-1)}\| = \lim_{k\to\infty} \left| \omega \frac{r^{(k-1)}}{a_{jj}} \right| = 0. \tag{2–38}$$

But $\boldsymbol{r}^{(k)} - \boldsymbol{r}^{(k-1)} = A(\boldsymbol{v}^{(k)} - \boldsymbol{v}^{(k-1)})$ and thus, for compatible norms,

$$\|\boldsymbol{r}^{(k)} - \boldsymbol{r}^{(k-1)}\| \leq \|A\| \cdot \|\boldsymbol{v}^{(k)} - \boldsymbol{v}^{(k-1)}\|.$$

Here $\|A\|$ is a constant and thus, from (2–38), we have

$$\lim_{k\to\infty} \|\boldsymbol{r}^{(k)} - \boldsymbol{r}^{(k-1)}\| = 0. \tag{2–39}$$

In particular, if we replace k by $k + 1$, *every* component must obey the relation

$$\lim_{k\to\infty} |r_i^{(k+1)} - r_i^{(k)}| = 0 \qquad i = 1, 2, \ldots, n. \tag{2–40}$$

According to Eq. (2–40), for any $\epsilon > 0$, there exists a number $K_2(\epsilon)$ such that

$$|r_i^{(k+1)} - r_i^{(k)}| < \epsilon \quad \text{for} \quad k > K_2(\epsilon) \qquad i = 1, 2, \ldots, n. \tag{2–41}$$

For i fixed and arbitrary integer $p \geq 1$, the triangle inequality yields, through (2–41),

$$|r_i^{(k+p)} - r_i^{(k)}| < p\epsilon \quad \text{for} \quad k > K_2(\epsilon) \qquad p \geq 1,\ i = 1, 2, \ldots, n. \tag{2–42}$$

Now combine (2–37) and (2–42). Let $K = \max(K_1(\epsilon), K_2(\epsilon))$. Then for the index $j \equiv k + p + 1 \pmod{n}$, (2–42) yields

$$|r_j^{(k+p)} - r_j^{(k)}| < p\epsilon, \quad k > K \quad j \equiv k + p + 1 \pmod{n}. \tag{2–43}$$

Since $|r_j^{(k+p)} - r_j^{(k)}| \geq |r_j^{(k)}| - |r_j^{(k+p)}|$, Eq. (2–43) yields

$$|r_j^{(k)}| < p\epsilon + |r_j^{(k+p)}|, \quad k > K \qquad j \equiv k + p + 1 \pmod{n}. \tag{2–44}$$

If we replace k by $k + p + 1$ in (2–37), through (2–44) we arrive at

$$|r_j^{(k)}| < (p + 1)\epsilon, \quad k > K \qquad j \equiv k + p + 1 \pmod{n}. \tag{2–45}$$

With the help of (2–45), the individual components of the residual vector $\boldsymbol{r}^{(k)}$ can be estimated for k sufficiently large. Fix the index k and let p take on the values $0, 1, 2, \ldots, n - 1$. In virtue of the congruence for j, j takes on all the values from 1 to n. With the Euclidean norm, we thus arrive at the estimate

$$\|\boldsymbol{r}^{(k)}\| = \left\{\sum_{j=1}^{n} |r_j^{(k)}|^2\right\}^{1/2} < (1^2 + 2^2 + \cdots + n^2)^{1/2}\,\epsilon = \text{const}\ \epsilon,\ k > K. \tag{2–46}$$

The convergence of the residual vector $\boldsymbol{r}^{(k)}$ to zero is assured by (2–46) for every value of the factor ω in the interval $0 < \omega < 2$. Then by the result

$$A(\boldsymbol{v}^{(k)} - \boldsymbol{x}) = \boldsymbol{r}^{(k)},$$

or

$$\|\boldsymbol{v}^{(k)} - \boldsymbol{x}\| = \|\boldsymbol{A}^{-1}\boldsymbol{r}^{(k)}\| \leq \|\boldsymbol{A}^{-1}\|\cdot\|\boldsymbol{r}^{(k)}\|,$$

the trial vectors $\boldsymbol{v}^{(k)}$ also converge to the solution $\boldsymbol{x}$.

With Theorem 2–6, the convergence of the successive displacement method ($\omega = 1$) is now also proven.

Theorem 2–6 gives no information on the convergence of overrelaxation by using different possible values of ω, and tells nothing whatsoever about the optimum choice of the relaxation parameter for achieving the best possible convergence. For this, additional information about the structure of matrix $\boldsymbol{A}$ is necessary. Nonetheless, we can say that the worse the condition of matrix $\boldsymbol{A}$ of the system of equations, the closer the optimum value of ω lies to 2. In such a case, at each relaxation step we jump far beyond the minimal point to a new approximation which leaves the quadratic function $F(\boldsymbol{v})$ almost as large as it was before. The strategy of making the best improvement in each individual relaxation step (going to the minimal point) is not the best one for achieving the optimum long-term result.

2-2-4 Optimum Choice of Relaxation Parameter

This section will give the theory for the determination of the optimum relaxation parameter ω for a special class of matrices that turn up in the discretization of elliptic boundary value problems (see Chapter 5 and Ref. 13).

Analogous to the investigation of convergence of the successive displacement method, we examine the approximation vectors $\boldsymbol{v}^{(0)}, \boldsymbol{v}^{(1)}, \boldsymbol{v}^{(2)}, \ldots, \boldsymbol{v}^{(m)}, \ldots$ that result after completed cycles of n individual successive relaxation steps. The overrelaxation method may be described as follows. In each individual step, the successive displacement method is used to find a component of an auxiliary vector—call it $\boldsymbol{v}^{(m+1/2)}$—from which the new component $\boldsymbol{v}^{(m+1)}$ is found by adding to $\boldsymbol{v}^{(m)}$ the product of ω and the difference between the corresponding components of $\boldsymbol{v}^{(m)}$ and $\boldsymbol{v}^{(m+1/2)}$. Use of the decomposition (2–19) of the symmetric, positive definite matrix $\boldsymbol{A} = \boldsymbol{E} + \boldsymbol{D} + \boldsymbol{F}$ with $\boldsymbol{F} = \boldsymbol{E}^{\mathrm{T}}$ yields the two steps described above:

$$\boldsymbol{E}\boldsymbol{v}^{(m+1)} + \boldsymbol{D}\boldsymbol{v}^{(m+1/2)} + \boldsymbol{F}\boldsymbol{v}^{(m)} + \boldsymbol{b} = 0 \tag{2–47}$$

and

$$\left.\begin{aligned} \boldsymbol{v}^{(m+1)} &= \boldsymbol{v}^{(m)} + \omega(\boldsymbol{v}^{(m+1/2)} - \boldsymbol{v}^{(m)}) \\ &= (1-\omega)\boldsymbol{v}^{(m)} + \omega\boldsymbol{v}^{(m+1/2)} \end{aligned}\right\}. \tag{2–48}$$

An iteration rule for the approximations is found by elimination of the auxiliary vector $\boldsymbol{v}^{(m+1/2)}$ from Eqs. (2–47) and (2–48), and the result is

$$(\omega \boldsymbol{E} + \boldsymbol{D})\boldsymbol{v}^{(m+1)} + [\omega \boldsymbol{F} + (\omega - 1)\boldsymbol{D}]\boldsymbol{v}^{(m)} + \omega \boldsymbol{b} = 0. \qquad (2\text{–}49)$$

By Theorem 2–6, only the values of ω in the interval $0 < \omega < 2$ enter into consideration. After division by $\omega \neq 0$, the following equation replaces Eq. (2–49):

$$(\boldsymbol{E} + \omega^{-1}\boldsymbol{D})\boldsymbol{v}^{(m+1)} + [\boldsymbol{F} + (1 - \omega^{-1})\boldsymbol{D}]\boldsymbol{v}^{(m)} + \boldsymbol{b} = 0. \qquad (2\text{–}50)$$

The iteration rule (2–50) reduces to (2–20) of the Gauss-Seidel technique in the case of $\omega = 1$. Since $(\boldsymbol{E} + \omega^{-1}\boldsymbol{D})$ is a nonsingular lower triangular matrix, Eq. (2–50) becomes

$$\boldsymbol{v}^{(m+1)} = -(\boldsymbol{E} + \omega^{-1}\boldsymbol{D})^{-1}[\boldsymbol{F} + (1 - \omega^{-1})\boldsymbol{D}]\boldsymbol{v}^{(m)} - (\boldsymbol{E} + \omega^{-1}\boldsymbol{D})^{-1}\boldsymbol{b}. \qquad (2\text{–}51)$$

Thereby ω-dependency of the iteration matrix $\boldsymbol{M}(\omega)$ of overrelaxation is given by

$$\boldsymbol{M}(\omega) = -(\boldsymbol{E} + \omega^{-1}\boldsymbol{D})^{-1}[\boldsymbol{F} + (1 - \omega^{-1})\boldsymbol{D}]. \qquad (2\text{–}52)$$

To achieve optimal convergence, the spectral radius $\rho(\boldsymbol{M}(\omega))$, which is a function of ω, must be made as small as possible. The eigenvalues of $\boldsymbol{M}(\omega)$ can be related to those of a more elementary matrix in the following special case.

Definition 2–2

A matrix $\boldsymbol{A}$ is called block tridiagonal if it exhibits the following structure:

$$\boldsymbol{A} = \begin{bmatrix} \boldsymbol{D}_1 & \boldsymbol{F}_1 & & & & & \\ \boldsymbol{E}_1 & \boldsymbol{D}_2 & \boldsymbol{F}_2 & & & 0 & \\ & \boldsymbol{E}_2 & \boldsymbol{D}_3 & \boldsymbol{F}_3 & & & \\ & & \cdot & \cdot & \cdot & & \\ & & & \cdot & \cdot & \cdot & \\ & 0 & & & \boldsymbol{E}_{m-2} & \boldsymbol{D}_{m-1} & \boldsymbol{F}_{m-1} \\ & & & & & \boldsymbol{E}_{m-1} & \boldsymbol{D}_m \end{bmatrix}. \qquad (2\text{–}53)$$

The matrices $\boldsymbol{D}_i$ indicate square matrices of various orders, while the matrices $\boldsymbol{E}_k$ and $\boldsymbol{F}_k$ are generally rectangular.[1] *Outside of the band of submatrices, all*

[1]Matrices $\boldsymbol{F}_k$ have as many rows as $\boldsymbol{D}_k$ and as many columns as $\boldsymbol{D}_{k+1}$; similarly for $\boldsymbol{E}_k$.

elements of A *are zero. If in addition all the submatrices* D_i *along the diagonal are themselves diagonal matrices,* A *is called diagonally block tridiagonal.*

In particular, the tridiagonal matrices such as in (2–17) are diagonally block tridiagonal, since in this case all submatrices in (2–53) are of order one.

Theorem 2-7. *The optimal relaxation parameter* ω_{opt} *for a symmetric definite system of equations whose coefficient matrix* A *is diagonally block tridiagonal is given by*

$$\omega_{\text{opt}} = \frac{2}{1 + \sqrt{1 - \lambda_1^2}}. \qquad (2\text{–}54)$$

Here λ_1 *is the largest eigenvalue of the matrix* $-D^{-1}(E + F)$ *obtained from the decomposition* (2–19) *of* A.[1]

Proof: Because of the requirements imposed on A, D is a purely diagonal matrix with diagonal elements which are greater than zero. Define $D^{1/2}$ as the matrix satisfying $D^{1/2}D^{1/2} = D$, containing on its diagonal the positive square roots of the corresponding elements of D, and define $D^{-1/2}$ as its inverse. The matrix $-D^{-1}(E + F)$ is thus similar to $B = -D^{-1/2}(E + F)$ $D^{-1/2}$ through the transformation matrix $D^{1/2}$. Matrix B is symmetric with $(E + F)$, so that in consequence of the invariance of the eigenvalues in a similarity transformation all eigenvalues λ_i of the matrix $-D^{-1}(E + F)$ are real. The eigenvalues λ_i are the zeros of the polynomial $P(\lambda)$

$$P(\lambda) = |E + \lambda D + F| = \begin{vmatrix} \lambda D_1 & F_1 & & & & \\ E_1 & \lambda D_2 & F_2 & & & \\ & E_2 & \lambda D_3 & F_3 & & \\ & & \cdot & \cdot & \cdot & \\ & & & E_{m-2} & \lambda D_{m-1} & F_{m-1} \\ & & & & E_{m-1} & \lambda D_m \end{vmatrix}. \qquad (2\text{–}55)$$

The value of the determinant (2–55) is unchanged in multiplying by -1 all rows and columns crossing at the diagonal matrices D_{2k-1} having odd indices. This alteration leaves

$$P(\lambda) = \begin{vmatrix} \lambda D_1 & -F_1 & & & & \\ -E_1 & \lambda D_2 & -F_2 & & & \\ & -E_2 & \lambda D_3 & -F_3 & & \\ & & \cdot & \cdot & \cdot & \\ & & & -E_{m-2} & \lambda D_{m-1} & -F_{m-1} \\ & & & & -E_{m-1} & \lambda D_m \end{vmatrix}. \qquad (2\text{–}56)$$

[1] Matrix $-D^{-1}(E + F)$ will turn up again in Sec. 2–3–3 as the iteration matrix of the simultaneous displacements method, but only in the special case $D = I$.

If λ in (2–56) is replaced everywhere by $-\lambda$, and if the factor -1 is taken out of the determinant, we gain the result

$$P(-\lambda) = (-1)^n P(\lambda). \tag{2–57}$$

The polynomial $P(\lambda)$ is an even or odd function of λ, depending on the parity of its degree. Therefore, the nonzero real eigenvalues λ_i appear in pairs of identical magnitude but opposite sign, and the multiplicity of the zero eigenvalue is even or odd.

After these preparatory remarks about the eigenvalues λ_i, the eigenvalues μ_i of the iteration matrix $\boldsymbol{M}(\omega)$ as defined in Eq. (2–52) will now be related to λ_i. The eigenvalues μ_i are the zeros of the polynomial

$$\begin{aligned} Q(\mu) &= |(\boldsymbol{E} + \omega^{-1}\boldsymbol{D})\mu + \boldsymbol{F} + (1 - \omega^{-1})\boldsymbol{D}| \\ &= |\boldsymbol{E}\mu + \omega^{-1}\cdot(\mu + \omega - 1)\boldsymbol{D} + \boldsymbol{F}|. \end{aligned} \tag{2–58}$$

For brevity's sake, set

$$\omega^{-1}(\kappa + \omega - 1) = \zeta, \tag{2–59}$$

so that, explicitly, Eq. (2–58) becomes

$$Q(\mu) = \begin{vmatrix} \zeta\boldsymbol{D}_1 & \boldsymbol{F}_1 & & & & \\ \mu\boldsymbol{E}_1 & \zeta\boldsymbol{D}_2 & \boldsymbol{F}_2 & & & \\ & \mu\boldsymbol{E}_2 & \zeta\boldsymbol{D}_3 & \boldsymbol{F}_3 & & \\ & & \cdot & \cdot & \cdot & \\ & & & \mu\boldsymbol{E}_{m-2} & \zeta\boldsymbol{D}_{m-1} & \boldsymbol{F}_{m-1} \\ & & & & \mu\boldsymbol{E}_{m-1} & \zeta\boldsymbol{D}_m \end{vmatrix}. \tag{2–60}$$

In order to convert the determinant (2–60) into a form analogous to the one of (2–55), multiply each of the columns belonging to the diagonal matrix $\boldsymbol{D}_k$ $(k = 1, 2, \ldots, m)$ by the factor $\mu^{(k/2-1)}$. Then multiply each of the rows belonging to $\boldsymbol{D}_k$ by the factor $\mu^{(1-k)/2}$. The coefficients of the submatrices $\boldsymbol{E}_k$ and $\boldsymbol{F}_k$ thus disappear, and at the same time all the diagonal matrices $\boldsymbol{D}_k$ acquire identical factors $\zeta\cdot\mu^{-1/2}$. With reference to an arbitrary jth row and jth column crossing at $\boldsymbol{D}_k$, this alteration causes the determinant (2–60) to be multiplied by the product of the respective factors; in other words,

$$\mu^{(k/2-1)}\mu^{(1-k)/2} = \mu^{-1/2}.$$

This factor is constant for every index $j = 1, 2, \ldots, n$, and thus (2–60) yields

$$Q(\mu) = \mu^{n/2} \cdot \begin{vmatrix} \zeta\mu^{-1/2}\boldsymbol{D}_1 & \boldsymbol{F}_1 & & & & \\ \boldsymbol{E}_1 & \zeta\mu^{-1/2}\boldsymbol{D}_2 & \boldsymbol{F}_2 & & & \\ & \boldsymbol{E}_2 & \zeta\mu^{-1/2}\boldsymbol{D}_3 & \boldsymbol{F}_3 & & \\ & & \cdot & \cdot & \cdot & \\ & & & \boldsymbol{E}_{m-2} & \zeta\mu^{-1/2}\boldsymbol{D}_{m-1} & \boldsymbol{F}_{m-1} \\ & & & & \boldsymbol{E}_{m-1} & \zeta\mu^{-1/2}\boldsymbol{D}_m \end{vmatrix}. \tag{2–61}$$

Comparison with (2–55) reveals the relation

$$Q(\mu) = \mu^{n/2}P(\zeta\mu^{-1/2}) = \mu^{n/2}P(\mu^{-1/2}\omega^{-1}(\mu + \omega - 1)). \tag{2–62}$$

Equation (2–62) shows the desired connection between the eigenvalues λ_i and μ_i. Recall that $0 < \omega < 2$. If λ be an arbitrary zero of $P(\lambda)$, and if μ should satisfy the equation

$$\frac{(\mu + \omega - 1)^2}{\mu} = \omega^2\lambda^2, \tag{2–63}$$

then (2–62) says that μ is a zero of $Q(\mu)$, and thus also an eigenvalue of $\boldsymbol{M}(\omega)$. Conversely, λ is always a zero of $P(\lambda)$ if not only μ is a zero of $Q(\mu)$, but also λ fulfills Eq. (2–63). This invertible, unique correlation of the eigenvalues λ_i and μ_i is the key to the discussion of convergence behavior dependent on ω.

Let us take first the successive displacements method with $\omega = 1$. Equation (2–63) then reduces to

$$\mu = \lambda^2 \qquad (\omega = 1). \tag{2–64}$$

Nonzero eigenvalues μ_i are the squares of the pairs of eigenvalues $\pm\lambda_i \neq 0$ and are thus real and positive. A value $\mu_i = 0$ corresponds to $\lambda_i = 0$. By Theorems 2–4 and 2–6, it follows that $0 \leq \mu_i < 1$ $(i = 1, 2, \ldots, n)$. Therefore, the values λ_i all have magnitude smaller than unity and lie on the real axis of a complex λ-plane within the interval $(-1, +1)$ (see Fig. 3).

Now let us turn to overrelaxation, with $1 < \omega < 2$. The relation (2–63) depicts a unique mapping of the eigenvalues λ_i onto the eigenvalues μ_i depending on the value ω. If the λ_i are given, the μ_i are the solutions of the quadratic equation

$$\mu^2 + [2(\omega - 1) - \omega^2\lambda_i^2]\mu + (\omega - 1)^2 = 0. \tag{2–65}$$

To each eigenvalue $\lambda_i = 0$ there now corresponds $\mu_i = -(\omega - 1)$. To each

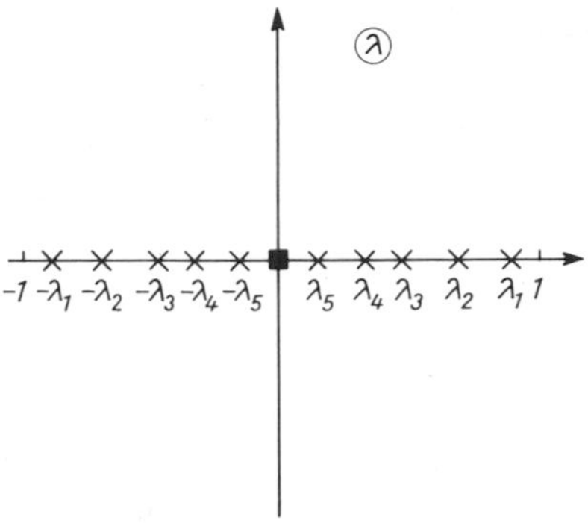

Figure 3. The eigenvalues λ.

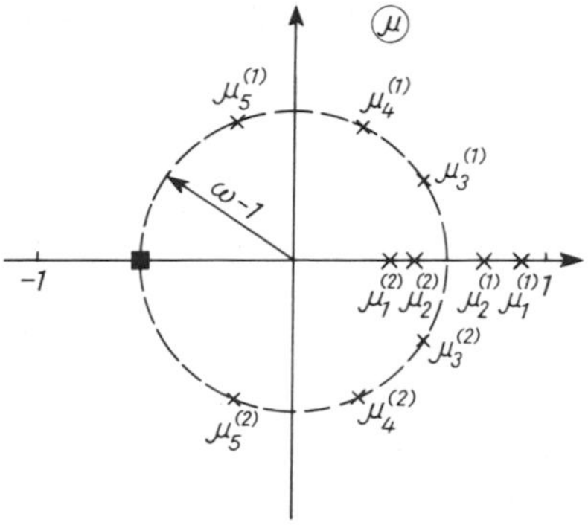

Figure 4. The eigenvalues μ_i for $\omega > 1$.

pair of eigenvalues $\pm\lambda_i \neq 0$ there correspond two values $\mu_i^{(1)}$ and $\mu_i^{(2)}$ whose product, by Eq. (2–65), equals $(\omega - 1)^2$. If the two values are complex and thus complex conjugates, they are of magnitude $(\omega - 1)$. The magnitude of the complex eigenvalues μ_i is independent of λ_i; thus they, together with the particular real eigenvalue $\mu_i = -(\omega - 1)$, lie on a circle of radius $(\omega - 1)$ with its center at the origin of a complex μ-plane (see Fig. 4). If the pair of solutions $\mu_i^{(1)}$ and $\mu_i^{(2)}$ is real, both lie inversely to the said circle (in Fig. 4), that is, their product is $(\omega - 1)^2$. Unless the two paired solutions are equal to one another, the larger of the two will be greater than $(\omega - 1)$. In this case, the spectral radius $\rho(\boldsymbol{M}(\omega))$ is established by the largest real eigenvalue $\mu_1^{(1)}$. It is easily shown that $\mu_1^{(1)} > \mu_1^{(2)}$ corresponds to the pair of eigenvalues of greatest magnitude $\pm\lambda_i$.

Next, we will show that $\mu_1^{(1)}$, which is a function of ω, is monotonically decreasing with increasing ω, provided that it is real and larger than $(\omega - 1)$. The coefficients of the quadratic equation (2–65) are continuous functions of ω; thus the solution $\mu_1^{(1)}$ is also continuous in ω. Differentiation of Eq. (2–65) with respect to ω yields

$$\frac{d\mu_1^{(1)}}{d\omega} = -\frac{\mu_1^{(1)} - \mu_1^{(1)}\omega\lambda_1^2 + \omega - 1}{\mu_1^{(1)} - \frac{1}{2}\omega^2\lambda_1^2 + \omega - 1}. \tag{2–66}$$

Since $\mu_1^{(1)} = \lambda_1^2 < 1$, for $\omega = 1$ we get from Eq. (2–66) the value

$$\frac{d\mu_1^{(1)}}{d\omega} = -\frac{\mu_1^{(1)} - \mu_1^{(1)}\cdot\lambda_1^2}{\mu_1^{(1)} - \frac{1}{2}\mu_1^{(1)}} = -2(1 - \lambda_1^2) < 0.$$

The derivative of $\mu_1^{(1)}$ with respect to ω is thus negative at $\omega = 1$, and retains the same sign in the ω region under consideration. In fact, the derivative $d\mu_1^{(1)}/d\omega$ is a continuous function of ω just like $\mu_1^{(1)}$ as long as the denominator in Eq. (2–66) is not zero. The numerator in Eq. (2–66) certainly cannot change sign for $\omega > 1$, since its zero appears at

$$\omega = \frac{1 - \mu_1^{(1)}}{1 - \lambda_1^2 \mu_1^{(1)}} < 1.$$

The denominator vanishes at

$$\mu_1^{(1)} = \tfrac{1}{2}\omega^2\lambda_1^2 - (\omega - 1). \tag{2-67}$$

But this value of $\mu_1^{(1)}$ corresponds precisely with the value of ω for which the two solutions $\mu_1^{(1)}$ and $\mu_1^{(2)}$ coincide. For this value of ω, the derivative becomes infinite, and the function $\mu_1^{(1)}(\omega)$ has a branch point there.

Therefore, the spectral radius $\rho(\boldsymbol{M}(\omega)) = \mu_1^{(1)}(\omega)$ decreases monotonically with increasing $\omega \geq 1$ up to the critical value of ω for which $\mu_1^{(1)} = \mu_1^{(2)}$. The discriminant of the quadratic equation (2–65)

$$[2(\omega - 1) - \omega^2\lambda_1^2]^2 - 4(\omega - 1)^2 = 0$$

disappears for this value. This determining equation defining the critical value ω_{crit} yields, with the limits $0 < \omega < 2$, the only possible solution,

$$\omega_{\text{crit}} = \frac{2}{1 + \sqrt{1 - \lambda_1^2}}. \tag{2-68}$$

For this value of ω_{crit}, all eigenvalues μ_i lie on the circle with radius $(\omega_{\text{crit}} - 1)$. For the values of ω in the interval $\omega_{\text{crit}} < \omega < 2$, all eigenvalues μ_i lie on a similar circle of increased radius $(\omega - 1)$. The spectral radius $\rho(\boldsymbol{M}(\omega))$ thus becomes truly minimal in the interval $1 \leq \omega < 2$ for $\omega_{\text{opt}} = \omega_{\text{crit}}$, in accord with Eq. (2–68). An analogous examination in the interval $0 < \omega < 1$ shows that $\rho(\boldsymbol{M}(\omega)) > \rho(\boldsymbol{M}(1))$. Thus ω_{opt} represents, by Eq. (2–68), the optimal relaxation parameter, and the optimum spectral radius is

$$\rho(\boldsymbol{M}(\omega_{\text{opt}})) = \omega_{\text{opt}} - 1. \tag{2-69}$$

By (2–64), the spectral radius $\rho(\boldsymbol{M})$ of the successive displacements method equals λ_1^2 in (2–54) for a diagonally blockwise tridiagonal matrix A. Table 4 shows the gain in overrelaxation through optimum choice of relaxation factor, as compared to the successive displacement method, and it tabulates the spectral radii and corresponding iteration steps k which are necessary, by Eq. (2–30). The great superiority of overrelaxation is especially clear when the spectral radius of the successive displacements method lies close to unity.

Figure 5 shows the spectral radius $\rho(\boldsymbol{M}(\omega))$ as a function of ω in the interval $1 \leq \omega \leq 2$, based on the discussion associated with the proof of Theorem 2–7, with $\lambda_1^2 = 0.95$. With ω approaching ω_{opt} from below,

Table 4 OVERRELAXATION COMPARED TO SUCCESSIVE DISPLACEMENT METHOD

Successive Displacement			*Overrelaxation*	
$\rho(M) = \lambda_1^2$	k	ω_{opt}	$\rho(M(\omega_{\text{opt}})) = \omega_{\text{opt}} - 1$	k
0.500	4	1.1716	0.1716	2
0.700	7	1.2922	0.2922	2
0.800	11	1.3820	0.3820	3
0.900	18	1.5195	0.5195	4
0.950	45	1.6345	0.6345	5
0.990	230	1.8182	0.8182	12
0.999	2300	1.9387	0.9387	37

$\rho(M(\omega))$ exhibits a vertical tangent, while $\rho(M(\omega))$ increases linearly toward unity for $\omega > \omega_{\text{opt}}$. If ω is chosen slightly too small, the convergence behavior is very sensitive in the vicinity of ω_{opt}; if, however, it is slightly too

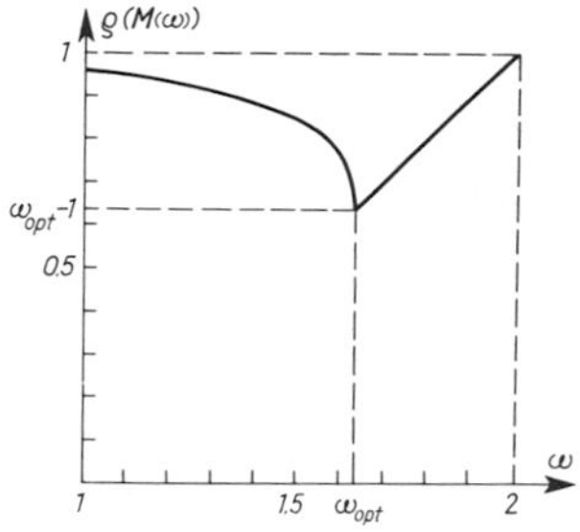

Figure 5. Spectral radius in overrelaxation as a function of ω.

large, the slowing of convergence is not so serious. It is far better to overestimate the value of ω_{opt} than to underestimate it. For a practical choice of ω_{opt}, see Sec. 5–2–2.

Example 2-5. System (2–17) exhibits a tridiagonal coefficient matrix $\boldsymbol{A}$; thus Theorem 2–7 may be invoked. Eigenvalues λ_i of the $-\boldsymbol{D}^{-1}(\boldsymbol{E} + \boldsymbol{F})$ matrix are the zeros of the polynomial

$$P(\lambda) = \begin{vmatrix} 2\lambda & -1 & & \\ -1 & 3\lambda & -1 & \\ & -1 & 3\lambda & -1 \\ & & -1 & 2\lambda \end{vmatrix} = 36\lambda^4 - 16\lambda^2 + 1 = 0.$$

$P(\omega)$ is an even polynomial, and its zeros are

$$\lambda_i = \pm\sqrt{\frac{4 \pm \sqrt{7}}{18}}.$$

They are all real and paired. The largest is $\lambda_1 = 0.6076$, and thus, by (2–54), $\omega_{opt} = 1.115$. The corresponding optimum radius of convergence is $\rho(\boldsymbol{M}(\omega_{opt})) = 0.115$, and thus $k \cong 1$. If $\omega = 1.1$, the optimal value is underestimated very little. By solving the quadratic equation (2–65), we find $\rho(\boldsymbol{M}(1.1)) = \mu_1^{(1)} = 0.195$ and thus $k \cong 1.4$. The far larger overestimate $\omega = 1.2$ is less consequential, since with $\omega > \omega_{opt}$, $\rho(\boldsymbol{M}(1.2)) = 0.2$ and, consequently, $k \cong 1.4$. The two convergence radii are essentially identical, and therein lies the explanation for the equally rapid convergence observed in Table 3. Since we gain quantitatively two decimals for every three cycles, and since the error vector magnitude at the start is smaller than 6, the deviation of the vector $\boldsymbol{v}^{(7)}$ from the exact solution by at most two units in the fourth decimal is thus explained for both $\omega = 1.1$ and $\omega = 1.2$.

PROBLEMS

2-2. For the system of equations $\boldsymbol{Ax} + \boldsymbol{b} = 0$ with

$$\boldsymbol{A} = \begin{bmatrix} 2 & -1 & 0 & 0 \\ -1 & 4 & -1 & 0 \\ 0 & -1 & 4 & -1 \\ 0 & 0 & -1 & 2 \end{bmatrix}, \quad \boldsymbol{b} = \begin{bmatrix} -3 \\ -5 \\ 15 \\ -7 \end{bmatrix} \text{ and the solution } \boldsymbol{x} = \begin{bmatrix} 2 \\ 1 \\ -3 \\ 2 \end{bmatrix},$$

carry out several steps, both with the method of hand relaxation and of successive displacements, and comment on convergence properties. What is the convergence rate in the successive displacement method here?

2-3. For the two matrices of Problem 2-1,

$$\boldsymbol{A}_1 = \begin{bmatrix} 9 & -1 \\ -1 & 9 \end{bmatrix} \text{ and } \boldsymbol{A}_2 = \begin{bmatrix} 31 & 29 \\ 29 & 31 \end{bmatrix},$$

find the convergence rates of the successive displacement method.

2-4. Find the optimal relaxation parameters for the three matrices of Problems 2-2 and 2-3 and the corresponding convergence rates in overrelaxation.

2-5. For the system of equations

$$31x_1 + 29x_2 - 33 = 0$$
$$29x_1 + 31x_2 - 27 = 0$$

examine numerically the convergence for various relaxation factors in the vicinity of the optimal value, and compute the corresponding convergence radii. Compare the convergence rates found numerically with theoretical predictions.

2-3 GRADIENT METHODS

2-3-1 The Principle

The general relaxation principle of Sec. 2–1–2 permits a free choice of direction vectors $\boldsymbol{p}$, as long as $\boldsymbol{p}$ is not orthogonal to residual vector $\boldsymbol{r}$ of trial vector $\boldsymbol{v}$. As gradient of the quadratic function $F(\boldsymbol{v})$ of the trial vector $\boldsymbol{v}$ to be minimized, the residual vector $\boldsymbol{r}$ points in the direction which increases $F(\boldsymbol{v})$ locally in the most rapid manner. It is thus only natural to use the gradient of $F(\boldsymbol{v})$ at the approximation point in order to establish the relaxation direction. Relaxation methods using the current or even the past residual vectors are called *gradient methods.* The successive displacement method and the method of overrelaxation are not gradient methods.

To derive the gradient methods, we are substantially guided by geometric arguments. In the case $n = 2$, the contour lines of the quadratic function $F(\boldsymbol{v}) =$ constant in a Cartesian coordinate system $(\boldsymbol{v}_1, \boldsymbol{v}_2)$ form concentric ellipses which are similar to one another, and whose common center coincides with the minimum of $F(\boldsymbol{v})$. Starting from a trial point $\boldsymbol{v}^{(0)}$, we arrive at the common center on the orthogonal trajectory through $\boldsymbol{v}^{(0)}$ (see Fig. 6). This identifies a gradient method representing a special case of the *principle of steepest descent* (Ref. 64). In principle, the method described leads to the problem of solving the system of differential equations

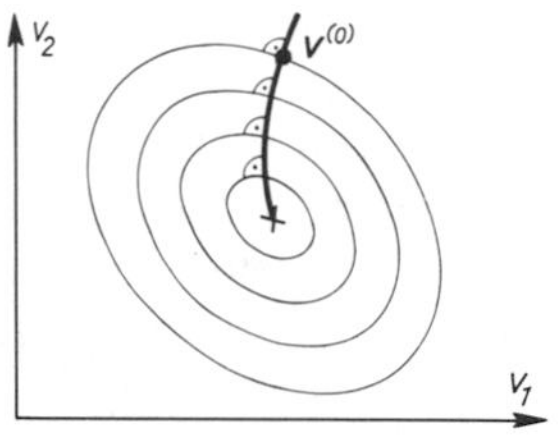

Figure 6. Principle of steepest descent.

$$\frac{d\boldsymbol{v}}{dt} = -\text{grad } F(\boldsymbol{v}) = -\boldsymbol{A}\boldsymbol{v} - \boldsymbol{b}, \qquad (2\text{–}70)$$

where t is a parameter on the trajectory. The system (2–70) can be solved by any classical numerical technique. The t-axis is divided up and the trajectory replaced by a sequence of contiguous straight-line segments. The choice of subdivision is crucial for the resultant gradient method. With a fine subdivision, the line segments remain close to the true trajectory, but we reach the vicinity of the solution point only after many steps. A coarse subdivision raises the danger that the segments diverge from the true trajectory.

2-3-2 Method of Steepest Descent

In the *method of steepest descent*, the relaxation direction $\boldsymbol{p}$ is defined by the negative of the residual vector.

$$\boldsymbol{p}^{(k)} = -\boldsymbol{r}^{(k-1)}, \qquad k = 1, 2, \ldots \qquad (2\text{–}71)$$

This direction is followed to the minimum point. By (2–8), the parameter $t_{\min}$ is determined by

$$t_{\min} = \frac{(\boldsymbol{r}^{(k-1)}, \boldsymbol{r}^{(k-1)})}{(A\boldsymbol{r}^{(k-1)}, \boldsymbol{r}^{(k-1)})}. \tag{2–72}$$

The value $R(\boldsymbol{x}) = (A\boldsymbol{x}, \boldsymbol{x})/(\boldsymbol{x}, \boldsymbol{x})$ for $\boldsymbol{x} \neq 0$ is called the *Rayleigh quotient* of the vector $\boldsymbol{x}$. It plays a role in certain methods of determining eigenvalues (see Chapter 4). In this terminology, the value of the parameter $t_{\min}$ in Eq. (2–72) equals the reciprocal of the Rayleigh quotient for the residual vector $\boldsymbol{r}^{(k-1)}$.

A direct consequence of Theorem 2–2 is the following.

Theorem 2-8. *In the method of steepest descent, the new residual vector* $\boldsymbol{r}^{(k)}$ *after the completion of a relaxation step is orthogonal to the previous* $\boldsymbol{r}^{(k-1)}$.

From a geometric point of view, relaxation involves describing a piecewise linear path with right-angled corners in an n-dimensional Euclidean space, with the path terminating at the minimum of the quadratic function $F(\boldsymbol{v})$. Unfortunately, it turns out that despite the choice of the best local direction along with the largest reduction of $F(\boldsymbol{v})$ in each relaxation displacement, convergence is not very good in general. Once again the tactic of seeking the most efficient goal by choosing the best local option does not lead to the best overall strategy.

2-3-3 The Simultaneous Displacement Method

Instead of selecting a new parameter t by Eq. (2–72) with each step as the reciprocal of the Rayleigh quotient, we can reduce the calculation time with the *simultaneous displacement method* by choosing t constant (and positive) for all iteration steps, and thereby refraining from going to the minimum at every step. The calculation rule is

$$\boldsymbol{v}^{(k)} = \boldsymbol{v}^{(k-1)} - t\boldsymbol{r}^{(k-1)}, \qquad t = \text{const.} > 0, \qquad k = 1, 2, \ldots. \tag{2–73}$$

The constant chosen for t is not entirely arbitrary. For the convergence of the simultaneous displacement method, it is necessary and sufficient that $0 < t < 2/\lambda_1$, where λ_1 indicates the largest eigenvalue of matrix A (see Ref. 61).

In the literature the simultaneous displacement method is generally taken to mean a special case of the method described above. The symmetric definite system of equations $A\boldsymbol{x} + \boldsymbol{b} = 0$ is altered by a trivial substitution of variables and a multiplication of the rows while maintaining symmetry, such that the diagonal elements of $\boldsymbol{A}$ are unity. By this procedure the decomposition of $\boldsymbol{A}$ is reduced to

$$\boldsymbol{A} = \boldsymbol{E} + \boldsymbol{I} + \boldsymbol{F},$$

where the identity matrix $\boldsymbol{I}$ has replaced the customary diagonal matrix $\boldsymbol{D}$. Furthermore, the parameter t is chosen to be 1, so that in each relaxation step the approximation vector $\boldsymbol{v}^{(k-1)}$ is corrected by the negative of the residual vector $\boldsymbol{r}^{(k-1)}$:

$$\boldsymbol{v}^{(k)} = \boldsymbol{v}^{(k-1)} - \boldsymbol{r}^{(k-1)}, \qquad k = 1, 2, \ldots. \tag{2–74}$$

Under the given suppositions for $\boldsymbol{A}$, the computation rule (2–74) has a straightforward interpretation. Each individual component $\boldsymbol{v}^{(k-1)}$ is altered such that the residual of the jth equation is zero, without regard to the correction of the other components. The alterations, that is, the negative residuals r_j, are all added onto the trial vector. Therein lies the difference between the simultaneous displacement and successive displacement methods. The former method was proposed by *Jacobi*; thus it is also known as the *Jacobi iteration method.*

The convergence conditions and convergence rate of the simultaneous displacement method (2–74) must be examined anew, since the relaxation direction does not lead to the minimal point. With the decomposition $\boldsymbol{A} = \boldsymbol{E} + \boldsymbol{I} + \boldsymbol{F}$ and the definition of $\boldsymbol{r}^{(k-1)} = \boldsymbol{A}\boldsymbol{v}^{(k-1)} + \boldsymbol{b}$, Eq. (2–74) becomes

$$\boldsymbol{v}^{(k)} = -(\boldsymbol{E} + \boldsymbol{F})\boldsymbol{v}^{(k-1)} - \boldsymbol{b} \qquad k = 1, 2, \ldots. \tag{2–75}$$

The procedure (2–75) is again a general iteration method, as in (2–22). The iteration matrix of the simultaneous displacement method is accordingly given by

$$\boldsymbol{M} = -(\boldsymbol{E} + \boldsymbol{F}). \tag{2–76}$$

Theorem 2-9. *The simultaneous displacement method always converges if the matrix* $\boldsymbol{A} = \boldsymbol{E} + \boldsymbol{I} + \boldsymbol{F}$ *is strongly diagonal dominant* (see Definition 1–5 in Sec. 1–3–2).

Proof: The matrix norm

$$\|\boldsymbol{M}\| = \|-(\boldsymbol{E} + \boldsymbol{F})\| = \max_i \sum_{\substack{k=1 \\ k \neq i}}^{n} |a_{ik}| \tag{2–77}$$

is smaller than unity because of the strong diagonal dominance of $\boldsymbol{A}$. From this observation follows convergence according to Theorem 2–3.

The norm $\|M\|$ of Eq. (2–77) yields an upper bound for the spectral radius $\rho(M)$, whence we can derive a lower bound for the convergence rate $R(M) = -\log_{10} \rho(M)$, and with it an underestimation of the speed of convergence.

A consequence of Theorem 2–4 is

Theorem 2-10. *The simultaneous displacement method for* $A = E + I + F$ *converges if and only if the magnitudes of the eigenvalues of the iteration matrix* $M = -(E + F)$ *are all smaller than unity.*

Tantamount to the convergence condition of Theorem 2–10 is the assertion that the eigenvalues of the given symmetric coefficient matrix A lie in the interval 0 to 2. The convergence of the simultaneous displacement method is thus directly linked to the eigenvalues of the symmetric coefficient matrix, which are in themselves of interest from other considerations. For the successive displacement and overrelaxation methods, however, the eigenvalues of consequence were those of quite another matrix, which was related to the equation matrix in a complicated fashion.

In contrast to the successive displacement method, the simultaneous displacement method need not converge for a symmetric definite system of equations, as we can observe in the following counterexample from Ref. 17. The symmetric matrix

$$A = \begin{bmatrix} 1 & a & a \\ a & 1 & a \\ a & a & 1 \end{bmatrix}, \quad a \text{ real}, \tag{2–78}$$

is seen to be positive definite for all values of a in the interval $-0.5 < a < 1$ through the formal Cholesky decomposition, with radicands 1, $(1 - a^2)$, $(2a + 1)(a - 1)^2/(1 - a^2)$. In addition, the eigenvalues of the iteration matrix $M = -(E + F)$ are the zeros of

$$P(\lambda) = \begin{vmatrix} -\lambda & -a & -a \\ -a & -\lambda & -a \\ -a & -a & -\lambda \end{vmatrix} = -(\lambda + 2a)(\lambda - a)^2.$$

The magnitude of the dominant eigenvalue of M is $|\lambda_1| = 2|a|$. The magnitude is less than unity if and only if $-0.5 < a < 0.5$. The simultaneous displacement method diverges with $|\lambda_1| \geq 1$ for the positive definite matrix A of Eq. (2–78) for all a values in the interval $0.5 \leq a < 1$.

For the simultaneous displacement method there is no general criterion for convergence analogous to Theorem 2–5. Nevertheless, we have access to this result:

Theorem 2-11. *The simultaneous displacement method always converges for a symmetric definite system of equations if the coefficient matrix A is diagonally block tridiagonal.*

Proof: The eigenvalues λ_i of the matrix $\boldsymbol{M} = -(\boldsymbol{E} + \boldsymbol{F})$ are the same ones that turn up in the proof of Theorem 2–7 in the special case $\boldsymbol{D} = \boldsymbol{I}$. By (2–64), they are related to the eigenvalues of the successive displacement method by $\lambda_i^2 = \mu_i$. For successive displacements, $0 \leq \mu_i < 1$, thus $|\lambda_i| < 1$, and convergence follows for simultaneous displacements by Theorem 2–4.

In the special case of diagonally block tridiagonal matrices $\boldsymbol{A}$, the correlation between the eigenvalues λ_i and μ_i of the iteration matrices of the simultaneous and successive displacement methods yields a concrete statement about the relationship of their convergence. The spectral radii are given by

$$\rho(\boldsymbol{M}_{\text{simul}}) = \lambda_1, \quad \rho(\boldsymbol{M}_{\text{succ}}) = \mu_1 = \lambda_1^2 = \rho^2(\boldsymbol{M}_{\text{simul}}). \tag{2–79}$$

The convergence rate of the successive displacement method is twice as great as that of the simultaneous displacement method, according to Eq. (2–79). In this special case, the former converges twice as fast as the latter. Thus when compared to successive displacement, or even to the overrelaxation method, simultaneous displacement is not even competitive in this particular case.

PROBLEMS

2-6. Solve the two symmetric definite systems of equations in Problem 2-1 both by the method of steepest descent and by the simultaneous displacement method. By a graphic representation of approximation points and relaxation directions, depict the convergence of the two methods.

2-7. Using the spectral radius, prove that the simultaneous displacement method always converges for a symmetric definite system of equations in two unknowns. (*Note:* This general result does not hold for systems having more variables.)

2-8. Show that the simultaneous displacement method diverges for all values of a that make the matrix

$$\boldsymbol{A} = \begin{bmatrix} 1 & \frac{\sqrt{2}}{2} & a \\ \frac{\sqrt{2}}{2} & 1 & \frac{\sqrt{2}}{2} \\ a & \frac{\sqrt{2}}{2} & 1 \end{bmatrix}$$

positive definite.

2-4 THE METHOD OF CONJUGATE GRADIENTS

2-4-1 Derivation

The first relaxation step of the *Hestenes-Stiefel method* (Ref. 31) is the same as in the method of steepest descents. We start by selecting some trial vector $\boldsymbol{v}^{(0)}$ and then choosing the negative residual vector as the relaxation direction, that is, $\boldsymbol{p}^{(1)} = -\boldsymbol{r}^{(0)} = -\text{grad } F(\boldsymbol{v}^{(0)}) = -(A\boldsymbol{v}^{(0)} + \boldsymbol{b})$. We continue in this direction to the minimum point. The resulting value of the parameter $t_{\min}$ will be designated q_1.

$$\boldsymbol{v}^{(1)} = \boldsymbol{v}^{(0)} - q_1\boldsymbol{r}^{(0)}, \quad q_1 = \frac{(\boldsymbol{r}^{(0)}, \boldsymbol{r}^{(0)})}{(A\boldsymbol{r}^{(0)}, \boldsymbol{r}^{(0)})} = -\frac{(\boldsymbol{r}^{(0)}, \boldsymbol{p}^{(1)})}{(A\boldsymbol{p}^{(1)}, \boldsymbol{p}^{(1)})}. \tag{2–80}$$

In the general kth relaxation step ($k \geq 2$), we seek the minimum not merely in the direction of the residual vector $\boldsymbol{r}^{(k-1)}$, but in the two-dimensional plane passing through $\boldsymbol{v}^{(k-1)}$ and spanned by the previous relaxation direction $\boldsymbol{p}^{(k-1)}$ and the new residual vector $\boldsymbol{r}^{(k-1)}$. This generalization is motivated by the geometric fact that the two-dimensional plane intersects in an ellipse the contour surface of the quadratic function $F(\boldsymbol{v}) = F(\boldsymbol{v}^{k-1})$ passing through the approximation point $\boldsymbol{v}^{(k-1)}$. The ellipse of the intersection passes through the point $\boldsymbol{v}^{(k-1)}$, where it is tangent to the former relaxation direction $\boldsymbol{p}^{(k-1)}$, since $\boldsymbol{v}^{(k-1)}$ is a minimum point (see Fig. 7). The ellipses of intersection of the plane in question with various contour surfaces $F(\boldsymbol{v}) =$ constant are thus concentric and similar. The minimum of $F(\boldsymbol{v})$ in this plane is therefore taken at the common center of the ellipses. In order to arrive at the ellipse center point $\boldsymbol{M}$ by leaving $\boldsymbol{v}^{(k-1)}$ in the relaxation direction $\boldsymbol{p}^{(k)}$, $\boldsymbol{p}^{(k)}$ and $\boldsymbol{p}^{(k-1)}$ must be *conjugate directions* with respect to the ellipse of intersection. Thus clearly they must be conjugate directions with respect to every ellipsoid $F(\boldsymbol{v}) =$ constant. They must consequently fulfill the relation (2–81).

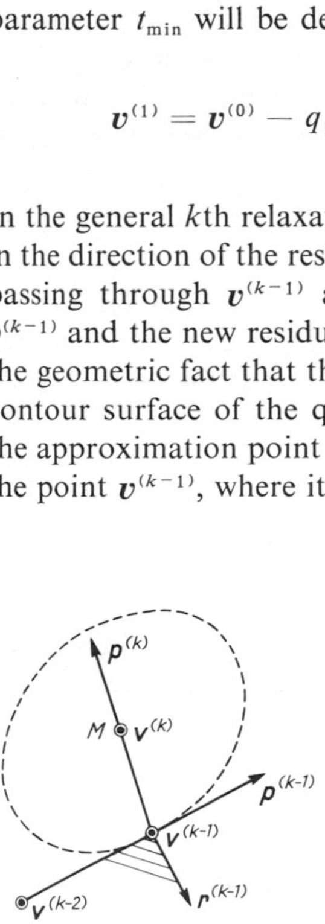

Figure 7. Method of conjugate gradients.

$$(A\boldsymbol{p}^{(k)}, \boldsymbol{p}^{(k-1)}) = (\boldsymbol{p}^{(k)}, A\boldsymbol{p}^{(k-1)}) = 0. \tag{2–81}$$

The direction vector $\boldsymbol{p}^{(k)}$ is taken as a linear combination of $\boldsymbol{r}^{(k-1)}$ and $\boldsymbol{p}^{(k-1)}$, where the coefficient of $\boldsymbol{r}^{(k-1)}$ is certainly nonzero and can be taken to be -1 for normalizing.

$$\boldsymbol{p}^{(k)} = -\boldsymbol{r}^{(k-1)} + e_{k-1}\boldsymbol{p}^{(k-1)} \qquad k = 2, 3, \ldots. \tag{2–82}$$

From Eq. (2–81), coefficient e_{k-1} is found to be

$$e_{k-1} = \frac{(\boldsymbol{r}^{(k-1)}, \boldsymbol{Ap}^{(k-1)})}{(\boldsymbol{p}^{(k-1)}, \boldsymbol{Ap}^{(k-1)})} \qquad k = 2, 3, \ldots. \tag{2–83}$$

With the direction $\boldsymbol{p}^{(k)}$ so identified, we proceed to the minimum point

$$\boldsymbol{v}^{(k)} = \boldsymbol{v}^{(k-1)} + q_k \boldsymbol{p}^{(k)} \quad \text{with} \quad q_k = -\frac{(\boldsymbol{r}^{(k-1)}, \boldsymbol{p}^{(k)})}{(\boldsymbol{Ap}^{(k)}, \boldsymbol{p}^{(k)})} \qquad k = 2, 3, \ldots. \tag{2–84}$$

Note that the denominators of the expressions for q_1, e_{k-1}, and q_k are greater than zero in consequence of the positive definiteness of $\boldsymbol{A}$ for every nonzero direction vector $\boldsymbol{p}^{(k)}$.

The *method of conjugate gradients* (Ref. 31) is essentially defined by the formulas (2–80), (2–82), (2–83), and (2–84), once a starting vector $\boldsymbol{v}^{(0)}$ has been chosen.

2-4-2 Properties and Simplifications

If we use (2–84), we find the residual vector $\boldsymbol{r}^{(k)}$ to be given after the kth relaxation step by

$$\boldsymbol{r}^{(k)} = \boldsymbol{Av}^{(k)} + \boldsymbol{b} = \boldsymbol{Av}^{(k-1)} + q_k \boldsymbol{Ap}^{(k)} + \boldsymbol{b} = \boldsymbol{r}^{(k-1)} + q_k(\boldsymbol{Ap}^{(k)}). \tag{2–85}$$

This represents a recursion formula for the residual vectors. Since the vector $\boldsymbol{Ap}^{(k)}$ in Eq. (2–84) must be calculated in any case, $\boldsymbol{r}^{(k)}$ is found with less effort than through actual substitution of the trial vector in the equations.

After completion of the kth relaxation step, $\boldsymbol{v}^{(k)}$ is by its structure the minimum point in the two-dimensional plane spanned by $\boldsymbol{r}^{(k-1)}$ and $\boldsymbol{p}^{(k-1)}$, or alternatively by $\boldsymbol{r}^{(k-1)}$ and $\boldsymbol{p}^{(k)}$. The residual vector $\boldsymbol{r}^{(k)}$, being the gradient of $F(\boldsymbol{v})$, is orthogonal to this plane in consequence of a generalization of Theorem 2–2, and thus there follow the relations

$$(\boldsymbol{r}^{(k)}, \boldsymbol{r}^{(k-1)}) = 0, \tag{2–86}$$

$$(\boldsymbol{r}^{(k)}, \boldsymbol{p}^{(k-1)}) = 0, \qquad (\boldsymbol{r}^{(k)}, \boldsymbol{p}^{(k)}) = 0. \tag{2–87}$$

Analogous to the method of steepest descent, two successive residual vectors are orthogonal. The relations (2–86) and (2–87) permit a modification of the formulas (2–84) and (2–83) for the calculation of the values q_k and e_{k-1}. By (2–82), it follows, using (2–87) for $k - 1$ in place of k, that

$$(\boldsymbol{r}^{(k-1)}, \boldsymbol{p}^{(k)}) = -(\boldsymbol{r}^{(k-1)}, \boldsymbol{r}^{(k-1)}) + e_{k-1}(\boldsymbol{r}^{(k-1)}, \boldsymbol{p}^{(k-1)}) = -(\boldsymbol{r}^{(k-1)}, \boldsymbol{r}^{(k-1)}),$$

so that from it comes the symmetrically constructed expression for q_k, replacing (2–84),

$$q_k = \frac{(r^{(k-1)}, r^{(k-1)})}{(Ap^{(k)}, p^{(k)})} \qquad k = 1, 2, \ldots. \tag{2–88}$$

As long as $r^{(k-1)}$ is nonzero, the positive definiteness of A assures that q_k is greater than zero. The recursion formula (2–85) for the residual vectors, using $k - 1$ in place of k, gives

$$Ap^{(k-1)} = \frac{1}{q_{k-1}}(r^{(k-1)} - r^{(k-2)}). \tag{2–89}$$

The numerator of e_{k-1} in Eq. (2–83), with reference to Eq. (2–86), is

$$\begin{aligned}(r^{(k-1)}, Ap^{(k-1)}) &= \frac{1}{q_{k-1}}[(r^{(k-1)}, r^{(k-1)}) - (r^{(k-1)}, r^{(k-2)})] \\ &= \frac{1}{q_{k-1}}(r^{(k-1)}, r^{(k-1)}).\end{aligned}$$

The combination of (2–83) and (2–88) yields, after the last equation,

$$e_{k-1} = \frac{(r^{(k-1)}, r^{(k-1)})}{(r^{(k-2)}, r^{(k-2)})} \qquad k = 2, 3, \ldots. \tag{2–90}$$

The values e_{k-1} of Eq. (2–90) appear as inner products, formed from consecutive residual vectors. As long as $r^{(k-1)}$ is nonzero, the values e_{k-1} are greater than zero.

Theorem 2-12. *In the method of conjugate gradients, the relaxation directions* $p^{(k)}$ $(k = 1, 2, \ldots)$ *form a system of conjugate directions, and the residual vectors* $r^{(k)}(k = 0, 1, 2, \ldots)$ *form an orthogonal system.*

Proof: We prove the properties by complete induction on the index k of relaxation steps.

Induction premise: After the kth step with $k \geq 1$,

$$(r^{(i)}, r^{(j)}) = 0 \quad \text{for} \quad i \neq j \qquad 0 \leq i, j \leq k; \tag{2–91}$$

$$(p^{(i)}, Ap^{(j)}) = 0 \quad \text{for} \quad i \neq j \qquad 1 \leq i, j \leq k. \tag{2–92}$$

Furthermore, the residual vectors $r^{(0)}, r^{(1)}, \ldots, r^{(k)}$ are nonzero.

Induction assertion: The vectors $\boldsymbol{r}^{(k+1)}$ and $\boldsymbol{p}^{(k+1)}$ fulfill

$$(\boldsymbol{p}^{(k+1)}, \boldsymbol{A}\boldsymbol{p}^{(j)}) = 0 \qquad j = 1, 2, \ldots, k; \tag{2–93}$$

$$(\boldsymbol{r}^{(k+1)}, \boldsymbol{r}^{(j)}) = 0 \qquad j = 0, 1, \ldots, k. \tag{2–94}$$

Induction proof: For $j = k$, (2–93) is fulfilled because of the construction of $\boldsymbol{p}^{(k+1)}$. By (2–82), it follows for $1 \leq j < k$, via (2–92), that

$$(\boldsymbol{p}^{(k+1)}, \boldsymbol{A}\boldsymbol{p}^{(j)}) = -(\boldsymbol{r}^{(k)}, \boldsymbol{A}\boldsymbol{p}^{(j)}) + e_k(\boldsymbol{p}^{(k)}, \boldsymbol{A}\boldsymbol{p}^{(j)}) = -(\boldsymbol{r}^{(k)}, \boldsymbol{A}\boldsymbol{p}^{(j)}).$$

By (2–89), with j in place of $k - 1$, this becomes

$$(\boldsymbol{p}^{(k+1)}, \boldsymbol{A}\boldsymbol{p}^{(j)}) = -\frac{1}{q_j}[(\boldsymbol{r}^{(k)}, \boldsymbol{r}^{(j)}) - (\boldsymbol{r}^{(k)}, \boldsymbol{r}^{(j-1)})] = 0$$

as a result of (2–91) and of q_j being greater than zero.

The assertion (2–94) holds for $j = k$ by (2–86). It remains to be proven for $0 \leq j < k$. Because of the recursion formula for the residual vectors (2–85), it follows, in accordance with the assertion (2–91), that

$$(\boldsymbol{r}^{(k+1)}, \boldsymbol{r}^{(j)}) = (\boldsymbol{r}^{(k)}, \boldsymbol{r}^{(j)}) + q_{k+1}(\boldsymbol{A}\boldsymbol{p}^{(k+1)}, \boldsymbol{r}^{(j)}) = q_{k+1}(\boldsymbol{A}\boldsymbol{p}^{(k+1)}, \boldsymbol{r}^{(j)}).$$

By (2–82), with j in place of $k - 1$,

$$\boldsymbol{r}^{(j)} = -\boldsymbol{p}^{(j+1)} + e_j\boldsymbol{p}^{(j)} \qquad j = 1, 2, \ldots, k - 1.$$

Thus

$$(\boldsymbol{r}^{(k+1)}, \boldsymbol{r}^{(j)}) = q_{k+1}[-(\boldsymbol{A}\boldsymbol{p}^{(k+1)}, \boldsymbol{p}^{(j+1)}) + e_j(\boldsymbol{A}\boldsymbol{p}^{(k+1)}, \boldsymbol{p}^{(j)})] \qquad 1 \leq j < k.$$

Because of the self-adjointness of $\boldsymbol{A}$ and the property (2–93) just proven, both inner products of the last expression vanish. Finally, for $j = 0$, $\boldsymbol{r}^{(0)} = -\boldsymbol{p}^{(1)}$ and thus, through (2–93),

$$(\boldsymbol{r}^{(k+1)}, \boldsymbol{r}^{(0)}) = -q_{k+1}(\boldsymbol{A}\boldsymbol{p}^{(k+1)}, \boldsymbol{p}^{(1)}) = 0.$$

Induction check: For $k = 1$, (2–91) is fulfilled through (2–86), while the second part of the induction premise (2–92) is inapplicable.

Theorem 2–12 expresses a remarkable property of the method of conjugate gradients, namely, that the residual vectors form an orthogonal system. They belong to an n-dimensional vector space, and thus the orthogonal system can contain at most n nonzero vectors. At the very latest, the $(n + 1)$th residual vector $\boldsymbol{r}^{(n)}$ must vanish in the sequence of the mutually orthogonal vectors $\boldsymbol{r}^{(0)}, \boldsymbol{r}^{(1)}, \boldsymbol{r}^{(2)}, \ldots$. Thus, at the very latest, the trial vector

$\boldsymbol{v}^{(n)}$ gives the solution because of $\boldsymbol{A}\boldsymbol{v}^{(n)} + \boldsymbol{b} = \boldsymbol{r}^{(n)} = 0$. We formulate this important result in

Theorem 2-13. *The method of conjugate gradients yields the solution in, at most, n steps.*

The conjugate gradients method exhibits the remarkable property that, in contrast to the simultaneous displacement method, no eigenvalues need be calculated in order to assure convergence. In addition, the iteratively constructed method leads theoretically to a *finite* process by Theorem 2–13. The numerical calculation, however, deviates from the theory, since the mutual orthogonality of the residual vectors cannot be maintained exactly, so that in general $\boldsymbol{r}^{(n)}$ differs from zero. The worse the condition of the coefficient matrix A, the greater the deviation. This observation, however, is not troublesome and we simply continue the calculation well past the nth step. Such a decision is justified, since the technique is decidedly a relaxation technique which reduces the value of the quadratic function at every step. Compare this with Sec. 5–2–5 and the detailed analysis of Ref. 13, where the results of bigger numerical problems are given and where refinements of the gradient methods are shown.

The method of conjugate gradients is usable in principle for the solution of an arbitrary symmetric definite system of equations. It is especially advantageous if the matrix $\boldsymbol{A}$ is not full but contains numerous zero elements and if, in addition, the individual equations show an internal regularity. Difference equations resulting from discretization of a boundary value problem possess these properties. The matrix $\boldsymbol{A}$ is then defined through *operator equations*, and the computation of the auxiliary vector $\boldsymbol{z} = \boldsymbol{A}\boldsymbol{p}^{(k)}$ in Eq. (2–88) can take place through a sequence of explicit expressions for the individual components, whereby the many zeros of $\boldsymbol{A}$ are entirely taken into account (see Chapter 5 and Ref. 13). For instance, the determining equations for the matrix of Eq. (2–17) are

$$
\begin{aligned}
z_1 &= 2p_1 - p_2 \\
z_i &= -p_{i-1} + 3p_i - p_{i+1}, \qquad i = 2, 3 \\
z_4 &= -p_3 + 2p_4.
\end{aligned}
$$

For general systems of equations not stemming from boundary value problems, the method of conjugate gradients is not recommended.

2-4-3 The Computational Procedure

The computation steps of the method of conjugate gradients will be grouped in the appropriate sequence for better visualization.

Start:

$$\boxed{\begin{array}{l} \text{Choice of } \boldsymbol{v}^{(0)} \\ \boldsymbol{r}^{(0)} = \boldsymbol{A}\boldsymbol{v}^{(0)} + \boldsymbol{b}, \quad \boldsymbol{p}^{(1)} = -\boldsymbol{r}^{(0)} \end{array}} \tag{2-95}$$

Relaxation Step $(k = 1, 2, \ldots)$:

$$\boxed{\begin{array}{l} \left.\begin{array}{l} e_{k-1} = \dfrac{(\boldsymbol{r}^{(k-1)}, \boldsymbol{r}^{(k-1)})}{(\boldsymbol{r}^{(k-2)}, \boldsymbol{r}^{(k-2)})} \\ \boldsymbol{p}^{(k)} = -\boldsymbol{r}^{(k-1)} + e_{k-1}\boldsymbol{p}^{(k-1)} \end{array}\right\} \quad k \geq 2 \\ q_k = \dfrac{(\boldsymbol{r}^{(k-1)}, \boldsymbol{r}^{(k-1)})}{(\boldsymbol{A}\boldsymbol{p}^{(k)}, \boldsymbol{p}^{(k)})} \\ \boldsymbol{v}^{(k)} = \boldsymbol{v}^{(k-1)} + q_k\boldsymbol{p}^{(k)}, \quad \boldsymbol{r}^{(k)} = \boldsymbol{r}^{(k-1)} + q_k(\boldsymbol{A}\boldsymbol{p}^{(k)}) \end{array}} \tag{2-96}$$

ALGOL Procedure for the Method of Conjugate Gradients. In the vector sequences $\boldsymbol{p}^{(k)}$, $\boldsymbol{v}^{(k)}$, $\boldsymbol{r}^{(k)}$, only the most recent is of interest, and the previous are totally unimportant. The index k may thus be omitted in the program. Nonetheless, we must bear in mind that the value of the inner product $(\boldsymbol{r}^{(k-1)}, \boldsymbol{r}^{(k-1)})$ is still used for the computation of the next e value. The values e_{k-1} and q_k, which are only of transitory significance, are carried along as complete sequences and summarized as results, because they can be used for the computation of the smallest eigenvalues of $\boldsymbol{A}$ (see Sec. 4–6–6). In accord with the last observation of Sec. 2–4–2, the computation of the auxiliary vector $\boldsymbol{z} = \boldsymbol{A}\boldsymbol{p}$ for an arbitrary vector $\boldsymbol{p}$ is defined by a procedure called *op*, which turns up as a formal parameter. For further simplification, the method is inevitably begun with $\boldsymbol{v}^{(0)} = 0$, so that $\boldsymbol{r}^{(0)} = \boldsymbol{b}$. The procedure is terminated as soon as $\boldsymbol{r}^{(k)} = 0$, or else, at latest, after some preselected number $n1$ of steps. A more refined criterion for termination is developed in Ref. 22 and employed in a correspondingly elaborate problem.

The program parameters are defined as:

- n — Order of the system of equations $\boldsymbol{A}\boldsymbol{x} + \boldsymbol{b} = 0$
- $n1$ — Maximum number of relaxation steps allowed
- b — Elements of the constant vector
- *op* — The procedure $op(n, p, z)$ computes the vector $\boldsymbol{z} = \boldsymbol{A}\boldsymbol{p}$ from an n-dimensional vector $\boldsymbol{p}$
- x — Elements of the solution vector after at most $n1$ steps
- q — Values q_k as given in (2–96)
- e — Values e_k as given in (2–96)
- *ncg* — Number of steps carried out so far

ALGOL Procedure No. 5

```
procedure cg (n, n1, b, op, x, q, e, ncg);
          value n, n1;
          integer n, n1, ncg;  array b, x, q, e;
          procedure op;
begin integer i, k;  real rr, rr1, h;  array z, p, r[1 : n];
    for i: = 1 step 1 until n do
    begin x[i] : = 0;  r[i] : = b[i];  p[i] : = -r[i] end;
    for k: = 1 step 1 until n1 do
cgstep:
    begin rr: = 0;  ncg: = k - 1;
        for i: = 1 step 1 until n do
            rr: = rr + r[i] ↑ 2;
        if rr = 0 then goto terminate;
        if k > 1 then
        begin e[k - 1]: = rr/rr1;
            for i: = 1 step 1 until n do
                p[i]: = e[k - 1] × p[i] - r[i]
        end;
        op(n, p, z);
        h: = 0;
        for i = 1 step 1 until n do
            h: = h + p[i] × z[i];
        q[k]: = rr/h;
        for i: = 1 step 1 until n do
        begin x[i] : = x[i] + q[k] × p[i];
                r[i] : = r[i] + q[k] × z[i]
        end i;
        rr1 : = rr
    end k;
    ncg: = n1;
terminate:
end cg
```

Example 2-6. The system of equations (2–17) serves as an illustration. Table 5 summarizes the vectors and individual quantities using the notation of the ALGOL program. In accord with the order of matrix A, four steps are carried out. A computer having six significant places has been simulated. Fewer places have been listed where significant places have been lost. The round-off errors here are not serious.

Table 5 METHOD OF CONJUGATE GRADIENTS

$k =$	0	1	2	3	4
$e_{k-1} =$	—	—	0.218864	0.257925	0.017032
$p^{(k)} =$	—	−1.00000	1.43886	1.63321	0.18044
	—	4.00000	2.21774	−0.61095	0.11736
	—	7.00000	1.00185	1.11467	0.03222
	—	0	3.10067	0.78351	−0.13232
$z^{(k)} = Ap^k =$	—	−6.00000	0.65998	3.87737	0.24352
	—	6.00000	4.21251	−4.58073	0.20386
	—	17.0000	−2.31286	3.17145	0.20386
	—	−7.00000	5.19949	0.45235	−0.23242
$h = (z^{(k)}, p^{(k)}) =$	—	149.000	24.0966	13.0207	0.101252
$q^{(k)} =$	—	0.442953	0.599462	0.286139	0.626723
$v^{(k)} =$	0	−0.442953	0.419589	0.886914	1.00000
	0	1.77181	3.10126	2.92644	2.99999
	0	3.10067	3.70124	4.02019	4.00000
	0	0	1.85873	2.08292	1.99999
$r^{(k)} =$	1	−1.65772	−1.26209	−0.15262	0.00000
	−4	−1.34228	1.18296	−0.12777	−0.00001
	−7	0.53020	−0.85627	0.05121	0.00001
	0	−3.10067	0.01623	0.14566	−0.00000
$rr = (r^{(k)}, r^{(k)}) =$	66	14.4450	3.72573	0.063457	$2_{10}-10$

PROBLEMS

2-9. Solve the systems of equations of Problems 2-2 and 2-5 by the method of conjugate gradients. Show the orthogonality of the residual vectors (as a check on calculations).

2-10. For the system of Problem 2-5 carry out the computations consistently to four significant figures. What is the approximation after two steps, starting from the origin? How large is the deviation from the exact solution, and how large is the residual vector?

3 DATA FITTING

3-1 FORMULATION OF THE PROBLEM

Data fitting involves a set of n known quantities l_i which exhibit the character of measured quantities, from which m other unknowns x_j are to be found. The fitting problem exists whenever the number of available measured quantities is greater than the number needed to determine uniquely the unknowns sought. For the unknowns, a number of determining equations are given. Because of the inaccuracies of the measurements, these equations contain contradictions which are eliminated by applying corrections to the measurements. The corrections must be kept as small as possible in a manner to be discussed, so that the measurements can retain their integrity as much as possible. For instance, we could require that the *sum of the squares* of the corrections be made a minimum (the *Gaussian principle*), or else that the *maximum of the magnitudes* of the corrections be minimized (the *Chebyshev principle*). The Chebyshev principle can be reduced to a problem in linear programming (Ref. 64); this is not as easy to handle as the Gaussian principle. The Gaussian principle, also known as the *method of least squares*, is based on probability theory with an assumption of a Gaussian distribution of random errors in the measurements. The behavior and propagation of observation errors can then be studied. From both a mathematical and a numerical standpoint, the method of least squares is the simplest and easiest to manipulate among the logical alternatives. It is used in most applications, and subsequent to its discovery by *Gauss* in 1794, the method and its applications in calculations have been thoroughly explored.

For a more thorough discussion of the formulation and of the problems encountered in data fitting, the reader is referred to Ref. 26.

3-1-1 Unconstrained Data Fitting

In unconstrained fitting the m unknown quantities x_j ($j = 1, 2, \ldots, m$) are related to the n known, observed quantities l_i ($i = 1, 2, \ldots, n$) via n given functions $f_i(x_1, x_2, \ldots, x_m)$ ($i = 1, 2, \ldots, n$), so that the unknowns have to be found from the system of equations

$$f_i(x_1, x_2, \ldots, x_m) - l_i = 0 \qquad i = 1, 2, \ldots, n; \quad n > m. \tag{3–1}$$

The system (3–1), which is commonly nonlinear, is usually overdetermined and thus not solvable. In order to resolve this contradiction, deviator quantities or *residuals* r_i are introduced into each equation, so that (3–1) is replaced by the system (3–2):

$$f_i(x_1, x_2, \ldots, x_m) - l_i = r_i \qquad i = 1, 2, \ldots, n; \quad n > m. \tag{3–2}$$

The residuals r_i can be interpreted as corrections to the corresponding measurements l_i. For the discussion that follows, assume that all the measurements were carried out with equal accuracy. Otherwise, corresponding weight factors may, in fact must, be utilized (Refs. 26, 64).

The requirement of the Gaussian principle that $\sum_{i=1}^{n} r_i^2$ be minimized can be directly applied only for *linear* functions $f_i(x_1, x_2, \ldots, x_m)$ of the unknowns x_j. In the general case involving nonlinear functions, the equations (3–2) must be *linearized* by choosing approximate values $\bar{x}_j$ of the unknowns x_j and calculating according to the correction principle (Ref. 64) with the substitution $x_j = \bar{x}_j + \xi_j$. The variables ξ_j represent small quantities which turn up as the new unknowns in place of x_j.

In consequence of the linearization, we can write approximately[1]

$$f_i(x_1, x_2, \ldots, x_m) \sim f_i(\bar{x}_1, \bar{x}_2, \ldots, \bar{x}_m) + \sum_{j=1}^{m} \frac{\partial f_i(\bar{x}_1, \bar{x}_2, \ldots, \bar{x}_m)}{\partial x_j} \xi_j. \tag{3–3}$$

Given that

$$\frac{\partial f_i(\bar{x}_1, \bar{x}_2, \ldots, \bar{x}_m)}{\partial x_j} = c_{ij}, \quad f_i(\bar{x}_1, \bar{x}_2, \ldots, \bar{x}_m) - l_i = d_i \tag{3–4}$$

$$i = 1, 2, \ldots, n; \quad j = 1, 2, \ldots, m,$$

the linearized system of Eq. (3–2) becomes

$$\sum_{j=1}^{m} c_{ij}\xi_j + d_i = r_i \qquad i = 1, 2, \ldots, n; \quad n > m. \tag{3–5}$$

[1]The linearization is possible only if the functions $f_i(x_1, x_2, \ldots, x_m)$ are sufficiently differentiable.

The system of *error equations* (3–5) must be solved with the requirement $\sum_{i=1}^{n} r_i^2 =$ minimum. In Eq. (3–5) we find an overdetermined linear system of equations, with more equations than unknowns. In what follows, only linear equations are considered.

Example 3-1. *The Intersection Method in Surveying.* At known fixed points P_i with coordinates (x_i, y_i), azimuth angles are measured to a new point P with unknown coordinates (x, y). For simplicity, assume that at each fixed point P_i the northerly direction is known exactly, so that the azimuth angle ϕ_i between north and the direction from P_i to P can be measured (see Fig. 8).[1] For solving the problem, the actual angles are expressed

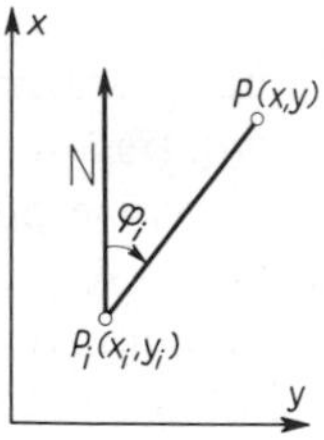

Figure 8. Intersection method.

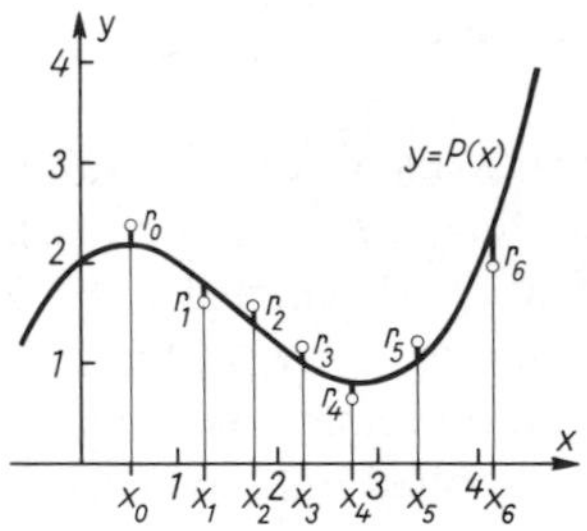

Figure 9. Adjustment to a polynomial.

formally through the known and unknown coordinates, and these are related to the measured quantities. This approach yields the nonlinear error equations

$$\arctan \frac{y - y_i}{x - x_i} - \phi_i = r_i \qquad i = 1, 2, \ldots, n.$$

Choice of approximate coordinates x and y with the correction statement $x = \bar{x} + \xi$, $y = \bar{y} + \eta$ leads us to the linearization, which yields, after elementary computations, the linearized error equations

$$\frac{-(\bar{y} - y_i)\xi + (\bar{x} - x_i)\eta}{(\bar{y} - y_i)^2 + (\bar{x} - x_i)^2} + \arctan\left(\frac{\bar{y} - y_i}{\bar{x} - x_i}\right) - \phi_i = r_i \qquad i = 1, 2, \ldots, n.$$

Example 3-2. Consider the problem of constructing a polynomial $P(x) = \sum_{i=1}^{n} p_j x^j$ of degree $n < N$ giving the closest fit to $N + 1$ points with given Cartesian coordinates (x_i, y_i) $(i = 1, 2, \ldots, N)$, as shown in Fig. 9. This

[1]The coordinate system of Fig. 8 is commonly used in geodesy.

leads directly to the linear system of error equations

$$\sum_{j=0}^{n} p_j x_i^j - y_i = r_i \qquad i = 0, 1, 2, \ldots, N.$$

Here the unknowns are the polynomial coefficients p_j whose number is less than the number of equations. Figure 9 illustrates qualitatively the case $n = 3$ and $N = 6$. The residuals r_i signify the deviation in the y coordinate of the curve $y = P(x)$ from the given points.

The *fitting of direct observations*, in which various measurements l_i for a single sought-after quantity x are known, can be construed to be a special case of unconstrained data fitting, since the error equations are

$$x - l_i = r_i \qquad i = 1, 2, \ldots, n.$$

3-1-2 Constrained Data Fitting

In problems of *constrained fitting* there exist measurements l_j for the n unknowns x_j, and the unknowns x_j must fulfill exactly a sequence of m *constraints* (3–6):

$$f_i(x_1, x_2, \ldots, x_n) = 0 \qquad i = 1, 2, \ldots, m < n \tag{3–6}$$

The values x_j are to be found such that on one hand they satisfy the conditions (3–6), and that on the other hand the sum of the squares of the corrections $v_j = x_j - l_j$ is a minimum according to the Gaussian principle:

$$\sum_{j=1}^{n} v_j^2 = \sum_{j=1}^{n} (x_j - l_j)^2 = \text{minimum}. \tag{3–7}$$

The nonlinear equations (3–6) are linearized according to the *correction principle*. The measured values l_j are available as approximations to the desired quantities x_j. The statement $x_j = l_j + v_j$ with the corrections v_j as the new unknowns puts in place of (3–6) the linearized constraints

$$\sum_{j=1}^{n} \frac{\partial f_i(l_1, l_2, \ldots, l_n)}{\partial x_j} v_j + f_i(l_1, l_2, \ldots, l_n) = 0 \qquad i = 1, 2, \ldots, m < n. \tag{3–8}$$

Utilization of definitions

$$\frac{\partial f_i(l_1, l_2, \ldots, l_n)}{\partial x_j} = p_{ij}, \qquad f_i(l_1, l_2, \ldots, l_n) = q_i \tag{3–9}$$

produces the linearized constraints

$$\sum_{j=1}^{n} p_{ij} v_j + q_i = 0 \qquad i = 1, 2, \ldots, m < n. \tag{3–10}$$

They must be fulfilled while $\sum_{j=1}^{n} v_j^2$ is being minimized. The use of constrained fitting with small corrections $v_j = x_j - l_j$ as unknowns is the most common method used in *geodesy*.

Below, only linear constraints enter into consideration, so that the task of constrained fitting represents a classical extremum problem with linear side conditions.

Example 3-3. In a triangle the three interior angles x_1, x_2, x_3 are measured, yielding the measured quantities l_1, l_2, l_3. Since the sum of the angles of a triangle is 180 degrees, the linear constraint to be fulfilled is

$$x_1 + x_2 + x_3 - 180 = 0.$$

The substitution $x_j = l_j + v_j$ $(j = 1, 2, 3)$ yields

$$v_1 + v_2 + v_3 + (l_1 + l_2 + l_3 - 180) = 0. \qquad (3\text{–}11)$$

The problem then involves making $v_1^2 + v_2^2 + v_3^2$ a minimum under the side condition (3–11).

PROBLEMS

3-1. In a trapezoid *ABCD* having right angles at corners *B* and *C* (see sketch), measurements yield *CD* to be 6 cm long, *AB* 4 cm, *BC* 2.5 cm, *BD* 6.5 cm, and *AC* 4.7 cm. Formulate the problem using both constrained and unconstrained models. Linearize both formulations, using the correction principle. (*Hint:* Note that the unconstrained problem has three unknowns for the five measurements, while in the other case two constraints for five unknowns must be formulated.)

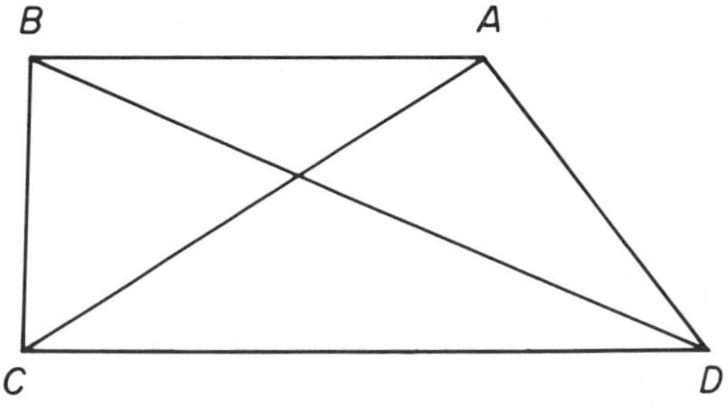

Trapezoid for Problem 3-1.

3-2. In order to find the height x of a tower standing vertically on a horizontal plane (see sketch), vertical angles α, β, and γ were measured from points A, B, and C lying at distances a, b, and c from the base of the tower, respectively.

Formulate the problem in the form both of unconstrained and constrained data fitting. (*Hint:* In the first case, three error equations based on the measured angles are to be set up; in the second, two conditions for the three adjusted angles must be found.)

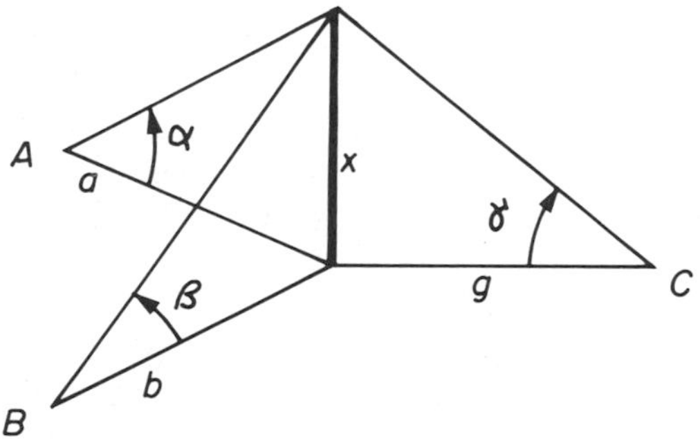

Tower for Problem 3-2.

3-2 UNCONSTRAINED FITTING

3-2-1 Gauss' Normal Equations

The system of n *linear error equations* in the unconstrained fitting of $m < n$ unknowns x_j

$$\sum_{j=1}^{m} c_{ij}x_j + d_i = r_i \qquad i = 1, 2, \ldots, n \tag{3–12}$$

is to be solved by the method of least squares. The elements c_{ij} ($i = 1, 2, \ldots, n$; $j = 1, 2, \ldots, m$) form a "tall" rectangular matrix $\boldsymbol{C} = (c_{ij})$ with more rows than columns. The constants d_i and the residuals r_i form "long" vectors $\boldsymbol{d} = (d_1, d_2, \ldots, d_n)^{\mathrm{T}}$ and $\boldsymbol{r} = (r_1, r_2, \ldots, r_n)^{\mathrm{T}}$, while the unknowns x_j form a "short" vector $\boldsymbol{x} = (x_1, x_2, \ldots, x_m)^{\mathrm{T}}$. Thus the error equations (3–12) may be written

$$\boldsymbol{Cx} + \boldsymbol{d} = \boldsymbol{r}. \tag{3–13}$$

The variety of vectors involved in the error equations of (3–13) may be visualized through Fig. 10. Gauss' principle requires that the square of the length of the residual vector $\boldsymbol{r}$ is minimized. This square $(\boldsymbol{r}, \boldsymbol{r})$ can be expressed by the matrix $\boldsymbol{C}$ and the vectors $\boldsymbol{x}$ and $\boldsymbol{d}$, using (3–13):

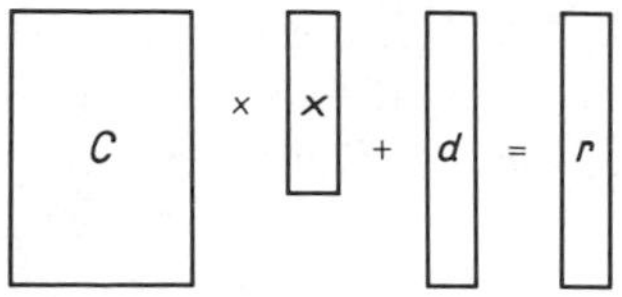

Figure 10. Structure of the error equations.

$$(\boldsymbol{r}, \boldsymbol{r}) = (\boldsymbol{Cx} + \boldsymbol{d}, \boldsymbol{Cx} + \boldsymbol{d}) = (\boldsymbol{Cx}, \boldsymbol{Cx}) + (\boldsymbol{Cx}, \boldsymbol{d}) + (\boldsymbol{d}, \boldsymbol{Cx}) + (\boldsymbol{d}, \boldsymbol{d}).$$

Elementary rearrangement yields the representation

$$(\boldsymbol{r}, \boldsymbol{r}) = (\boldsymbol{C}^{\mathrm{T}}\boldsymbol{Cx}, \boldsymbol{x}) + 2(\boldsymbol{C}^{\mathrm{T}}\boldsymbol{d}, \boldsymbol{x}) + (\boldsymbol{d}, \boldsymbol{d}). \tag{3–14}$$

We can readily check that all the matrix and vector operations in (3–14) make sense. Expression (3–14) represents a quadratic function of the vector $\boldsymbol{x}$ which must be minimized through suitable choice of $\boldsymbol{x}$. Since this is independent of the constant summand $(\boldsymbol{d}, \boldsymbol{d})$, the minimum is found, invoking Theorem 2–1, through the solution of the linear system of equations

$$\boldsymbol{C}^{\mathrm{T}}\boldsymbol{Cx} + \boldsymbol{C}^{\mathrm{T}}\boldsymbol{d} = 0. \tag{3–15}$$

Equations (3–15) are called *Gauss' normal equations*. Their solution yields the values of the unknowns x_j. The elements a_{jk} of the matrix $\boldsymbol{A} = \boldsymbol{C}^{\mathrm{T}}\boldsymbol{C}$ and the components b_j of the constant vector $\boldsymbol{b} = \boldsymbol{C}^{\mathrm{T}}\boldsymbol{d}$ of the normal equations (3–15) $\boldsymbol{Ax} + \boldsymbol{b} = 0$ are computed from the *column vectors* $\boldsymbol{c}_i$ of the matrix $\boldsymbol{C}$ and from the constant vector $\boldsymbol{d}$ as inner products in accord with (3–16) and (3–17).

$$a_{jk} = (\boldsymbol{c}_j, \boldsymbol{c}_k) = \sum_{i=1}^{n} c_{ij}c_{ik} \qquad j, k = 1, 2, \ldots, m. \tag{3–16}$$

$$b_j = (\boldsymbol{c}_j, \boldsymbol{d}) = \sum_{i=1}^{n} c_{ij}d_i \qquad j = 1, 2, \ldots, m. \tag{3–17}$$

Theorem 3-1. *The coefficient matrix* $\boldsymbol{A} = \boldsymbol{C}^{\mathrm{T}}\boldsymbol{C}$ *of Gauss' normal equations is symmetric and its number of rows corresponds to the number of unknowns. In the event that the matrix* $\boldsymbol{C}$ *is of maximal rank* m, *the system of normal equations is symmetric definite.*

Proof: The symmetry of $\boldsymbol{A}$ is evident, by (3–16), because of the commutativity of the inner product. The quadratic form $Q(\boldsymbol{x}) = (\boldsymbol{Ax}, \boldsymbol{x}) = (\boldsymbol{C}^{\mathrm{T}}\boldsymbol{Cx}, \boldsymbol{x}) = (\boldsymbol{Cx}, \boldsymbol{Cx})$ is nonnegative for any $\boldsymbol{x}$, being an inner product of the vector $\boldsymbol{Cx}$ with itself. It vanishes only if $\boldsymbol{Cx} = 0$. The vector $\boldsymbol{Cx}$ represents a linear combination of column vectors of $\boldsymbol{C}$. Because of the given maximum rank of $\boldsymbol{C}$, they are linearly independent. Thus $\boldsymbol{Cx} = 0$ requires that $\boldsymbol{x} = 0$, and with it the positive definiteness of $\boldsymbol{A}$ is established.

Note: If $\boldsymbol{C}$ is not of maximum rank, the system of error equations (3–13) displays a multiplicity of possible solutions such that each solution makes the square of the residual a minimum. This exceptional case calls for special methods which are dependent on the precise statement of the problem. In what follows we consider only the conventional case of positive definite normal equations.

3-2-2 The Solution of the Normal Equations

The symmetric definite property of Gauss normal equations makes possible their solution by the Cholesky method, or by some relaxation scheme. The desirable solution procedure is thereby indicated in principle. But we should be wary of a frequently encountered phenomenon. The normal equations of a larger system of error equations often exhibit an unfavorable condition, so that the unknowns of the normal equations cannot be determined with great accuracy. The condition number κ of a symmetric, positive definite matrix A is defined as the ratio of the largest to the smallest eigenvalue. The largest and smallest eigenvalues, respectively, of a symmetric matrix A are, by Theorem 4–3 (see Sec. 4–3), the maximum and minimum, respectively, of the Rayleigh quotient of all the nonzero vectors $\boldsymbol{x}$:

$$\lambda_{\max} = \max_{x \neq 0} \frac{(A\boldsymbol{x}, \boldsymbol{x})}{(\boldsymbol{x}, \boldsymbol{x})}, \quad \lambda_{\min} = \min_{x \neq 0} \frac{(A\boldsymbol{x}, \boldsymbol{x})}{(\boldsymbol{x}, \boldsymbol{x})}. \tag{3–18}$$

Restricting the normalized vectors $\boldsymbol{x}$ to the m unit vectors $\boldsymbol{e}_k$, Eq. (3–18) yields the bounds

$$\lambda_{\max} = \max_{(x,x)=1} \frac{(A\boldsymbol{x}, \boldsymbol{x})}{(\boldsymbol{x}, \boldsymbol{x})} \geq \max_{k=1,\ldots,m} (A\boldsymbol{e}_k, \boldsymbol{e}_k) = \max_k a_{kk},$$

$$\lambda_{\min} = \min_{(x,x)=1} \frac{(A\boldsymbol{x}, \boldsymbol{x})}{(\boldsymbol{x}, \boldsymbol{x})} \leq \min_{k=1,\ldots,m} (A\boldsymbol{e}_k, \boldsymbol{e}_k) = \min_k a_{kk}.$$

Theorem 3-2. *The largest (smallest) eigenvalue of a symmetric matrix A is at least (at most) as great as the largest (smallest) diagonal element of A.* This gives the useful estimate of the condition number κ of symmetric definite normal equations

$$\kappa \geq \frac{\max_k a_{kk}}{\min_k a_{kk}}. \tag{3–19}$$

Frequently, the smallest eigenvalue is greatly overestimated by the upper bound, thus the bound (3–19) for the condition number would be correspondingly too small.

A more qualitative statement about the smallest eigenvalue of the normal equations is given by the following intuitive argument. Because of $A = C^{\mathrm{T}}C$, Eq. (3–18) signifies the same as $\lambda_{\min} = \min_{(x,x)=1} (C\boldsymbol{x}, C\boldsymbol{x})$. The product $\boldsymbol{y} = C\boldsymbol{x}$ can be interpreted as a linear combination of column vectors $\boldsymbol{c}_i$ of C with the coefficients x_i. Thus $\lambda_{\min}$ is the minimum of the square of the length of all linear combinations of column vectors

$$\boldsymbol{y} = \sum_{i=1}^{m} x_i \boldsymbol{c}_i$$

subject to the constraint

$$\sum_{i=1}^{m} x_i^2 = 1.$$

Under the supposition that the column vectors c_i of C are linearly dependent (that is, C is not of maximum rank m), values x_i exist, with $\sum_{i=1}^{m} x_i^2 = 1$, such that $\sum_{i=1}^{m} x_i c_i = 0$. Then $\lambda_{\min} = 0$ corresponds to A being singular. If the column vectors c_i are not exactly linearly dependent, but only approximately so, it is revealing to note that the zero vector can almost be represented as a linear combination of the c_i's, that is to say, the vector $y = \sum_{i=1}^{m} x_i c_i$ with $\sum_{i=1}^{m} x_i^2 = 1$ can be small though not arbitrarily small. Accordingly, the condition number will be large.

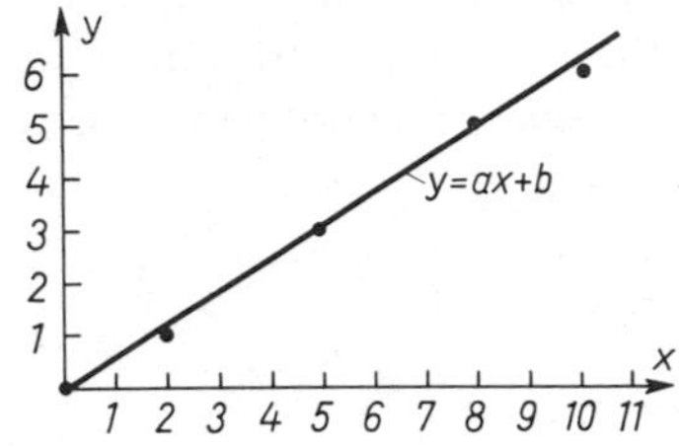

Figure 11. Adjustment to a straight line.

Example 3-4. A straight line with equation $y = ax + b$ is to pass through five measurement points from a physical experiment (see Fig. 11). Their Cartesian coordinates are (0, 0), (2, 1), (5, 3), (8, 5), (10, 6). The unknown coefficients a and b are to be specified in such a manner that the residuals in the error equations (3–20) are minimal in the sense of Gauss' principle:

$$\begin{aligned} b &= r_1 \\ 2a + b - 1 &= r_2 \\ 5a + b - 3 &= r_3 \\ 8a + b - 5 &= r_4 \\ 10a + b - 6 &= r_5 \end{aligned} \qquad C = \begin{bmatrix} 0 & 1 \\ 2 & 1 \\ 5 & 1 \\ 8 & 1 \\ 10 & 1 \end{bmatrix}, \quad d = \begin{bmatrix} 0 \\ -1 \\ -3 \\ -5 \\ -6 \end{bmatrix}. \tag{3–20}$$

The normal equations and their solutions are

$$\begin{aligned} 193a + 25b - 117 &= 0 \\ 25a + 5b - 15 &= 0 \end{aligned} \qquad a = \frac{21}{34}, \quad b = -\frac{3}{34}.$$

The two eigenvalues of the matrix $A = C^T C$ are $\lambda_{\max} = 196.268$ and $\lambda_{\min} = 1.732$, so that the condition number $\kappa \cong 113$ is not too bad. The estimate (3–19) yields $\kappa \geq 38.6$.

Example 3-5. In order to illustrate a poor condition of normal equations, the equation of a straight line $y = ax + b$ will be found from an experiment involving a large number of measurements. The ordinates y_i are measured at n equidistant integer values of abcissa $x_l = l$ ($l = 1, 2, \ldots, n$). The error equations are

$$la + b - y_l = r_l \qquad l = 1, 2, \ldots, n.$$

The coefficient matrix A of the normal equations is

$$A = \begin{bmatrix} \sum_{l=1}^{n} l^2 & \sum_{l=1}^{n} l \\ \sum_{l=1}^{n} l & n \end{bmatrix} = \begin{bmatrix} \frac{1}{6}n(n+1)(2n+1) & \frac{1}{2}n(n+1) \\ \frac{1}{2}n(n+1) & n \end{bmatrix}.$$

By Theorem 3–2, $\lambda_{\max} \geq n(n+1)(2n+1)/6 > n^3/3$. Furthermore, the product of the two eigenvalues of A equals determinant $|A| = n^2(n^2 - 1)/12$, so that for $\lambda_{\min}$ we have the estimate

$$\lambda_{\min} = \frac{n^2(n^2-1)}{12\lambda_{\max}} < \frac{n^4}{12\lambda_{\max}} < \frac{n}{4}.$$

For the condition number κ we have the inequality

$$\kappa = \frac{\lambda_{\max}}{\lambda_{\min}} > \frac{4}{3}n^2.$$

The condition number κ grows with the square of n and can thus be arbitrarily large. For $n = 100$, $\kappa > 13{,}333$, so that in a numerical solution of the normal equations, we can count on losing four significant figures.

Example 3-6. A poor condition of normal equations is often especially prominent when a polynomial of degree $n < N$ is to pass through $N + 1$ points (see Example 3–2). For instance, suppose that $N + 1 = 8$ points are given, with equidistant abcissas $x_l = l$ ($l = 1, 2, \ldots, 8$), and polynomial of degree $n = 4$ is to be found. The elements c_{ij} of the error equation matrix $\boldsymbol{C}$ are given by the values $c_{ij} = x_i^{j-1}$ ($i = 1, 2, \ldots, 8$; $j = 1, 2, 3, 4, 5$). With the use of the diagonal elements of the pertinent normal equation matrix, the estimate (3–19) yields $\kappa \geq 3.08 \cdot 10^6$. The condition number is significantly greater if we are given $N + 1 = 12$ points with abcissae $x_1 = -6$, $x_2 = -5, \ldots, x_6 = -1$, $x_7 = 1, \ldots, x_{12} = 6$ to establish a polynomial of degree $n = 8$. The estimate (3–19) gives the result $\kappa \geq 4.96 \cdot 10^{11}$. A digital computer with 11-place mantissa will quite possibly give an absurd result, since the matrix $\boldsymbol{A} = \boldsymbol{C}^T\boldsymbol{C}$ is numerically singular.

PROBLEMS

3-3. The system of error equations

$$\begin{aligned} x_1 + x_2 - 1 &= r_1 \\ x_1 + 2x_2 - 3 &= r_2 \\ x_1 + 3x_2 - 6 &= r_3 \\ x_1 + 4x_2 - 10 &= r_4 \\ x_1 + 5x_2 - 15 &= r_5 \end{aligned}$$

is to be solved, using normal equations. What is the condition of the matrix of the normal equations?

3-4. In order to determine the coefficients of a quadratic function $y = ax^2 + bx + c$, n ordinates y_i are measured at equidistant points $x_i = i$ $(i = 1, 2, \ldots, n)$. What is the minimum growth of the condition number of the corresponding matrix of the normal equations, as a function of n? More generally, what is the estimate for a polynomial of pth degree $(p \ll n)$?

3-3 CONSTRAINED FITTING

3-3-1 The Correlate Equations

The task of constrained data fitting consists of determining the values of n unknowns $x_1, x_2, \ldots, x_n$, for which there exist measurements $l_1, l_2, \ldots, l_n$, such that the unknowns satisfy m linear constraints

$$\sum_{j=1}^{n} p_{ij}x_j + q_i = 0 \qquad i = 1, 2, \ldots, m < n \tag{3–21}$$

and at the same time minimize the quantity

$$\sum_{j=1}^{n} (x_j - l_j)^2 = \sum_{j=1}^{n} v_j^2. \tag{3–22}$$

This extremum problem with side conditions will be solved, using the method of *Lagrange multipliers.* The multipliers will be designated $-2t_i$ $(i = 1, 2, \ldots, m)$. This problem is equivalent to that of finding the stationary value of the *Lagrange function* (3–23):

$$L = \sum_{j=1}^{n} (x_j - l_j)^2 - 2\sum_{i=1}^{m} t_i \left\{ \sum_{j=1}^{n} p_{ij}x_j + q_i \right\}. \tag{3–23}$$

For this, the first derivative of L with respect to the unknowns x_j must necessarily vanish.

$$\frac{\partial L}{\partial x_j} = 2(x_j - l_j) - 2\sum_{i=1}^{n} t_i p_{ij} = 0, \qquad j = 1, 2, \ldots, n. \tag{3–24}$$

Together with the m constraint equations (3–21), there are $n + m$ linear equations for the unknown quantities $x_1, x_2, \ldots, x_n; t_1, t_2, \ldots, t_m$ which can, in addition, be greatly simplified.

The elements p_{ij} $(i = 1, 2, \ldots, m;\ j = 1, 2, \ldots, n)$ form a "wide" rectangular matrix $\boldsymbol{P} = (p_{ij})$ with fewer rows than columns. The unknowns x_j, the measurements l_j, and the corrections v_j are "long" vectors $\boldsymbol{x} = (x_1, x_2, \ldots, x_n)^T$, $\boldsymbol{l} = (l_1, l_2, \ldots, l_n)^T$, $\boldsymbol{v} = (v_1, v_2, \ldots, v_n)^T$, each with n components. The constants q_i and the *correlates* t_i form "short" vectors, each with m components, namely, $\boldsymbol{q} = (q_1, q_2, \ldots, q_m)^T$ and the *correlate vector* $\boldsymbol{t} = (t_1, t_2, \ldots, t_m)^T$. Thus the constraints (3–21) are

$$\boldsymbol{Px} + \boldsymbol{q} = 0, \tag{3–25}$$

and the requirement (3–22) is

$$(\boldsymbol{v}, \boldsymbol{v}) = (\boldsymbol{x} - \boldsymbol{l}, \boldsymbol{x} - \boldsymbol{l}) = \text{minimum}. \tag{3–26}$$

The variety of vectors involved in the equations of condition (3–25) are shown in Fig. 12.

Solution of the necessary conditions (3–24) for vector $\boldsymbol{x}$ gives the *correlate equations*

$$\boldsymbol{x} = \boldsymbol{P}^T\boldsymbol{t} + \boldsymbol{l}. \tag{3–27}$$

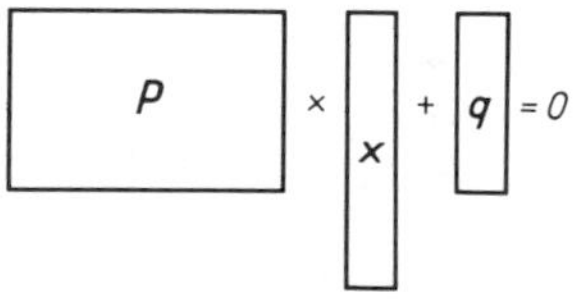

Figure 12. Structure of the equations of condition.

Equations (3–27) relate the correlates t_i, which are introduced as multipliers, to the vector $\boldsymbol{x}$ to be determined, by way of the given matrix $\boldsymbol{P}$ and the measurement vector $\boldsymbol{l}$. Substitution of the correlate equations (3–27) into the equations of condition (3–25) yields a linear system of *normal equations* for the correlate vector $\boldsymbol{t}$ alone:

$$\boldsymbol{PP}^T\boldsymbol{t} + (\boldsymbol{Pl} + \boldsymbol{q}) = 0. \tag{3–28}$$

Thus we arrive at a system of linear equations not in the unknowns themselves, but in the correlates which were introduced as auxiliary quantities. If the correlates in (3–28) are computed, the required unknowns x_j are found from the correlate equations (3–27).

The elements of the coefficient matrix $\boldsymbol{B} = \boldsymbol{PP}^T$ and the components of the constant vector $\boldsymbol{d} = \boldsymbol{Pl} + \boldsymbol{q}$ of the normal equations (3–28) are computed from the *row vectors* $\boldsymbol{p}_i$ of the matrix $\boldsymbol{P}$ and the vectors $\boldsymbol{l}$ and $\boldsymbol{q}$ as inner products, according to (3–29) and (3–30):

$$b_{ik} = (\boldsymbol{p}_i, \boldsymbol{p}_k) = \sum_{j=1}^{n} p_{ij} p_{kj} \qquad i, k = 1, 2, \ldots, m, \tag{3–29}$$

$$d_i = (\boldsymbol{p}_i, \boldsymbol{l}) + q_i = q_i + \sum_{j=1}^{n} p_{ij} l_j \qquad i = 1, 2, \ldots, m. \tag{3–30}$$

Analogous to unconstrained data fitting, we have

Theorem 3-3. *The matrix* $\boldsymbol{B} = \boldsymbol{P}\boldsymbol{P}^{\mathrm{T}}$ *of the normal equation of constrained data fitting is symmetric, and its number of rows equals the number of constraints. It is positive definite if the constraints are linearly independent; that is, if* $\boldsymbol{P}$ *is of maximal rank* m.

We require the linear independence of the equations of constraint here, because otherwise the number of equations could be reduced. Therefore, the normal equations for the correlates are symmetric definite.

Note: In the event that the fitting problem is formulated in the corrections $v_j = x_j - l_j$, as is indispensable for constraints that are nonlinear in origin, the measurements l_j are taken into account in the coefficients of the constraint equations by (3–9), and the applicable measurement vector for the corrections must, of course, be set equal to zero. Accordingly, (3–27) and (3–28) simplify to

$$\boldsymbol{v} = \boldsymbol{P}^{\mathrm{T}}\boldsymbol{t} \qquad \text{(correlate equations)}, \tag{3–31}$$

$$\boldsymbol{P}\boldsymbol{P}^{\mathrm{T}}\boldsymbol{t} + \boldsymbol{q} = 0 \qquad \text{(normal equations)}. \tag{3–32}$$

This is the conventional form for surveying problems.

Example 3-7. The three angles and three sides of a triangle have been measured, with the values given in Table 6. The measurements have dimensions of angles and of lengths. The sum of the squares of the measurement errors are to be minimized under the working hypothesis that the angles are measured in degrees and the lengths in millimeters.

A triangle is uniquely determined by three suitable quantities. The six unknowns must fulfill three constraints, such as the one specifying that the

Table 6 MEASUREMENTS OF THE TRIANGLE

Unknowns x_i	$\alpha = x_1$	$\beta = x_2$	$\gamma = x_3$	$a = x_4$	$b = x_5$	$c = x_6$
Measured values l_i	67°30′	52°	60°	172 mm	146 mm	160 mm

sum of the angles must be 180 degrees, or ones giving the law of sines formulated for two combinations of pairs of sides. Naturally, any other set of three mutually independent relations of the triangle would also be admissible:

$$\left.\begin{aligned} x_1 + x_2 + x_3 - 180 &= 0 \\ x_4 \sin x_2 - x_5 \sin x_1 &= 0 \\ x_5 \sin x_3 - x_6 \sin x_2 &= 0 \end{aligned}\right\} . \tag{3–33}$$

Linearization with the substitution $x_j = l_j + v_j$ $(j = 1, 2, \ldots, 6)$ converts (3–33) into linear constraints:

$$\left.\begin{aligned} v_1 + v_2 + v_3 + (-180 + l_1 + l_2 + l_3) &= 0 \\ -l_5 v_1 \cos l_1 + l_4 v_2 \cos l_2 + v_4 \sin l_2 - v_5 \sin l_1 & \\ + (l_4 \sin l_2 - l_5 \sin l_1) &= 0 \\ -l_6 v_2 \cos l_2 + l_5 v_3 \cos l_3 + v_5 \sin l_3 - v_6 \sin l_2 & \\ + (l_5 \sin l_3 - l_6 \sin l_2) &= 0 \end{aligned}\right\} . \tag{3–34}$$

Considering that the linearization formula $\sin(x_0 + \Delta x) \approx \sin x_0 + \Delta x \cos x_0$ is correct for arguments in arc measure, we can write (3–34) by Table 6 in the form

$$\left.\begin{aligned} v_1 + v_2 + v_3 - 0.5000 &= 0 \\ -0.9752 v_1 + 1.8483 v_2 + 0.7880 v_4 - 0.9239 v_5 + 0.6466 &= 0 \\ -1.7194 v_1 + 1.2741 v_3 + 0.8660 v_5 - 0.7880 v_6 + 0.3560 &= 0 \end{aligned}\right\} . \tag{3–35}$$

The matrix $\boldsymbol{P}$ and the vector $\boldsymbol{q}$ of the constraints, and the matrix $\boldsymbol{B}$ and the constant vector $\boldsymbol{d}$ of the normal equations $\boldsymbol{Bt} + \boldsymbol{d} = 0$, with $\boldsymbol{B} = \boldsymbol{PP}^{\mathrm{T}}$ and $\boldsymbol{d} = \boldsymbol{q}$, are given by

$$\boldsymbol{P} = \begin{bmatrix} 1 & 1 & 1 & 0 & 0 & 0 \\ -0.9752 & 1.8483 & 0 & 0.7880 & -0.9239 & 0 \\ -1.7194 & 0 & 1.2741 & 0 & 0.8660 & -0.7880 \end{bmatrix},$$

$$\boldsymbol{q} = \begin{bmatrix} -0.5000 \\ 0.6466 \\ 0.3560 \end{bmatrix},$$

$$\boldsymbol{B} = \begin{bmatrix} 3.0000 & 0.8731 & -0.4453 \\ 0.8731 & 5.8418 & 0.8767 \\ -0.4453 & 0.8767 & 5.9506 \end{bmatrix}, \quad \boldsymbol{d} = \begin{bmatrix} -0.5000 \\ 0.6466 \\ 0.3560 \end{bmatrix}.$$

The Cholesky method yields the upper triangular matrix $\boldsymbol{R}$; forward substitution yields the vector $\boldsymbol{y}$; and backward substitution yields the correlate vector $\boldsymbol{t}$:

$$\boldsymbol{R} = \begin{bmatrix} 1.7321 & 0.5041 & -0.2571 \\ & 2.3638 & 0.4257 \\ & & 2.3882 \end{bmatrix}, \quad \boldsymbol{y} = \begin{bmatrix} 0.2887 \\ -0.3351 \\ -0.0582 \end{bmatrix},$$

$$\boldsymbol{t} = \begin{bmatrix} 0.2030 \\ -0.1374 \\ -0.0244 \end{bmatrix}. \tag{3–36}$$

From this result, using (3–31), we arrive at the correction vector $\boldsymbol{v}$ and the vector $\boldsymbol{x}$ of the desired adjusted values

$$\boldsymbol{v} = (0.379, \quad -0.051, \quad 0.172, \quad -0.108, \quad 0.106, \quad 0.019)^{\mathrm{T}},$$
$$\boldsymbol{x} = \boldsymbol{l} + \boldsymbol{v} = (67.879, \quad 51.949, \quad 60.172, \quad 171.892, \quad 146.106, \quad 160.019)^{\mathrm{T}}.$$

The first three are in units of degrees of angle, the last three in millimeters.

3-3-2 Duality of Data Fitting

The formulation of the problems of fitting by the unconstrained and constrained method can be geometrically interpreted in an n-dimensional space V_n in consequence of their linear character. Thus a mutual relationship between the two can be derived.

In accord with the error equations (3–13) of unconstrained fitting $\boldsymbol{Cx} + \boldsymbol{d} = \boldsymbol{r}$, the residual vector $\boldsymbol{r}$ must be representable as a sum of $\boldsymbol{d}$ and a linear combination of the column vectors $\boldsymbol{c}_i$ of the matrix $\boldsymbol{C}$ with the coefficients x_i. If $\boldsymbol{C}$ has maximum rank m, the residual vector $\boldsymbol{r}$ can be considered as a point in the m-dimensional subspace F', spanned by the m column vectors of $\boldsymbol{C}$, but displaced to the point $\boldsymbol{d}$. Furthermore, since the normal equations require that $\boldsymbol{C}^{\mathrm{T}}\boldsymbol{Cx} + \boldsymbol{C}^{\mathrm{T}}\boldsymbol{d} = \boldsymbol{C}^{\mathrm{T}}(\boldsymbol{Cx} + \boldsymbol{d}) = \boldsymbol{C}^{\mathrm{T}}\boldsymbol{r} = 0$ as a result of Gauss's principle, the residual vector $\boldsymbol{r}$ must be orthogonal to all column vectors of $\boldsymbol{C}$, so that it occupies the foot of the perpendicular from the origin to the subspace F'. Figure 13 shows the case $n = 3$, $m = 2$. The space F' is a two-dimensional plane passing through $\boldsymbol{d}$.

The linear constraints (3–21) $\boldsymbol{Px} + \boldsymbol{q} = 0$ require that the solution point $\boldsymbol{x}$ to be found lie in the intersection of the m hyperplanes defined by the m linear equations. If the constraints are linearly independent, the intersection of the m hyperplanes E_i is an $(n - m)$-dimensional plane F'. The normal vectors of the hyperplanes E_i are given by the row vectors $\boldsymbol{p}_i$ of $\boldsymbol{P}$. In addition,

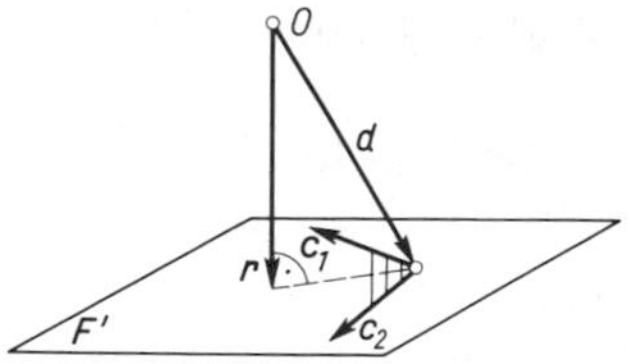

Figure 13. Adjustment by unconstrained fitting.

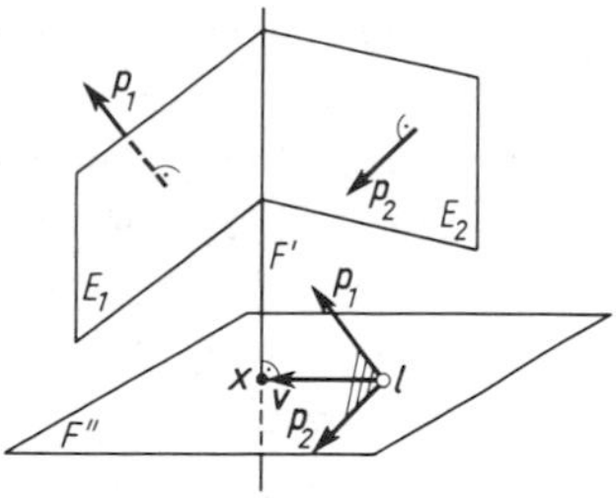

Figure 14. Adjustment by constrained fitting.

as a result of Gauss' principle, the correlate equations $\boldsymbol{x} = \boldsymbol{P}^{\mathrm{T}}\boldsymbol{t} + \boldsymbol{l}$ confine the solution point $\boldsymbol{x}$ to the m-dimensional plane F'' spanned by the m row vectors $\boldsymbol{p}_i$ of $\boldsymbol{P}$, but displaced to the point $\boldsymbol{l}$. It is thus characterized as the intersection point of the two subspaces F' and F'' which are perpendicular to one another, since each vector of one space is entirely orthogonal to every vector in the other space, and the sum of the dimensions of the two spaces is n. The solution point $\boldsymbol{x}$ is the foot of the perpendicular from point $\boldsymbol{l}$ to the space F'.

Figure 14 shows the case $n = 3$, $m = 2$. The intersection of the planes E_1 and E_2 is a one-dimensional line F'. The two-dimensional space F'' defined by the correlate equations is a plane through the point $\boldsymbol{l}$. The solution point $\boldsymbol{x}$ is the intersection of F' and F'' and simultaneously the perpendicular of $\boldsymbol{l}$ on the line F'.

The solution point $\boldsymbol{x}$ of the constrained fitting problem can be characterized as an intersection of two mutually perpendicular subspaces F' and F''. The same is true for unconstrained fitting, since we add to the subspace F', in which the residual vector $\boldsymbol{r}$ must lie, a space F'' perpendicular to it. F$''$ is the intersection of m hyperplanes through the origin, whose normal directions are given by the column vectors $\boldsymbol{c}_i$. Therefore, fitting with and without constraints are both equivalent to the same problem of finding the intersection point of two mutually perpendicular planes F' and F''.

This result can be expressed algebraically as follows. The m unknowns x_i of the unconstrained fit occur as coefficients of the linear combination of n-dimensional column vectors $\boldsymbol{c}_i$ in the error equations, while the n unknowns x_i of the constrained fit represent the coordinates of the solution point in V_n directly. Conversely, the m correlates t_i of constrained fitting are coefficients of the linear combination of n-dimensional row vectors $\boldsymbol{p}_i$ in the constraint equations, while the n components r_i of the residual vector in the unconstrained fit represent the solution point in V_n. Thus a unique or *dual* interrelationship exists between the quantities of the two forms of data fitting. The error equations and the correlate equations are *dual* to one another.

Duality Principle of Data Fitting. *For every problem of fitting data without constraints there exists a corresponding dual problem of constrained data fitting, and vice versa.*

The duality of the calculus of observations permits the linking of one problem to the other algebraically. Frequently, this can be carried out easily; it is even possible at times to formulate the same fitting problem in two ways. For instance, fitting of three measured angles of a triangle can be formulated not merely as a constrained fitting, as in Example 3–3, but also such that the two adjusted values x_1 and x_2 of two angles to be evaluated can be chosen as unknowns, and the adjusted value of the third angle can be expressed with the help of the sum of the angles as $(180 - x_1 - x_2)$. This procedure is the source of a linear system of error equations of unconstrained data fitting:

$$\begin{aligned} x_1 \qquad\qquad &- l_1 = r_1 \\ x_2 \qquad &- l_2 = r_2 \\ -x_1 - x_2 + (180 &- l_3) = r_3. \end{aligned}$$

The transition from one problem to the dual one and the simultaneous treatment of both problems have practical applications in regard to considerations of *hypercircles* (Refs. 36, 65), which yield upper and lower bounds for the magnitude of the residual vector.

PROBLEMS

3-5. For a function $y = f(x)$, the values $y_1 = 270$, $y_2 = 260$, $y_3 = 248$, $y_4 = 235$, and $y_5 = 213$ are measured at equidistant locations $x_1 = 1$, $x_2 = 2, \ldots,$ $x_5 = 5$. The ordinates y_i are to be given the smallest possible alteration, in the sense of least squares, such as to fit exactly a parabola of second degree. (*Hint:* For any four consecutive ordinates, the third difference must be $y_i - 3y_{i-1} + 3y_{i-2} - y_{i-3} = 0$. This leads to two constraint equations.) What is the corresponding dual problem?

3-4 THE METHOD OF ORTHOGONALIZATION IN DATA FITTING

Section 3–2–2 pointed out the difficulties of solving normal equations because of their poor condition. In this section we develop a method for solution of error equations avoiding the normal equations, leading to greater numerical stability. The method can also be applied to constrained fitting.

3-4-1 Schmidt Orthogonalization

We are given p linearly independent vectors $\boldsymbol{a}_i$ $(i = 1, 2, \ldots, p)$ of an n-dimensional vector space $(p \leq n)$. In the p-dimensional subspace spanned by them, a basis of p orthonormal vectors $\boldsymbol{b}_i$ is to be found. These are methodically determined by the *Schmidt orthogonalization process*, using a suitable linear combination of the given vectors.

First step: The first given vector $\boldsymbol{a}_1$, which is nonzero as a consequence of the linear independence, is normalized by division by its own norm, which will be designated for later use as

$$r_{11} = \sqrt{(\boldsymbol{a}_1, \boldsymbol{a}_1)} \neq 0.$$

Then

$$\boldsymbol{b}_1 = \boldsymbol{a}_1 / r_{11}.$$

General kth step: $\boldsymbol{b}_1, \boldsymbol{b}_2, \ldots, \boldsymbol{b}_{k-1}$ are the orthonormal vectors already found by the *Schmidt* technique, constructed from the $(k-1)$ linearly independent vectors $\boldsymbol{a}_1, \boldsymbol{a}_2, \ldots, \boldsymbol{a}_{k-1}$, such that

$$(\boldsymbol{b}_i, \boldsymbol{b}_j) = \begin{Bmatrix} 1 & i = j \\ 0 & i \neq j \end{Bmatrix} \qquad i, j = 1, 2, \ldots, k-1. \tag{3–37}$$

In order to find $\boldsymbol{b}_k$ with the help of $\boldsymbol{a}_k$, a vector $\boldsymbol{x}$ must be constructed, orthogonal to the orthonormal vectors $\boldsymbol{b}_1, \boldsymbol{b}_2, \ldots, \boldsymbol{b}_{k-1}$, through the substitution

$$\boldsymbol{x} = \boldsymbol{a}_k - \sum_{j=1}^{k-1} r_{jk} \boldsymbol{b}_j. \tag{3–38}$$

The coefficient of $\boldsymbol{a}_k$ in (3–38) can be normalized to unity, since it is certainly nonzero in accord with our procedure. The required orthogonality of $\boldsymbol{x}$ means that

$$(\boldsymbol{b}_i, \boldsymbol{x}) = (\boldsymbol{b}_i, \boldsymbol{a}_k) - \sum_{j=1}^{k-1} r_{jk}(\boldsymbol{b}_i, \boldsymbol{b}_j) = 0 \qquad i = 1, 2, \ldots, k-1,$$

and yields, with invocation of (3–37), the *explicit* formulas for r_{ik}

$$r_{ik} = (\boldsymbol{b}_i, \boldsymbol{a}_k) \qquad i = 1, 2, \ldots, k-1. \tag{3–39}$$

Here we see the advantage of the substitution (3–38) as a linear combination of $\boldsymbol{a}_k$ and the vectors $\boldsymbol{b}_1, \boldsymbol{b}_2, \ldots, \boldsymbol{b}_{k-1}$ which have already been orthonormalized. In contrast, a representation of $\boldsymbol{x}$ as a linear combination of given $\boldsymbol{a}_i$ leads at every step to the problem of solving a general system of linear equations.

The coefficients r_{ik} (3–39) are computed as inner products of the orthonormalized vectors $\boldsymbol{b}_i$ with the newly added vector $\boldsymbol{a}_k$. The vector $\boldsymbol{x}$ found by Eq. (3–38) is always nonzero in consequence of the required linear independence of the $\boldsymbol{a}_i$'s and can be normalized as

$$\boldsymbol{b}_k = \left(\boldsymbol{a}_k - \sum_{j=1}^{k-1} r_{jk}\boldsymbol{b}_j\right)\Big/ r_{kk} \tag{3-40}$$

with the normalizing constant

$$r_{kk} = \sqrt{\left(\boldsymbol{a}_k - \sum_{j=1}^{k-1} r_{jk}\boldsymbol{b}_j,\ \boldsymbol{a}_k - \sum_{j=1}^{k-1} r_{jk}\boldsymbol{b}_j\right)}. \tag{3-41}$$

In the case of linear independence of the p vectors $\boldsymbol{a}_i$ given, the constructive method ends up with a system of p orthonormal vectors $\boldsymbol{b}_i$. Linear dependence of the given vectors $\boldsymbol{a}_i$ results in the disappearance of a normalizing constant r_{kk}. This phenomenon may be used as a test for linear independence of the given vectors.

It is important to note in the Schmidt method that in arriving at an additional linearly independent vector $\boldsymbol{a}_{p+1}$ and in extending the system, the p orthonormal vectors $\boldsymbol{b}_1, \boldsymbol{b}_2, \ldots, \boldsymbol{b}_p$ remain unaltered. The procedure can easily be continued. The sequence in which the given vectors $\boldsymbol{a}_i$ are fed into the procedure significantly affects the system of orthonormal vectors $\boldsymbol{b}_i$. Since the $\boldsymbol{a}_i$'s are normally used in ascending order, $\boldsymbol{a}_1, \boldsymbol{a}_2, \ldots, \boldsymbol{a}_p$ are frequently said to be orthonormalized from left to right.

In order to clarify the connection between the given and the orthonormalized vectors, the relations (3–40) will be solved for the $\boldsymbol{a}_k$'s:

$$\left.\begin{aligned}
\boldsymbol{a}_1 &= r_{11}\boldsymbol{b}_1 \\
\boldsymbol{a}_2 &= r_{12}\boldsymbol{b}_1 + r_{22}\boldsymbol{b}_2 \\
\boldsymbol{a}_3 &= r_{13}\boldsymbol{b}_1 + r_{23}\boldsymbol{b}_2 + r_{33}\boldsymbol{b}_3 \\
&\cdots \\
\boldsymbol{a}_p &= r_{1p}\boldsymbol{b}_1 + r_{2p}\boldsymbol{b}_2 + r_{3p}\boldsymbol{b}_3 + \cdots + r_{pp}\boldsymbol{b}_p
\end{aligned}\right\}. \tag{3-42}$$

Now consider the p vectors $\boldsymbol{a}_i$ as column vectors of a rectangular matrix $\boldsymbol{A}$ with n rows and p columns ($p \leq n$). Similarly, consider the $\boldsymbol{b}_i$'s as columns of an $(n \times p)$ matrix $\boldsymbol{B}$, and use the coefficients r_{ik}, which are defined only for $i \leq k$, to define an upper triangular matrix $\boldsymbol{R}$ of order p. Then (3–42) yields

$$\boldsymbol{A} = \boldsymbol{B}\cdot\boldsymbol{R}. \tag{3-43}$$

Theorem 3-4. *The Schmidt orthogonalization method applied to the column vectors of a generally tall $(n \times p)$ matrix $\boldsymbol{A}$ of rank p $(p \leq n)$ yields the product $\boldsymbol{A} = \boldsymbol{B}\cdot\boldsymbol{R}$. Here $\boldsymbol{B}$ is an $(n \times p)$ matrix with orthonormalized columns, and $\boldsymbol{R}$ is a nonsingular upper triangular matrix of order p. If $\boldsymbol{A}$ is a nonsingular square matrix, the method yields a decomposition of $\boldsymbol{A}$ into a product of an orthogonal matrix $\boldsymbol{B}$ and a nonsingular upper triangular matrix.*

A square matrix is called *orthogonal* if its row and column vectors each form a system of orthonormal vectors. If the column vectors of a square matrix are orthonormal, the row vectors are likewise.

ALGOL Procedure for the Orthogonalization Method. In the kth step, as soon as the r_{jk}'s (as inner products of $\boldsymbol{b}_j$ with $\boldsymbol{a}_k$) and the linear combination (3–38) are calculated, the vector $\boldsymbol{a}_k$ is no longer needed. Thus the given column vectors of $\boldsymbol{A}$ may be directly turned into the orthonormal column vectors of the same matrix. This will be done in the procedure that follows. Furthermore, in the kth step, for the sake of simplicity, r_{jk} times the jth column of $\boldsymbol{A}$ is subtracted from the kth column right after computation of r_{jk}. This does not influence the computation of the following r_{jk} of the same step because of the orthonormal property of the first $(k - 1)$ columns. For the application of the procedure in Sec. 3–4–2, the upper triangular matrix $\boldsymbol{R}$ is needed; thus it appears as a parameter. We neglect here the special case of matrix $\boldsymbol{A}$ not having maximal rank p.

The procedure parameters are defined as:

- n Number of rows of $\boldsymbol{A}$, dimension of the vectors
- p Number of columns of $\boldsymbol{A}$, number of vectors ($p \leq n$)
- a Elements of the matrix $\boldsymbol{A}$ whose column vectors are to be orthonormalized. At the end, the columns of $\boldsymbol{A}$ will represent the desired system of orthonormal vectors.
- r Elements of the upper triangular matrix $\boldsymbol{R} = (r_{ik})$ of order p. Only the elements on or above the diagonal are defined.

ALGOL Procedure No. 6

```
procedure orth (n, p, a, r);
          value n, p;  integer n, p;  array a, r;
begin integer i, j, k;  real s;
   for k: = 1 step 1 until p do
   begin comment Computation of r(j, k) and orthogonalization
            of kth column;
      for j: = 1 step 1 until k - 1 do
      begin r[j, k]: = 0;
         for j: = 1 step 1 until n do
            r[j, k]: = r[j, k] + a[i, j] × a[i, k];
         for i: = 1 step 1 until n do
            a[i, k]: = a[i, k] - r[j, k] × a[i, j]
      end j:
      comment Normalization of kth column;
      s: = 0:
      for i: = 1 step 1 until n do
         s: = s + a[i, k] ↑ 2;
```

```
            r[k, k] : = sqrt(s);
            for i: = 1 step 1 until n do
                a[i, k] : = a[i, k]/r[k, k]
        end k
    end orth
```

3-4-2 Application to Data Fitting Problems

With the use of Gauss' principle, the solution of the error equations of unconstrained data fitting

$$\boldsymbol{Cx} + \boldsymbol{d} = \boldsymbol{r} \tag{3–44}$$

is traced back to the solution of the normal equations

$$\boldsymbol{C}^{\mathrm{T}}\boldsymbol{Cx} + \boldsymbol{C}^{\mathrm{T}}\boldsymbol{d} = 0, \quad \text{or} \quad \boldsymbol{C}^{\mathrm{T}}(\boldsymbol{Cx} + \boldsymbol{d}) = \boldsymbol{C}^{\mathrm{T}}\boldsymbol{r} = 0. \tag{3–45}$$

The residual vector $\boldsymbol{r}$ must be, on one hand, a linear combination of $\boldsymbol{d}$ and the column vectors of $\boldsymbol{C}$ according to Eq. (3–44), and, on the other hand, orthogonal to all column vectors of $\boldsymbol{C}$ by (3–45). This observation makes possible the following attack whose motivation is based on the arguments of Sec. 3–2–2. Using the column vectors $\boldsymbol{c}_1, \boldsymbol{c}_2, \ldots, \boldsymbol{c}_m$ of $\boldsymbol{C}$, we must form the system of m orthonormal vectors $\boldsymbol{s}_1, \boldsymbol{s}_2, \ldots, \boldsymbol{s}_m$ by the Schmidt orthogonalization technique; that is, we must achieve the decomposition

$$\boldsymbol{C} = \boldsymbol{SR} \tag{3–46}$$

by Theorem 3–4.

Next, the residual vector $\boldsymbol{r}$ is found through the *orthogonalization* of vector $\boldsymbol{d}$ with respect to $\boldsymbol{s}_1, \boldsymbol{s}_2, \ldots, \boldsymbol{s}_m$, which span the same subspace as $\boldsymbol{c}_1, \boldsymbol{c}_2, \ldots, \boldsymbol{c}_m$. The subsequent normalization is to be omitted:

$$\boldsymbol{r} = \boldsymbol{d} - \sum_{i=1}^{m} (\boldsymbol{d}, \boldsymbol{s}_i)\boldsymbol{s}_i = \boldsymbol{d} - \sum_{i=1}^{m} f_i\boldsymbol{s}_i, \quad f_i = (\boldsymbol{d}, \boldsymbol{s}_i) \qquad i = 1, 2, \ldots, m. \tag{3–47}$$

The values f_i may be collected in a small vector $\boldsymbol{f}$, which by Eq. (3–47) may be defined as

$$\boldsymbol{f} = \boldsymbol{S}^{\mathrm{T}}\boldsymbol{d}. \tag{3–48}$$

Thus, in accord with Eq. (3–47), the residual vector is

$$\boldsymbol{r} = \boldsymbol{d} - \boldsymbol{Sf}. \tag{3–49}$$

In addition, by Eqs. (3–44) and (3–46),

$$\boldsymbol{r} = \boldsymbol{SRx} + \boldsymbol{d}, \tag{3–50}$$

such that from (3–49) and (3–50) the relation $\boldsymbol{SRx} + \boldsymbol{Sf} = 0$ follows, from which, in light of the maximal rank of $\boldsymbol{S}$, it must also follow that

$$\boldsymbol{Rx} + \boldsymbol{f} = 0. \tag{3–51}$$

Equation (3–51) represents a linear system of equations for the unknowns x_i, utilizing the upper triangular matrix $\boldsymbol{R}$ found in orthogonalization and the constant vector $\boldsymbol{f}$ given by Eq. (3–48). The solution is trivially simple, using the method of backward substitution. The residual vector $\boldsymbol{r}$ need not be calculated if we are interested only in the solution vector $\boldsymbol{x}$. Finally, for the solution of the error equations $\boldsymbol{Cx} + \boldsymbol{d} = \boldsymbol{r}$, the algorithm crystallizes:

(a)	$\boldsymbol{C} = \boldsymbol{SR}$	(Orthonormalization)
(b)	$\boldsymbol{f} = \boldsymbol{S}^{\mathrm{T}}\boldsymbol{d}$	
(c)	$\boldsymbol{Rx} + \boldsymbol{f} = 0$	(Backward Substitution)
(d)	$\boldsymbol{r} = \boldsymbol{d} - \boldsymbol{Sf}$	(Residual vector)

(3–52)

Equation (3–52) reveals that with a change of the constant vector $\boldsymbol{d}$ in the error equations, orthogonalization need not be repeated. Knowing $\boldsymbol{S}$ and $\boldsymbol{R}$ suffices to find $\boldsymbol{f}$, $\boldsymbol{x}$, and $\boldsymbol{r}$.

ALGOL Procedure for the Solution of Error Equations. The procedure *orth* given in Sec. 3–4–1 is used as a global quantity. The elements of the upper triangular matrix $\boldsymbol{R}$ are labeled rr_{ik} to avoid a conflict of terminology. The components of the residual vector are r_k, and they are given in the results along with the solutions x_i. The elements of the error equation matrix $\boldsymbol{C}$ are altered through the orthonormalization process. By the end, the matrix has the meaning of $\boldsymbol{S}$. Thus it is possible subsequently to solve another system of error equations having the identical matrix $\boldsymbol{C}$ but a different measurement vector $\boldsymbol{d}$ without orthonormalizing again. Reason enough for retaining elements of the upper triangular matrix $\boldsymbol{R}$!

The procedure parameters are defined as:

- n Number of error equations; rows in $\boldsymbol{C}$
- m Number of unknowns; columns of $\boldsymbol{C}$; order of $\boldsymbol{R}$
- d Elements d_i of error equations
- c Elements c_{ij} of matrix $\boldsymbol{C}$
- x Elements x_i of the solution $\boldsymbol{x}$
- r Elements r_j of the residual vector $\boldsymbol{r}$

rr Elements rr_{ik} of the upper triangular matrix $\boldsymbol{R}$, defined on and above the diagonal

ALGOL Procedure No. 7

```
procedure fehlerorth (n, m, d, c, x, r. rr);
          value n, m;  integer n, m;  array d, c, x, r, rr;
begin integer i, k;  real s;  array f[1 : m];
   orth (n, m, c, rr);
   comment Solution of the quantities f(k) and successive
           orthogonalization of the constant vector;
   for i: = 1 step 1 until n do r[i] : = d[i];
   for k: = 1 step 1 until m do
   begin f[k] : = 0;
      for i: = 1 step 1 until n do
         f[k] : = f[k] + c[i, k] × r[i];
      for i: = 1 step 1 until n do
         r[i] : = r[i] − f[k] × c[i, k]
   end k;
   comment Computation of solutions x[i] by backward substitution;
   for i: = m step −1 until 1 do
   begin s: = f[i];
      for k : = i + 1 step 1 until m do
         s: = s + x[k] × rr[i, k];
      x[i] : = −s/rr[i, i]
   end i
end fehlerorth
```

Example 3-8. The orthonormalization of the column vectors of matrix $\boldsymbol{C}$ (Eq. 3–20) yields the decomposition $\boldsymbol{C} = \boldsymbol{SR}$, with

$$\boldsymbol{S} = \begin{bmatrix} 0 & 0.753426 \\ 0.143964 & 0.558236 \\ 0.359909 & 0.265452 \\ 0.575854 & -0.0273329 \\ 0.719818 & -0.222523 \end{bmatrix}, \quad \boldsymbol{R} = \begin{bmatrix} 13.8924 & 1.79955 \\ 0 & 1.32727 \end{bmatrix}.$$

By Eq. (3–52), furthermore, it follows that

$$\boldsymbol{f} = \boldsymbol{S}^{\mathrm{T}}\boldsymbol{d} = \begin{bmatrix} -8.42187 \\ 0.117211 \end{bmatrix}, \quad \boldsymbol{x} = \begin{bmatrix} 0.617661 \\ -0.088310 \end{bmatrix},$$

$$\boldsymbol{r} = (-0.08831, \;\; 0.14702, \;\; 0, \;\; -0.14703, \;\; 0.08830)^{\mathrm{T}}.$$

Aside from round-off errors and deviations attributable to the condition number, the solution $\boldsymbol{x}$ agrees with the values found in Sec. 3–2–2.

The problem of constrained data fitting, namely, that of finding a vector $\boldsymbol{x}$ such that $(\boldsymbol{x} - \boldsymbol{l}, \boldsymbol{x} - \boldsymbol{l})$ is a minimum obeying the side condition $\boldsymbol{Px} + \boldsymbol{q} = 0$, can be solved equally well in the *homogeneous* special case of $\boldsymbol{q} = 0$ by the *method of orthogonalization*. The homogeneity of the equations of constraint can always be assured through an appropriate substitution of variables. Then the constraints $\boldsymbol{Px} = 0$ require that the solution vector $\boldsymbol{x}$ be orthogonal to all row vectors of $\boldsymbol{P}$. In addition, with reference to the correlate equations $\boldsymbol{x} = \boldsymbol{P}^{\mathrm{T}}\boldsymbol{t} + \boldsymbol{l}$, $\boldsymbol{x}$ must consist of a sum of $\boldsymbol{l}$ and a linear combination of the row vectors of $\boldsymbol{P}$. We are thus confronted by the situation *dual* to the unconstrained fitting method. The problem is solved by the algorithm (3–53) dual to (3–52).

$$\boxed{\begin{array}{llll} \text{(a)} & \boldsymbol{P}^{\mathrm{T}} = \boldsymbol{SR} & & \text{(Orthonormalization of} \\ \text{(b)} & \boldsymbol{f} \ \ = \boldsymbol{S}^{\mathrm{T}}\boldsymbol{l} & & \text{the columns of } \boldsymbol{P}^{\mathrm{T}}) \\ \text{(c)} & \boldsymbol{x} \ \ = \boldsymbol{l} - \boldsymbol{Sf} & & \end{array}} \qquad (3\text{–}53)$$

As an indication of duality, the solution vector $\boldsymbol{x}$ manifests itself through the orthogonalization of $\boldsymbol{l}$. The correlate vector $\boldsymbol{t}$, which is usually of no interest, no longer turns up in this method of solution.

Example 3-9. Let us examine Example 3-7. In order for the method of orthogonalization to be applicable, the constraints must first be made homogeneous through a suitable substitution of the type $\boldsymbol{w} = \boldsymbol{v} + \boldsymbol{z}$. The vector $\boldsymbol{z}$ can be found, for example, as a special solution from equations (3–35) with $z_2 = z_3 = z_5 = 0$, as

$$\boldsymbol{z} = (-0.5000, \quad 0, \quad 0, \quad 0.2018, \quad 0, \quad 0.6392)^{\mathrm{T}}.$$

In the problem considered earlier, $(\boldsymbol{v}, \boldsymbol{v})$ was to be minimized. Now $(\boldsymbol{w} - \boldsymbol{z}, \boldsymbol{w} - \boldsymbol{z})$ must be made a minimum, so that in the formulation for $\boldsymbol{w}$, the vector $\boldsymbol{z}$ plays the role of the measurement vector. Matrix $\boldsymbol{P}$ is unchanged. Table 7 portrays the column vectors of $\boldsymbol{P}^{\mathrm{T}}$, the vector $\boldsymbol{z}$ as well as the column

Table 7 METHOD OF ORTHOGONALIZATION

$\boldsymbol{P}^{\mathrm{T}}$				$\boldsymbol{S}$			
$p_1 =$	$p_2 =$	$p_3 =$	$z =$	$s_1 =$	$s_2 =$	$s_3 =$	$w =$
1	−0.9752	−1.7194	−0.5000	0.5774	−0.5357	−0.5623	−0.1211
1	1.8483	0	0	0.5774	0.6588	−0.0553	−0.0509
1	0	1.2741	0	0.5774	−0.1231	0.6176	0.1720
0	0.7880	0	0.2018	0	0.3333	−0.0594	0.0936
0	−0.9239	0.8660	0	0	−0.3909	0.4323	0.1058
0	0	−0.7880	0.6392	0	0	−0.3300	0.6584

vectors of the orthonormal matrix $\boldsymbol{S}$, and the solution vector $\boldsymbol{x}$ found through orthogonalization.

By (3–53), the upper triangular matrix $\boldsymbol{R}$ and the vector $\boldsymbol{f}$ are given by

$$\boldsymbol{R} = \begin{bmatrix} 1.7321 & 0.5041 & -0.2571 \\ & 2.3638 & 0.4257 \\ & & 2.3882 \end{bmatrix} \quad \boldsymbol{f} = \begin{bmatrix} -0.2887 \\ 0.3351 \\ 0.0582 \end{bmatrix}. \tag{3–54}$$

The desired vector $\boldsymbol{x}$ of adjusted values is

$$\boldsymbol{l} = \boldsymbol{x} + \boldsymbol{w} - \boldsymbol{z} = (67.879,\ 51.949,\ 60.172,\ 171.892,\ 146.106,\ 160.019)^{\mathrm{T}},$$

where the first three components are in degrees of angle, and the last three are in millimeters.

3-4-3 Numerical Comparison with the Cholesky Method

In Secs. 3–2–1 and 3–4–2 two different methods of solution for the evaluation of the error equations in the unconstrained data fit were developed, using the method of least squares (see Table 8).

Table 8 COMPARISON OF SOLUTION METHODS IN EVALUATING ERROR EQUATIONS

$\boldsymbol{Cx} + \boldsymbol{d} = \boldsymbol{r}$	
Normal Equations, Cholesky	*Orthogonalization*
$\boldsymbol{A} = \boldsymbol{C}^{\mathrm{T}}\boldsymbol{C}$, $\boldsymbol{b} = \boldsymbol{C}^{\mathrm{T}}\boldsymbol{d}$ $\boldsymbol{Ax} + \boldsymbol{b} = 0$ (Normal Equations) $\boldsymbol{A} = \boldsymbol{R}^{\mathrm{T}}\boldsymbol{R}$ (Cholesky) $\boldsymbol{R}^{\mathrm{T}}\boldsymbol{y} + \boldsymbol{b} = 0$ (Forward Substitution) $\boldsymbol{Rx} - \boldsymbol{y} = 0$ (Backward Substitution)	$\boldsymbol{C} = \boldsymbol{SR}$ (Orthonormalization) $\boldsymbol{f} = \boldsymbol{S}^{\mathrm{T}}\boldsymbol{d}$ $\boldsymbol{Rx} + \boldsymbol{f} = 0$ (Backward Substitution)

Theorem 3-5. *The two methods of solution of error equations, namely, the Cholesky method for the solution of normal equations and the method of orthogonalization, are mathematically equivalent.*

Proof: First, we show that the upper triangular matrices $\boldsymbol{R}$ are identical in both cases. Indices will be used to distinguish the two. Since the column vectors of $\boldsymbol{S}$ are orthonormal, it follows that

$$\boldsymbol{A} = \boldsymbol{C}^{\mathrm{T}}\boldsymbol{C} = \boldsymbol{R}_{\text{orth}}^{\mathrm{T}}\boldsymbol{S}^{\mathrm{T}}\boldsymbol{S}\boldsymbol{R}_{\text{orth}} = \boldsymbol{R}_{\text{orth}}^{\mathrm{T}}\boldsymbol{R}_{\text{orth}}. \tag{3–55}$$

Also, the Cholesky decomposition $\boldsymbol{A} = \boldsymbol{R}_{\text{Chol}}^{\mathrm{T}}\boldsymbol{R}_{\text{Chol}}$ is unique. Thus $\boldsymbol{R}_{\text{Chol}}$

$= R_{\text{orth}}$. Once this equivalence is established, it follows also that

$$y = -(R^{\mathrm{T}})^{-1}b = -(R^{\mathrm{T}})^{-1}C^{\mathrm{T}}d = -(R^{\mathrm{T}})^{-1}R^{\mathrm{T}}S^{\mathrm{T}}d = -S^{\mathrm{T}}d = -f.$$

Consequently, the two methods of backward substitution are also identical.

As a result of the mathematical equivalence of the two techniques, we can compute the inverse of the matrix of the normal equations (the so-called *weight matrix*) which is frequently of interest, even if the matrix equation does not appear explicitly. Through the identity of upper triangular matrices involved, Eq. (3–55) leads to

$$A^{-1} = R^{-1}(R^{-1})^{\mathrm{T}}. \tag{3–56}$$

Subsequent to inversion of R (see Sec. 1–4–1), the specific elements of interest in the weight matrix, such as the diagonal elements, may be evaluated individually. We should note here the comment at the end of Sec. 1–4–4.

Despite the mathematical equivalence of the two techniques, *numerically* there exists a significant distinction, which is particularly evident in error equations with poorly conditioned normal equations. The observation to follow clarifies only some fundamentals and is thus mostly of a qualitative nature.

Let us examine the origin of the diagonal elements of R, which play a key role. Because of the orthonormalization, r_{kk} is interpreted as the *length* of the kth column vector of C *after* its orthogonalization to the first $(k - 1)$ columns. The length of the kth vector consequently decreases, or at least fails to increase, thus a diminution of numerical accuracy is expected in the obliteration of significant figures. In the Cholesky method, in contrast, r_{kk} originates from the corresponding element a_{kk} in the normal equations. But the value $a_{kk} = (c_k, c_k)$ represents the *square of the length* of the kth column vector of C. This square of length is gradually reduced to the value r_{kk}^2 in the Cholesky method, since nonnegative quantities are successively subtracted from the original one. In this procedure, too, we observe an obliteration of significant figures, the maximum of which can be shown to be related to the condition number of the normal equations. This brings us to the decisive point: Upon the resolution of the normal equations by the Cholesky method, the loss of significant figures occurs as the *square of the length* of the column vectors, while upon orthogonalization, it occurs as the *length* itself. In magnitude, the loss in accuracy in the second case is only half as great as in the first.

The observation demonstrates that matrix R is numerically more accurate if found by the orthogonalization method. Similarly, the computation of $f = S^{\mathrm{T}}d$ is subject to smaller errors than the forward substitution for y, as the loss of significant figures in S is roughly half as great as in R of the Cholesky decomposition. Ultimately, backward substitution in the case of orthogonalization results not only in a more accurate vector $\boldsymbol{f}$, but also a more accurate matrix $\boldsymbol{R}$, making the solution $\boldsymbol{x}$ more accurate, too.

Conclusion: *The orthogonalization method is preferable to that of the normal equations because of its numerical advantages. The solution will be numerically more accurate.*

This formulated statement can be further supplemented by the following qualitative observation: Given that κ is the condition number of the normal equations, the inaccuracy of the largest component of $\boldsymbol{x}$ may be κ units in the last figure (see Sec. 1–2). The orthogonalization method would lead to a corresponding inaccuracy of $\sqrt{\kappa}$ units. Thus the orthogonalization method can yield usable results when the normal equations run up against a stone wall.

The explicit statement given here for unconstrained data fitting also applies analogously for constrained fitting. The mathematical equivalence is evident in Examples 3–7 and 3–9 above.

PROBLEMS

3-6. Solve the system of error equations of Problem 3-3, using the method of orthogonalization.

3-7. Solve the constrained fitting problem of Problem 3-5, using the method of orthogonalization.

3-8. Using accuracy to three figures, solve the system of error equations

$$1.07x_1 + 1.10x_2 - 2.80 = r_1$$
$$1.07x_1 + 1.11x_2 - 2.70 = r_2$$
$$1.07x_1 + 1.15x_2 - 2.50 = r_3$$

both by the method of normal equations and by orthogonalization. All quantities are to be rounded off to three significant figures; for instance, $3 \cdot (1.07^2) = 3 \cdot (1.14) = 3.42$. Show that the resultant matrix of normal equations is numerically indefinite! Orthogonalization, on the other hand, yields a usable result.

3-9. The solution $\boldsymbol{x}$ of error equations $\boldsymbol{Cx} + \boldsymbol{d} = \boldsymbol{r}$ may be expressed directly in terms of the constant vector $\boldsymbol{d}$ as

$$\boldsymbol{x} = -\boldsymbol{R}^{-1} \cdot \boldsymbol{f} = -\boldsymbol{R}^{-1} \cdot \boldsymbol{S}^{\mathrm{T}} \cdot \boldsymbol{d}$$

through the method of orthogonalization, by (3-51) and (3-48). The matrix $\boldsymbol{C}^{+} = \boldsymbol{R}^{-1} \cdot \boldsymbol{S}^{\mathrm{T}}$ is called the *pseudoinverse* of $\boldsymbol{C}$, since formally it yields the solution of the error equations. Verify that $\boldsymbol{C}^{+}\boldsymbol{C} = \boldsymbol{I}$ (identity matrix of order m). What is the pseudoinverse $\boldsymbol{C}^{+}$ for the system of error equations of Problem 3-3? Use it to evaluate the solution $\boldsymbol{x}$.

3-5 THE METHOD OF CONJUGATE GRADIENTS IN DATA FITTING

The fact that the normal equations of constrained and unconstrained data fitting are both symmetric definite by Theorems 3–1 and 3–3 means that relaxation methods may be invoked for the solution. Thus Gauss, for instance, used hand relaxation for the solution of normal equations. In particular, in ref. 62 the method of conjugate gradients is used, and it is shown that the matrix of normal equations can be eliminated because of its origin as a product of two mutually transposed matrices. There ensues a different numerical approach which dispenses with normal equations. For general fitting problems with normal equation matrices having few zero elements, the resultant method is not recommended, as suggested in Sec. 2–4–2. Nonetheless the resultant method has become useful in the solution of large systems of error equations in geodesy, where matrices of error equations do not have a large proportion of nonzero elements (see Ref. 85). The method is shown in Appendix A. Note too that every data-fitting problem can be formulated and computed in the unconstrained as well as the constrained form, in consequence of the duality property of data fitting. Through simlutaneous solution of dual problems by the method of conjugate gradients, constantly improving *upper* and *lower* bounds for the magnitude of the minimal residual vector are available at every iteration step (see Ref. 36).

3-5-1 Unconstrained Data Fitting by the Method of Conjugate Gradients

The concept of residuals was used in regard to both the relaxation technique and the error equations. In order to avoid a conflict in terminology with the momentary juxtaposition of the two topics, the following designation is used: Let $\boldsymbol{v}^{(k)}$ be an approximation vector to the solution $\boldsymbol{x}$. Take $\boldsymbol{f}^{(k)}$ to indicate the pertinent *error vector* with reference to the error equations

$$\boldsymbol{C}\boldsymbol{v}^{(k)} + \boldsymbol{d} = \boldsymbol{f}^{(k)} \tag{3–57}$$

whereas $\boldsymbol{r}^{(k)}$ will indicate the *residual vector* with reference to the normal equations

$$\boldsymbol{C}^{\mathrm{T}}\boldsymbol{C}\boldsymbol{v}^{(k)} + \boldsymbol{C}^{\mathrm{T}}\boldsymbol{d} = \boldsymbol{r}^{(k)} \tag{3–58}$$

in the conventional sense of relaxation methods. Then clearly it follows that

$$\boldsymbol{r}^{(k)} = \boldsymbol{C}^{\mathrm{T}}\boldsymbol{f}^{(k)}. \tag{3–59}$$

The elimination of the matrix of the normal equations $\boldsymbol{A} = \boldsymbol{C}^{\mathrm{T}}\boldsymbol{C}$ in the denominator of q_k in the computation rule (2–96) takes place through the alteration

$$(\boldsymbol{A}\boldsymbol{p}^{(k)}, \boldsymbol{p}^{(k)}) = (\boldsymbol{C}^{\mathrm{T}}\boldsymbol{C}\boldsymbol{p}^{(k)}, \boldsymbol{p}^{(k)}) = (\boldsymbol{C}\boldsymbol{p}^{(k)}, \boldsymbol{C}\boldsymbol{p}^{(k)}). \tag{3–60}$$

$\boldsymbol{A}$ is eliminated from the recursion formula for the residual vectors $\boldsymbol{r}^{(k)}$ in (2–96) by proceeding to the error equations. For this purpose, the recursion formula $\boldsymbol{v}^{(k)} = \boldsymbol{v}^{(k-1)} + q_k\boldsymbol{p}^{(k)}$ for the approximate vectors in (2–96) is multiplied by $\boldsymbol{C}$.

$$\boldsymbol{C}\boldsymbol{v}^{(k)} = \boldsymbol{C}\boldsymbol{v}^{(k-1)} + q_k(\boldsymbol{C}\boldsymbol{p}^{(k)}). \tag{3–61}$$

Adding the constant vector $\boldsymbol{d}$ to both sides of (3–61), we have, with recourse to (3–57), the recursion formula for the error vectors

$$\boldsymbol{f}^{(k)} = \boldsymbol{f}^{(k-1)} + q_k(\boldsymbol{C}\boldsymbol{p}^{(k)}). \tag{3–62}$$

The relation (3–59) actually represents the connection between the error vectors and the residual vectors. The residual vectors also need to be known along the way, since they are needed for the computation of e_{k-1}, $\boldsymbol{p}^{(k)}$, and q_k. From (2–95) and (2–96), we arrive at the following algorithm for the solution of error equations $\boldsymbol{C}\boldsymbol{x} + \boldsymbol{d} = \boldsymbol{f}$ with the help of the method of conjugate gradients.

Start:

$$\boxed{\text{Choice of } \boldsymbol{v}^{(0)};\ \boldsymbol{f}^{(0)} = \boldsymbol{C}\boldsymbol{v}^{(0)} + \boldsymbol{d}} \tag{3–63}$$

Relaxation step ($k = 1, 2, \ldots.$):

$$\boxed{\begin{aligned}
\boldsymbol{r}^{(k-1)} &= \boldsymbol{C}^{\mathrm{T}}\boldsymbol{f}^{(k-1)} \\
e_{k-1} &= \frac{(\boldsymbol{r}^{(k-1)}, \boldsymbol{r}^{(k-1)})}{(\boldsymbol{r}^{(k-2)}, \boldsymbol{r}^{(k-2)})} && k \geq 2 \\
\boldsymbol{p}^{(k)} &\begin{cases} = -\boldsymbol{r}^{(k-1)} & k = 1 \\ = -\boldsymbol{r}^{(k-1)} + e_{k-1}\boldsymbol{p}^{(k-1)} & k \geq 2 \end{cases} \\
q_k &= \frac{(\boldsymbol{r}^{(k-1)}\ \boldsymbol{r}^{(k-1)})}{(\boldsymbol{C}\boldsymbol{p}^{(k)}, \boldsymbol{C}\boldsymbol{p}^{(k)})} \\
\boldsymbol{v}^{(k)} &= \boldsymbol{v}^{(k-1)} + q_k\boldsymbol{p}^{(k)} \\
\boldsymbol{f}^{(k)} &= \boldsymbol{f}^{(k-1)} + q_k(\boldsymbol{C}\boldsymbol{p}^{(k)})
\end{aligned}} \tag{3–64}$$

In the computation procedure of (3–63) and (3–64), $\boldsymbol{f}^{(k)}$, $\boldsymbol{d}$, and $\boldsymbol{C}\boldsymbol{p}^{(k)}$ are all n-dimensional vectors, while $\boldsymbol{r}^{(k-1)}$, $\boldsymbol{p}^{(k)}$, and $\boldsymbol{v}^{(k)}$ are all m-dimensional.

It is clear that for each relaxation step there are two operations of multiplying a matrix by a vector. The multiplication with $\boldsymbol{A}$ required in the method of conjugate gradients is now carried out in two parts. As a result the computation time per iteration step is increased. Nonetheless the matrix A of the normal equations need not be explicitly calculated.

Some features of the calculation procedure which emanate from those of the method of conjugate gradients are important. One direct consequence of Theorem 2–13 is

Theorem 3-6. *The algorithm* (3–63) *and* (3–64) *yields the solution after, at most, m steps.*

Furthermore, we have

Theorem 3-7. *In the solution of the equations of unconstrained data fitting using the method of conjugate gradients, the magnitude of the error vectors $\boldsymbol{f}^{(k)}$ decreases monotonically, in the strict sense.*

Proof: If we solve the recursion formulas involving the error vectors $\boldsymbol{f}^{(k)}$ in (3–64) for $\boldsymbol{f}^{(k-1)}$, it follows for the scalar product that

$$\begin{aligned}(\boldsymbol{f}^{(k-1)}, \boldsymbol{f}^{(k-1)}) = (\boldsymbol{f}^{(k)}, \boldsymbol{f}^{(k)}) - 2q_k(\boldsymbol{f}^{(k)}, \boldsymbol{C}\boldsymbol{p}^{(k)}) \\ + q_k^2(\boldsymbol{C}\boldsymbol{p}^{(k)}, \boldsymbol{C}\boldsymbol{p}^{(k)}).\end{aligned} \tag{3–65}$$

Since, however, by (3–59) and also (2–87),

$$(\boldsymbol{f}^{(k)}, \boldsymbol{C}\boldsymbol{p}^{(k)}) = (\boldsymbol{C}^{\mathrm{T}}\boldsymbol{f}^{(k)}, \boldsymbol{p}^{(k)}) = (\boldsymbol{r}^{(k)}, \boldsymbol{p}^{(k)}) = 0,$$

then in Eq. (3–65) the middle term vanishes, and (3–65) yields

$$(\boldsymbol{f}^{(k)}, \boldsymbol{f}^{(k)}) = (\boldsymbol{f}^{(k-1)}, \boldsymbol{f}^{(k-1)}) - q_k^2(\boldsymbol{C}\boldsymbol{p}^{(k)}, \boldsymbol{C}\boldsymbol{p}^{(k)}). \tag{3–66}$$

As long as $\boldsymbol{v}^{(k-1)} \neq \boldsymbol{x}$, then the residual vector $\boldsymbol{r}^{(k-1)}$ does not vanish, and neither does the relaxation vector $\boldsymbol{p}^{(k)}$. Also because of the maximal rank of $\boldsymbol{C}$, the vector $\boldsymbol{C}\boldsymbol{p}^{(k)}$ is nonzero, as is q_k. Thus there follows from Eq. (3–66) the strict monotonic nature of $(\boldsymbol{f}^{(k)}, \boldsymbol{f}^{(k)})$ with increasing k.

The significance of Theorem 3–7 is that the iterative process can, if desired, be broken off before the completion of m steps, if the square of the error $(\boldsymbol{f}^{(k)}, \boldsymbol{f}^{(k)})$ is small enough. This is called for if we require merely that the value of $(\boldsymbol{f}^{(k)}, \boldsymbol{f}^{(k)})$ be under a certain limit, and if we are satisfied with a corresponding approximation $\boldsymbol{v}^{(k)}$ for the solution $\boldsymbol{x}$. In the solution of larger systems of error equations of geodesy, we observe that frequently the permissible magnitude of $(\boldsymbol{f}^{(k)}, \boldsymbol{f}^{(k)})$ is reached after relatively few steps, and that the pertinent approximation $\boldsymbol{v}^{(k)}$ resembles the exact solution with sufficient accuracy. This observation, almost needless to say, reduces the total necessary computational effort.

Next, let us consider the execution in practice and its implications. Utilization of the method of conjugate gradients for the solution of error equations is reasonable only if the matrix $\boldsymbol{C}$ has very few nonzero terms. Such is the case in most geodetic applications, where each error equation links but a very small number of unknowns. In order to take advantage of the many zero elements in the calculation of $\boldsymbol{C}\boldsymbol{p}^{(k)}$, it is useful to substitute for $\boldsymbol{C}$ two reference matrices identifying the nonzero elements and their values. We can proceed analogously for a time-saving computation of $\boldsymbol{C}^{\mathrm{T}}\boldsymbol{f}^{(k-1)}$. If, for instance, only 2 per cent of the elements of $\boldsymbol{C}$ are nonzero, then this compact representation of $\boldsymbol{C}$ and $\boldsymbol{C}^{\mathrm{T}}$ takes up only 10 per cent as many cells in computer memory as the complete matrix $\boldsymbol{C}$. Through further improvements, the memory occupancy can be reduced further.

Given the assumption that the number of error equations is roughly double the number of unknowns, a count of the multiplicative computation steps taking into account the many zeros in $\boldsymbol{C}$ produces a total computation time proportional to m^2. In comparison, the method of normal equations and the orthogonalization methods are both m^3 processes.

The fact that the method constantly utilizes the given quantities of the error equations is numerically advantageous. As a result, the round-off errors that are inherent in the derivation of the normal equations or in orthogonalization can be avoided.

Conclusion: *For the solution of systems of error equations with many unknowns and a matrix* $\boldsymbol{C}$ *having a great preponderance of zero elements, the method of conjugate gradients is preferable to other methods.*

3-5-2 Constrained Data Fitting Using the Method of Conjugate Gradients

For constrained data fitting, it is preferable to consider the correction vector (or improvement vector) $\boldsymbol{v} = \boldsymbol{x} - \boldsymbol{l}$ as unknown, rather than the solution vector $\boldsymbol{x}$. For this vector $\boldsymbol{v}$, the equations of condition (3–25) are

$$\boldsymbol{P}\boldsymbol{v} + \boldsymbol{w} = 0, \quad \text{with} \quad \boldsymbol{w} = \boldsymbol{P}\boldsymbol{l} + \boldsymbol{q}. \tag{3–67}$$

Consequently, the correlate equations (3–27) become

$$\boldsymbol{v} = \boldsymbol{P}^{\mathrm{T}}\boldsymbol{t}, \tag{3–68}$$

and the normal equations for the correlate vector $\boldsymbol{t}$ are

$$\boldsymbol{P}\boldsymbol{P}^{\mathrm{T}}\boldsymbol{t} + \boldsymbol{w} = 0. \tag{3–69}$$

For the solution of the normal equations (3–69), the method of conjugate gradients gives us a procedure not just for the iterative treatment of approximations $\boldsymbol{t}^{(k)}$ of correlation vector $\boldsymbol{t}$, but also for the computation of the approximations $\boldsymbol{v}^{(k)}$ of the correction vector $\boldsymbol{v}$ via the correlate equations

(3–68). Note carefully that $\boldsymbol{v}^{(k)}$ now indicates an approximation for $\boldsymbol{v}$. The residual vector $\boldsymbol{r}^{(k)}$ is given by

$$\boldsymbol{r}^{(k)} = \boldsymbol{P}\boldsymbol{P}^{\mathrm{T}}\boldsymbol{t}^{(k)} + \boldsymbol{w}. \tag{3–70}$$

With the help of the relation

$$\boldsymbol{v}^{(k)} = \boldsymbol{P}^{\mathrm{T}}\boldsymbol{t}^{(k)} \tag{3–71}$$

and utilizing (3–70) and (3–71), we arrive at

$$\boldsymbol{r}^{(k)} = \boldsymbol{P}\boldsymbol{v}^{(k)} + \boldsymbol{w}, \tag{3–72}$$

which relates the approximation $\boldsymbol{v}^{(k)}$ to the residual vector $\boldsymbol{r}^{(k)}$.

Elimination of the normal equation matrix $\boldsymbol{A} = \boldsymbol{P}\boldsymbol{P}^{\mathrm{T}}$ from the denominator of q_k of the specification (2–96) takes place analogous to the preceding. Also, there follows from the recursion formula in $\boldsymbol{t}^{(k)}$

$$\boldsymbol{t}^{(k)} = \boldsymbol{t}^{(k-1)} + q_k\boldsymbol{p}^{(k)} \tag{3–73}$$

a corresponding recursion formula for the approximations $\boldsymbol{v}^{(k)}$

$$\boldsymbol{v}^{(k)} = \boldsymbol{v}^{(k-1)} + q_k(\boldsymbol{P}^{\mathrm{T}}\boldsymbol{p}^{(k)}) \tag{3–74}$$

if we multiply by $\boldsymbol{P}^{\mathrm{T}}$ consistent with (3–71).

In summary, the method of conjugate gradients applied to constrained data fitting with $\boldsymbol{P}\boldsymbol{v} + \boldsymbol{w} = 0$ and $(\boldsymbol{v}, \boldsymbol{v}) =$ minimum involves the following algorithm:

Start:

$$\boxed{\text{Choice of } \boldsymbol{t}^{(0)} = 0; \quad \boldsymbol{v}^{(0)} = 0.} \tag{3–75}$$

Relaxation steps ($k = 1, 2, \ldots$):

$$\boxed{\begin{aligned}
\boldsymbol{r}^{(k-1)} &= \boldsymbol{P}\boldsymbol{v}^{(k-1)} + \boldsymbol{w} & \\
e_{k-1} &= \frac{(\boldsymbol{r}^{(k-1)}, \boldsymbol{r}^{(k-1)})}{(\boldsymbol{r}^{(k-2)}, \boldsymbol{r}^{(k-2)})} & k \geq 2 \\
\boldsymbol{p}^{(k)} &= \begin{cases} -\boldsymbol{r}^{(k-1)} & k = 1 \\ -\boldsymbol{r}^{(k-1)} + e_{k-1}\boldsymbol{p}^{(k-1)} & k \geq 2 \end{cases} \\
q_k &= \frac{(\boldsymbol{r}^{(k-1)}, \boldsymbol{r}^{(k-1)})}{(\boldsymbol{P}^{\mathrm{T}}\boldsymbol{p}^{(k)}, \boldsymbol{P}^{\mathrm{T}}\boldsymbol{p}^{(k)})} & \\
\boldsymbol{t}^{(k)} &= \boldsymbol{t}^{(k-1)} + q_k\boldsymbol{p}^{(k)} & \\
\boldsymbol{v}^{(k)} &= \boldsymbol{v}^{(k-1)} + q_k(\boldsymbol{P}^{\mathrm{T}}\boldsymbol{p}^{(k)}) &
\end{aligned}} \tag{3-76}$$

In this algorithm, the initial vector $t^{(0)}$ is taken as zero and, $v^{(0)} = 0$ very simply. This is a reasonable choice because, correction vector v will be quite small. The recursion formula for vectors $t^{(k)}$ has been included for the sake of completeness, in order to demonstrate the *duality* of algorithms (3–64) and (3–76). However, the sequence of approximation vectors $t^{(k)}$, may be omitted without hampering the general procedure. Thus we end up with an algorithm for the solution of constrained fitting problems where not only the matrix of the normal equations is eliminated, but also the generally uninteresting correlate vector.

Theorem 2–13 leads to

Theorem 3-8. *The algorithm* (3–75) *and* (3–76) *yields the correction vector* v *associated with constrained data fitting after, at most, m steps.*

In addition, we have

Theorem 3-9. *In constrained data fitting using the method of conjugate gradients, the magnitude of the correction vector* $v^{(k)}$ *increases in the strict sense.*

Proof: Consider the change in the correction vector in the ith relaxation step

$$\Delta v^{(i)} = v^{(i)} - v^{(i-1)} = q_i(P^{\mathrm{T}}p^{(i)}). \tag{3–77}$$

Since the pairs of relaxation vectors $p^{(i)}$ are mutually conjugate, the changes $\Delta v^{(i)}$, in accord with

$$\begin{aligned}(\Delta v^{(i)}, \Delta v^{(j)}) &= q_i q_j(P^{\mathrm{T}}p^{(i)}, P^{\mathrm{T}}p^{(j)}) \\ &= q_i q_j(p^{(i)}, PP^{\mathrm{T}}p^{(j)}) = 0 \quad \text{for} \quad i \neq j,\end{aligned} \tag{3–78}$$

are thus mutually orthogonal. The kth approximation $v^{(k)}$ may, however, be formally represented as a finite sum of $v^{(0)}$ (which vanishes) and the k alteration vectors $\Delta v^{(1)}, \Delta v^{(2)}, \ldots, \Delta v^{(k)}$, as follows:

$$v^{(k)} = v^{(0)} + \sum_{i=1}^{k} \Delta v^{(i)} = \sum_{i=1}^{k} \Delta v^{(i)}. \tag{3–79}$$

In light of the pair orthogonality relation (3–78), it follows that

$$\begin{aligned}(v^{(k)}, v^{(k)}) &= \left(\sum_{i=1}^{k} \Delta v^{(i)}, \sum_{i=1}^{k} \Delta v^{(i)}\right) = \sum_{i=1}^{k} (\Delta v^{(i)}, \Delta v^{(i)}) \\ &= \sum_{i=1}^{k} q_i^2(P^{\mathrm{T}}p^{(i)}, p^{\mathrm{T}}p^{(i)}).\end{aligned} \tag{3–80}$$

The value of $(\boldsymbol{v}^{(k)}, \boldsymbol{v}^{(k)})$ actually increases in the strict sense with increasing k, since the terms in (3–80) are quantities greater than zero as long as the solution $\boldsymbol{v}$ has not yet been reached, by an argument analogous to the one used above. The remarks made in Sec. 3–5–1 may also be applied here.

Note: The theoretical significance of Theorems 3–7 and 3–9 lies in the fact that because of the duality in data fitting every fitting problem can in principle be formulated as either constrained or unconstrained, and can be computed accordingly. If we do both, the simultaneous use of the method of conjugate gradients yields upper and lower bounds (that approach each other at each iteration step) for the magnitude of the minimum error vector, or else of the correction vector (see Ref. 36).

PROBLEMS

3-10. Solve the system of error equations below, which is quite typical for a net adjustment in geodesy, by the method of conjugate gradients. Note the monotonic aspect of the magnitude of the error vectors. What observation can be made as a result for the residual vectors? Practically speaking, how many steps are required to achieve results?

$$\begin{aligned}
x_1 \qquad\qquad\qquad\qquad & - 0.1 = f_1 \\
x_1 - x_2 \qquad\qquad\qquad & + 0.2 = f_2 \\
x_2 - x_3 \qquad\quad & - 0.3 = f_3 \\
x_3 - x_4 & + 0.4 = f_4 \\
x_3 \qquad\quad & + 0.1 = f_5 \\
x_4 & - 0.2 = f_6 \\
-x_1 \qquad\qquad\qquad + x_4 & - 0.3 = f_7
\end{aligned}$$

3-11. Fitting of gravitational field measurements of the earth leads, in its simplest form, to the following problem: In order to find the gravitation at a sequence of m points, n differences of gravitational attraction between pairs of points are measured. At a point A, the gravitational attraction is known. For the measured differences, use random quantities between -1 and $+1$. For the net shown on page 110, set up the error equations (29 equations in 15 unknowns) and solve them on a computer, using the method of conjugate gradients, where the many zero elements are taken into account through reference matrices. Iteration is to be broken off as soon as the quantity (f_k, f_k) no longer decreases numerically over three iterations. How many iterations are needed?

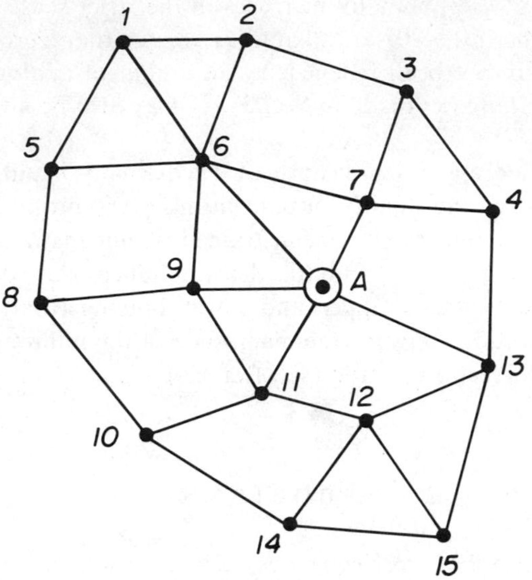

Net for Problem 3-11.

4 SYMMETRIC EIGENVALUE PROBLEMS

4-1 EIGENVALUE PROBLEMS OF PHYSICS

As a representative eigenvalue problem of physics, let us consider the problem of determining the *infinitesimal vibrations* of a general, conservative system capable of oscillations, with n *displacement coordinates* given by q_i $(i = 1, 2, \ldots, n)$. The goal is to find the motion of the system by determining the n displacement coordinates $q_i(t)$ as functions of time t. Assume q_i to be chosen such that $q_1 = q_2 = \cdots = q_n = 0$ corresponds to the equilibrium configuration. For a conservative system, the *potential energy* U is a function of q_i alone; that is, $U = U(q_1, q_2, \ldots, q_n)$. This function can be expanded in powers of q_i in the form

$$U(q_1, q_2, \ldots, q_n) = U_0 + \sum_{i=1}^{n} \alpha_i q_i + \sum_{i=1}^{n} \sum_{k=1}^{n} a_{ik} q_i q_k + \cdots.$$

The constant U_0 corresponding to the potential energy of equilibrium is irrelevant and can be set equal to zero. The linear expression in displacement coordinates q_i vanishes, since for equilibrium $q_i = 0$ the first partial derivatives of the potential energy with respect to each of the displacement coordinates are zero. The next expression in the expansion consists of squares of the coordinates, with which we will content ourselves, since we limit ourselves to motions close to the equilibrium position such that higher powers of the coordinates are of negligible importance. Therefore, with small motions about an equilibrium point, the potential energy can be represented as a quadratic form in q_i with constants a_{ik}, as in

$$U = \sum_{i=1}^{n} \sum_{k=1}^{n} a_{ik} q_i q_k, \qquad A = (a_{ik}) \text{ symmetric.} \tag{4–1}$$

For a stable equilibrium configuration, it is positive definite. The *kinetic energy* of the dynamic system is given by the sum of the squares of velocities v_i multiplied by half of each mass; that is,

$$T = \tfrac{1}{2}\sum_{i=1}^{n} m_i v_i^2.$$

The velocities v_i can be represented as functions of the time derivatives of the displacement coordinates as in $v_i = f_i\,(\dot{q}_1, \dot{q}_2, \ldots, \dot{q}_n)$. This function may be expanded in powers of $\dot{q}_i$, with a vanishing constant term, as in

$$v_i = \sum_{k=1}^{n} c_{ik}\dot{q} + \cdots.$$

We confine ourselves to motions such that second or higher powers of time derivatives of coordinates can be neglected when compared to the linear terms. Thus the expansion of velocities will be limited to the linear part. There results a positive definite quadratic form in the derivatives of the displacement coordinates for the kinetic energy

$$T = \sum_{i=1}^{n}\sum_{k=1}^{n} b_{ik}\dot{q}_i\dot{q}_k, \qquad \boldsymbol{B} = (b_{ik}) \text{ symmetric, positive definite.} \tag{4–2}$$

According to *Hamilton's Principle* (see Ref. 11), the line integral

$$W = \int_{t_0}^{t_1} \left\{\sum_{i=1}^{n}\sum_{k=1}^{n} [b_{ik}\dot{q}_i(t)\dot{q}_k(t) - a_{ik}q_i(t)q_k(t)]\right\} dt \tag{4–3}$$

over two arbitrary instants of time t_0 and t_1 takes on a stationary value. The movement is such that the functions $q_i(t)$ make the line integral stationary when compared with all admissible neighboring virtual motions $q_i^*(t)$; that is, motions having the same time interval, the same initial state $q_i^*(t_0) = q_i(t_0)$, and the same final state $q_i^*(t_1) = q_i(t_1)$. The *variation* of the integral W must vanish with arbitrary admissible functions $\delta q_k(t) = q_k^*(t) - q_k(t)$. The symmetry of the matrices $\boldsymbol{A}$ and $\boldsymbol{B}$ yields the result

$$\delta W = 2\int_{t_0}^{t_1} \left\{\sum_{i=1}^{n}\sum_{k=1}^{n} [b_{ik}\dot{q}_i(\delta\dot{q}_k) - a_{ik}q_i(\delta q_k)]\right\} dt.$$

The operations of variation and time derivative are interchangeable. Thus $(\delta\dot{q}_k) = (\delta q_k)^{\cdot}$. Integration by parts of the first double sum converts this expression into

$$\begin{aligned}\tfrac{1}{2}\delta W = {} & \sum_{i=1}^{n}\sum_{k=1}^{n} b_{ik}\dot{q}_i(\delta q_k)\Big|_{t_0}^{t_1} \\ & - \int_{t_0}^{t_1} \left\{\sum_{i=1}^{n}\sum_{k=1}^{n} [b_{ik}\ddot{q}_i + a_{ik}q_i](\delta q_k)\right\} dt.\end{aligned} \tag{4–4}$$

Since the variation among admissible functions vanishes for both $t = t_0$ and $t = t_1$, the first part of δW is zero. For variations $\delta q_k(t)$ which are otherwise arbitrary, the variation δW of Eq. (4–4) can vanish if and only if the second order differential equations with constant coefficients

$$\sum_{i=1}^{n} (b_{ik}\ddot{q}_i + a_{ik}q_i) = 0 \qquad k = 1, 2, \ldots, n \tag{4–5}$$

are fulfilled. These are the equations of motion for small vibrations about an equilibrium position. Their solution is given by the *normal vibrations*

$$q_i(t) = x_i \cos(\omega t) \qquad i = 1, 2, \ldots, n, \tag{4–6}$$

by which the position coordinates describe motions with the same *frequency* ω, and no phase differences, only with differences in *amplitude* x_i. If this form is substituted back into the equations of motion (4–5), and if the common factor $\cos \omega t$ is canceled out, we arrive at the *homogeneous system of equations* for the amplitudes x_i:

$$\sum_{i=1}^{n} (a_{ik} - \omega^2 b_{ik})x_i = 0 \qquad k = 1, 2, \ldots, n. \tag{4–7}$$

We seek the values $\lambda = \omega^2$ for which the system has a nontrivial solution. Through the symmetry of matrices $\boldsymbol{A}$ and $\boldsymbol{B}$, we have the *generalized symmetric eigenvalue problem*

$$(\boldsymbol{A} - \lambda \boldsymbol{B})\boldsymbol{x} = 0 \tag{4–8}$$

with the symmetric coefficient matrices $\boldsymbol{A}$ and $\boldsymbol{B}$ of the potential and kinetic energy, respectively. Because of the assumption of stable equilibrium, both matrices must be positive definite. The symmetry of the matrices is carried over to the eigenvalue problem, and it can and should be taken advantage of in the solution techniques (for example, see Sec. 4–8). The eigenvalue problem (4–8) has n eigenvalues λ_i yielding *n characteristic frequencies* ω_i of the vibrating system. The eigenvector $\boldsymbol{x}^{(i)}$ appropriate to λ_i has as components the amplitudes of vibration in the position coordinates and thus yields the form of *characteristic motions* of the system belonging to the characteristic frequency ω_i.

Suitable choice of position coordinates often serves to reduce the kinetic energy matrix $\boldsymbol{B}$ to an identity matrix $\boldsymbol{I}$. Then Eq. (4–8) simplifies to the particular eigenvalue problem

$$(\boldsymbol{A} - \lambda \boldsymbol{I})\boldsymbol{x} = 0 \tag{4–9}$$

with symmetric and positive definite matrix $\boldsymbol{A}$. In the following we will consider first the particular eigenvalue problem (4–9) before generalizing to the problem (4–8).

PROBLEMS

4-1. What is the eigenvalue problem of a system having three degrees of freedom whose potential and kinetic energies are given by the following quantities, as a function of coordinate positions q_k?

$$U = 9q_1^2 + 6q_1q_2 + 6q_1q_3 + 17q_2^2 + 10q_2q_3 + 27q_3^2$$
$$T = 4\dot{q}_1^2 + 9\dot{q}_2^2 + 16\dot{q}_3^2$$

4-2. Same as Problem 4-1 for

$$U = 8q_1^2 + 2q_1q_2 - 4q_1q_3 + 20q_2^2 + 6q_2q_3 + 15q_3^2$$
$$T = 4\dot{q}_1^2 + 2\dot{q}_1\dot{q}_2 + 10\dot{q}_2^2 + 4\dot{q}_2\dot{q}_3 + 15\dot{q}_3^2.$$

4-2 CRITICISM OF THE CHARACTERISTIC POLYNOMIAL

Suppose that the special eigenvalue problem $(\boldsymbol{A} - \lambda\boldsymbol{I})\boldsymbol{x} = 0$ is given and the eigenvalues λ_k and matching eigenvectors $\boldsymbol{x}^{(k)}$ are sought. The eigenvalues turn out to be the zeros of the *characteristic polynomial* of the matrix $\boldsymbol{A}$, defined as the determinant of the matrix $\boldsymbol{A} - \lambda\boldsymbol{I}$:

$$P(\lambda) = |\boldsymbol{A} - \lambda\boldsymbol{I}|. \tag{4–10}$$

If $\boldsymbol{A}$ is a matrix of order n, $P(\lambda)$ is a polynomial of true degree n. Its zeros make the matrix $\boldsymbol{A} - \lambda\boldsymbol{I}$ singular and, consequently, nontrivial solutions of the homogeneous system of equations $(\boldsymbol{A} - \lambda\boldsymbol{I})\boldsymbol{x} = 0$ exist.

Theoretically, the problem of determining the eigenvalues is solved when the characteristic polynomial is found and its zeros are computed by some familiar method. From the point of view of function theory, too, polynomials seem to be desirable, since operations with them are easy, they are bounded within any bounded interval, and they can be differentiated as often as we like. Numerically, however, they present difficulties which people in general virtually ignore. The existence of such difficulties in the case of multiple, or at least closely concentrated, zeros is widely known. Apart from this, in more general cases the roots of the polynomials react with great sensitivity to small changes in the polynomial coefficients. This means that the coefficients of the polynomial determine the zeros with a good deal of uncertainty.

In order to examine this phenomenon, consider the polynomial of true degree n with given coefficients a_k:

$$f(z) = a_n z^n + a_{n-1} z^{n-1} + \cdots + a_1 z + a_0, \qquad a_n \neq 0, \tag{4–11}$$

where the zeros $z_1, z_2, \ldots, z_n$ are to be found. *Perturbations* of the form ϵb_k will be applied to the coefficients a_k, where ϵ is to be a small parameter. The values b_k specify another polynomial

$$g(z) = b_n z^n + b_{n-1} z^{n-1} + \cdots + b_1 z + b_0, \tag{4–12}$$

which need no longer be of actual degree n. Besides the given polynomial of Eq. (4–11), consider the perturbed polynomial

$$h(z) = f(z) + \epsilon \cdot g(z), \tag{4–13}$$

whose zeros are given by $\zeta_1, \zeta_2, \ldots, \zeta_n$. We are interested in the differences $\zeta_k - z_k$ $(k = 1, 2, \ldots, n)$, and we assume for this problem that the zeros z_k are simple zeros. We want to compute the changes in zeros of $f(z)$ as compared to those of $h(z)$ in the first approximation, given the perturbation $\epsilon g(z)$, by using the *Newtonian formula* for the new zero ζ_k, with z_k as an approximation. Since $f(z_k) = 0$, we arrive at the approximation

$$\zeta_k \sim z_k - \frac{h(z_k)}{h'(z_k)} = z_k - \frac{f(z_k) + \epsilon g(z_k)}{f'(z_k) + \epsilon g'(z_k)} = z_k - \frac{\epsilon g(z_k)}{f'(z_k) + \epsilon g'(z_k)}. \tag{4–14}$$

Since the zeros z_k are simple, $f'(z) \neq 0$. The correction of z_k in Eq. (4–14) can be further linearized with ϵ sufficiently small, employing the relation

$$\frac{g(z_k)}{f'(z_k) + \epsilon g'(z_k)} \sim \frac{g(z_k)}{f'(z_k)} - \epsilon \frac{g(z_k) g'(z_k)}{[f'(z_k)]^2}. \tag{4–15}$$

From Eqs. (4–14) and (4–15) there follows a first approximation, with retention only of linear terms in ϵ for ζ_k,

$$\zeta_k \sim z_k - \epsilon \frac{g(z_k)}{f'(z_k)}. \tag{4–16}$$

By this formula, the magnitude of the change in z_k depends on the value of the perturbation polynomial $g(z)$ and the derivative of the given polynomial $f(z)$ at z_k.

Example 4-1. Suppose that the zeros of a polynomial of twelfth degree $(n = 12)$ are given as $z_1 = 1$, $z_2 = 2$, $z_3 = 3, \ldots, z_{11} = 11$, $z_{12} = 12$. Clearly these are simple, isolated, and widely spaced zeros. The polynomial is

$$f(z) = \prod_{k=1}^{12} (z - k) = z^{12} - 78z^{11} + 2717z^{10} -+ \cdots + 12!$$

with coefficients as given in Table 9. We are now to find the effect on the zeros of changing a *single* coefficient a_j in $f(z)$. For the sake of realism, assume that this coefficient has a *relative error* of the order of $\epsilon = 10^{-11}$. The perturbation polynomial is then

$$g(z) = a_j z^j \qquad (0 \leq j \leq 12)$$

for a fixed j. Furthermore,

$$f'(z_k) = f'(k) = (-1)^k (12 - k)!(k - 1)!.$$

Using the same approximation technique as in (4–16), we arrive at

$$\zeta_k \sim z_k - \epsilon \frac{a_j k^j}{(-1)^k (12 - k)!(k - 1)!},$$

or, for the magnitude of the change,

$$|\Delta z_k| = |\zeta_k - z_k| \sim \epsilon \frac{|a_j| k^j}{(12 - k)!(k - 1)!} = K_{k,j}\epsilon. \qquad (4\text{–}17)$$

Table 9 ILLUSTRATIVE POLYNOMIAL

k	a_k	$K_{k,6}$	$K_{k,7}$	$K_{k,8}$	ζ_k
0	479,001,600				
1	−1,486,442,880	$1.127 \cdot 10^0$	$1.735 \cdot 10^{-1}$	$1.878 \cdot 10^{-2}$	1.000 0000
2	1,931,559,552	$7.935 \cdot 10^2$	$2.443 \cdot 10^2$	$5.287 \cdot 10^1$	2.000 0000
3	−1,414,014,888	$4.519 \cdot 10^4$	$2.087 \cdot 10^4$	$6.775 \cdot 10^3$	2.999 9998
4	657,206,836	$7.617 \cdot 10^5$	$4.691 \cdot 10^5$	$2.030 \cdot 10^5$	4.000 0047
5	−206,070,150	$5.812 \cdot 10^6$	$4.474 \cdot 10^6$	$2.420 \cdot 10^6$	4.999 9553
6	44,990,231	$2.430 \cdot 10^7$	$2.244 \cdot 10^7$	$1.457 \cdot 10^7$	6.000 2245
7	−6,926,634	$6.126 \cdot 10^7$	$6.602 \cdot 10^7$	$5.001 \cdot 10^7$	6.999 3402
8	749,463	$9.750 \cdot 10^7$	$1.201 \cdot 10^8$	$1.040 \cdot 10^8$	8.001 2014
9	−55,770	$9.883 \cdot 10^7$	$1.370 \cdot 10^8$	$1.334 \cdot 10^8$	8.998 6303
10	2,717	$6.199 \cdot 10^7$	$9.544 \cdot 10^7$	$1.033 \cdot 10^8$	10.000 9538
11	−78	$2.196 \cdot 10^7$	$3.720 \cdot 10^7$	$4.427 \cdot 10^7$	10.999 6279
12	1	$3.366 \cdot 10^6$	$6.218 \cdot 10^6$	$8.073 \cdot 10^6$	12.000 0622

Table 9 shows the values of $K_{k,j}$ for $j = 6, 7, 8$ and $k = 1, 2, \ldots, 12$. The maximum of $K_{k,j}$ occurs at $k = 9$, $j = 7$ with $K_{9,7} = 1.370 \times 10^8$. If the coefficient $a_7 = -6{,}926{,}634$ is relatively altered by 10^{-11} to $a_7 = -6{,}926{,}634.0000692634$, then z_9 experiences a change of 0.00137. A computation to 20 places confirms the theory well (see Table 9), since the zeros ζ_k reflect the changes predicted by $K_{k,7}$ very accurately. The quantities $K_{k,j}$ also clearly show that the individual zeros have widely varying sensitivities.

In the use of a computer having an 11-place mantissa, a relative error of the order considered here is inevitable. If we develop a characteristic polynomial of matrix A with the eigenvalues as given above, the coefficients are saddled with relative errors of at least this size. Therefore, the zeros thus computed can lie far from the true eigenvalues. The situation deteriorates with increasing n. Complex zeros of a characteristic polynomial of a matrix with real eigenvalues can occur as well (see Ref. 74).

Conclusion: *The characteristic polynomial of a matrix is not well suited for computation of eigenvalues. If possible, it should be avoided. If not, we must compute by using a very large number of significant figures.*

In what follows, the characteristic polynomial is used for theoretical purposes, if at all. The numerical methods, however, never require its explicit formulation complete with coefficients.

PROBLEMS

4-3. In the first approximation, find the change in the zeros of the polynomial $z^4 - 130z^3 + 3129z^2 - 13100z + 10000$ (having zeros at $z = 1, 4, 25, 100$) if one of the coefficients undergoes a relative change of 10^{-5}. Which zero is the most sensitive?

4-4. Same as Problem 4-3 for the polynomial $z^8 - 130z^6 + 3129z^4 - 13100z^2 + 10000$ if one of the nonzero coefficients undergoes a relative change of 10^{-5}.

4-3 THE PRINCIPAL AXIS THEOREM

This section encompasses some facts and characteristics of eigenvalues and eigenvectors of a symmetric matrix A which together form the basis of several numerical methods. The eigenvalues are taken as the *real* solutions of an *extremum problem*, and the eigenvectors form an *orthonormal* system. From this follows the *principal axis theorem*, by which every symmetric matrix may be transformed to a diagonal form by an orthogonal similarity transformation.

Theorem 4-1. *Let A represent a symmetric matrix. The corresponding quadratic form $Q(\boldsymbol{x}) = (A\boldsymbol{x}, \boldsymbol{x})$ takes on a maximum λ_1 upon the unit sphere $(\boldsymbol{x}, \boldsymbol{x}) = 1$. The vector $\boldsymbol{x}$ corresponding to the maximum is an eigenvector, and λ_1 is its eigenvalue.*

Proof: The quadratic form $Q(\boldsymbol{x}) = (A\boldsymbol{x}, \boldsymbol{x})$ is a continuous function upon the unit sphere. The unit sphere is a closed, bounded set of the n-dimensional space. Thus there exists a point on the unit sphere, that is, a vector $\boldsymbol{x}_1$ with

$(\boldsymbol{x}_1, \boldsymbol{x}_1) = 1$, for which $Q(\boldsymbol{x})$, a continuous function, takes on its maximal value λ_1. It follows

$$(\boldsymbol{A}\boldsymbol{x}, \boldsymbol{x}) \leq \lambda_1 \text{ for all } \boldsymbol{x} \text{ with } (\boldsymbol{x}, \boldsymbol{x}) = 1 \tag{4–18}$$

and

$$(\boldsymbol{A}\boldsymbol{x}_1, \boldsymbol{x}_1) = \lambda_1 \quad \text{with} \quad (\boldsymbol{x}_1, \boldsymbol{x}_1) = 1. \tag{4–19}$$

For all vectors $\boldsymbol{x}$ of length unity, the inequality (4–18) means the same as

$$(\boldsymbol{A}\boldsymbol{x}, \boldsymbol{x}) \leq \lambda_1(\boldsymbol{x}, \boldsymbol{x}), \tag{4–20}$$

which is then trivially true for any nonzero vector, since every nonzero vector can be represented as a suitable multiple c of a vector of length unity, and the factor $c^2 > 0$ appears on both sides of (4–20). Because of the distributive property of the inner product, (4–20) leads to

$$(\boldsymbol{A}\boldsymbol{x} - \lambda_1\boldsymbol{x}, \boldsymbol{x}) = ((\boldsymbol{A} - \lambda_1\boldsymbol{I})\boldsymbol{x}, \boldsymbol{x}) \leq 0 \quad \text{for all} \quad \boldsymbol{x} \neq 0, \tag{4–21}$$

and by (4–19), it follows for the particular vector $\boldsymbol{x}_1$

$$(\boldsymbol{A}\boldsymbol{x}_1 - \lambda_1\boldsymbol{x}_1, \boldsymbol{x}_1) = ((\boldsymbol{A} - \lambda_1\boldsymbol{I})\boldsymbol{x}_1, \boldsymbol{x}_1) = 0 \quad \text{with} \quad (\boldsymbol{x}_1, \boldsymbol{x}_1) = 1. \tag{4–22}$$

It remains to show that (4–21) and (4–22) lead to $(\boldsymbol{A} - \lambda_1\boldsymbol{I})\boldsymbol{x}_1 = 0$. The matrix $\boldsymbol{B} = \boldsymbol{A} - \lambda_1\boldsymbol{I}$, like $\boldsymbol{A}$, is symmetric. Its corresponding quadratic form by (4–21) is $(\boldsymbol{B}\boldsymbol{x}, \boldsymbol{x}) \leq 0$ for all nonzero $\boldsymbol{x}$, but a vector $\boldsymbol{x}_1$ does exist such that $(\boldsymbol{B}\boldsymbol{x}_1, \boldsymbol{x}_1) = 0$, by (4–22). Let us now examine a one-parameter set of vectors $\boldsymbol{x} = \boldsymbol{x}_1 + t\boldsymbol{y}$ which contains the vector $\boldsymbol{x}_1$ when $t = 0$, and where $\boldsymbol{y}$ is an arbitrary vector. For the vectors $\boldsymbol{x}$ of this group we have

$$\begin{aligned}(\boldsymbol{B}\boldsymbol{x}, \boldsymbol{x}) &= (\boldsymbol{B}(\boldsymbol{x}_1 + t\boldsymbol{y}), \boldsymbol{x}_1 + t\boldsymbol{y}) = (\boldsymbol{B}\boldsymbol{x}_1, \boldsymbol{x}_1) + 2t(\boldsymbol{B}\boldsymbol{x}_1, \boldsymbol{y}) + t^2(\boldsymbol{B}\boldsymbol{y}, \boldsymbol{y}) \\ &= 2t(\boldsymbol{B}\boldsymbol{x}_1, \boldsymbol{y}) + t^2(\boldsymbol{B}\boldsymbol{y}, \boldsymbol{y}) \leq 0 \end{aligned} \tag{4–23}$$

for every arbitrary vector $\boldsymbol{y}$. But an expression of the form $at + bt^2$ with a nonzero changes sign when t passes the origin. In order for such an expression not to change sign, as desired in our case, necessarily $a = 0$. Carried over to our case, this observation applied to (4–23) leads to

$$(\boldsymbol{B}\boldsymbol{x}_1, \boldsymbol{y}) = ((\boldsymbol{A} - \lambda_1\boldsymbol{I})\boldsymbol{x}_1, \boldsymbol{y}) = 0 \tag{4–24}$$

for every vector $\boldsymbol{y}$. The equation (4–24) is valid only if

$$(\boldsymbol{A} - \lambda_1\boldsymbol{I})\boldsymbol{x}_1 = 0, \quad \text{or} \quad \boldsymbol{A}\boldsymbol{x}_1 = \lambda_1\boldsymbol{x}_1, \tag{4–25}$$

from which $\boldsymbol{x}_1$ emerges as an eigenvector and λ_1 as the corresponding eigenvalue of the matrix $\boldsymbol{A}$.

In order to prove the existence of another eigenvalue and eigenvector, all vectors $\boldsymbol{x}$ orthogonal to $\boldsymbol{x}_1$ will be examined. The totality of these vectors forms an $(n-1)$-dimensional subspace. This subspace is *invariant* with respect to the linear transformation $\boldsymbol{A}$, since it maps onto itself. Actually, given any vector $\boldsymbol{x}$ with $(\boldsymbol{x}, \boldsymbol{x}_1) = 0$, with $\boldsymbol{y} = \boldsymbol{A}\boldsymbol{x}$, and $\boldsymbol{A}$ symmetric,

$$(\boldsymbol{y}, \boldsymbol{x}_1) = (\boldsymbol{A}\boldsymbol{x}, \boldsymbol{x}_1) = (\boldsymbol{x}, \boldsymbol{A}\boldsymbol{x}_1) = (\boldsymbol{x}, \lambda_1\boldsymbol{x}_1) = \lambda_1(\boldsymbol{x}, \boldsymbol{x}_1) = 0.$$

That is, $\boldsymbol{y}$ is orthogonal to $\boldsymbol{x}_1$. The unit sphere in this $(n-1)$-dimensional subspace is a closed, bounded set such that on it, the continuous quadratic form $Q(\boldsymbol{x}) = (\boldsymbol{A}\boldsymbol{x}, \boldsymbol{x})$ takes on the maximum λ_2 for some vector $\boldsymbol{x}_2$ with the properties $(\boldsymbol{x}_1, \boldsymbol{x}_2) = 0$ and $(\boldsymbol{x}_2, \boldsymbol{x}_2) = 1$. Generalizing (4–21) and (4–22) leads to

$$((\boldsymbol{A} - \lambda_2\boldsymbol{I})\boldsymbol{x}, \boldsymbol{x}) \leq 0 \quad \text{for all} \quad \boldsymbol{x} \neq 0 \quad \text{and} \quad (\boldsymbol{x}, \boldsymbol{x}_1) = 0$$

and also

$$((\boldsymbol{A} - \lambda_2\boldsymbol{I})\boldsymbol{x}_2, \boldsymbol{x}_2) = 0.$$

We examine further the symmetric matrix $\boldsymbol{B}_1 = \boldsymbol{A} - \lambda_2\boldsymbol{I}$, whose corresponding transformation maps the $(n-1)$-dimensional subspace into itself. In addition, $\boldsymbol{y}$ is an arbitrary nonzero vector in this subspace with $(\boldsymbol{y}, \boldsymbol{x}_1) = 0$. Thus we form the one-parameter set of vectors $\boldsymbol{x} = \boldsymbol{x}_2 + t\boldsymbol{y}$ lying in the $(n-1)$-dimensional subspace, obeying the necessary conditions, and yielding the vector $\boldsymbol{x}_2$ when $t = 0$. For the vectors of this set, it follows for matrix $\boldsymbol{B}_1$ that

$$\begin{aligned}(\boldsymbol{B}_1\boldsymbol{x}, \boldsymbol{x}) &= (\boldsymbol{B}_1(\boldsymbol{x}_2 + t\boldsymbol{y}), \boldsymbol{x}_2 + t\boldsymbol{y}) = (\boldsymbol{B}_1\boldsymbol{x}_2, \boldsymbol{x}_2) + 2t(\boldsymbol{B}_1\boldsymbol{x}_2, \boldsymbol{y}) + t^2(\boldsymbol{B}_1\boldsymbol{y}, \boldsymbol{y}) \\ &= 2t(\boldsymbol{B}_1\boldsymbol{x}, \boldsymbol{y}_2) + t^2(\boldsymbol{B}_1\boldsymbol{y}, \boldsymbol{y}) \leq 0\end{aligned}$$

for every nonzero vector $\boldsymbol{y}$ with $(\boldsymbol{y}, \boldsymbol{x}_1) = 0$. In order for this last quantity not to change sign as t passes through the origin, necessarily $(\boldsymbol{B}_1\boldsymbol{x}_2, \boldsymbol{y}) = 0$. The vector $\boldsymbol{x}_2$ is a vector of the $(n-1)$-dimensional subspace. Because of the invariance of this subspace under the linear transformation $\boldsymbol{B}_1$, $\boldsymbol{B}_1\boldsymbol{x}_2$ belongs to this subspace, too. This vector must be orthogonal to any arbitrary vector of the subspace, thus it must vanish, and it follows that $\boldsymbol{B}_1\boldsymbol{x}_2 = 0$; that is, $\boldsymbol{A}\boldsymbol{x}_2 = \lambda_2\boldsymbol{x}_2$, so that $\boldsymbol{x}_2$ is an eigenvector of $\boldsymbol{A}$, and λ_2 the corresponding eigenvalue. As a result of the limitation of permissible comparison vectors $\boldsymbol{x}$ of length unity, evidently $\lambda_2 \leq \lambda_1$, since the new maximum cannot be greater.

The third eigenvalue λ_3 and corresponding eigenvector $\boldsymbol{x}_3$ result from the solution of the following extremum problem with side conditions: λ_3 is the maximum of $(\boldsymbol{Ax}, \boldsymbol{x})$ for all vectors $\boldsymbol{x}$ over the $(n - 2)$-dimensional unit sphere with $(\boldsymbol{x}, \boldsymbol{x}) = 1$, $(\boldsymbol{x}, \boldsymbol{x}_1) = 0$, and $(\boldsymbol{x}, \boldsymbol{x}_2) = 0$, and $\boldsymbol{x}_3$ is the vector for which the maximum occurs. In a consistent continuation of this constructive method, the existence of n eigenvectors and corresponding eigenvalues can be proven. The eigenvectors form an orthonormal system of real vectors as a result of the projections, and the eigenvalues are real since the quadratic form for real vectors can take on only real values. Of course, it is possible that several equal eigenvalues turn up in this constructive technique. If we have the case of

$$\lambda_1 > \lambda_2 > \cdots > \lambda_k = \lambda_{k+1} = \cdots = \lambda_{k+p-1} > \lambda_{k+p} > \cdots > \lambda_n,$$

then λ_k is called a p-tuple eigenvalue. In that case every linear combination of eigenvectors corresponding to the eigenvalues $\lambda_k, \lambda_{k+1}, \ldots, \lambda_{k+p-1}$ is also an eigenvector. The eigenvectors for a multiple eigenvalue are not unique. This result is given by

Theorem 4-2. *Every symmetric matrix $\boldsymbol{A}$ of order n has n real eigenvalues, and the corresponding n eigenvectors constitute an orthogonal system. In the case of a multiple eigenvalue of multiplicity p, the eigenvectors form a p-dimensional subspace.*

If we replace the word maximum in Theorem 4–1 by the word minimum, the theorem remains valid by way of an analogous proof. An immediate consequence thereof is

Theorem 4-3. *The maximum (minimum) of the Rayleigh quotient $\boldsymbol{R}(\boldsymbol{x}) = (\boldsymbol{Ax}, \boldsymbol{x})/(\boldsymbol{x}, \boldsymbol{x})$ for an arbitrary vector $\boldsymbol{x} \neq 0$ equals the largest (smallest) eigenvalue $\lambda_{\max}$ $(\lambda_{\min})$ of the symmetric matrix $\boldsymbol{A}$. The vector $\boldsymbol{x}$ for which the Rayleigh quotient attains its extremum is the corresponding eigenvector. The value of the Rayleigh quotient of a symmetric matrix $\boldsymbol{A}$ lies in the range $\lambda_{\min} \leq \boldsymbol{R}(\boldsymbol{x}) \leq \lambda_{\max}$.*

An important consequence of Theorem 4–3 is that the eigenvalues of a positive definite matrix $\boldsymbol{A}$ are positive, since the value of the Rayleigh quotient for all nonzero vectors $\boldsymbol{x}$ is greater than zero.

Theorem 4-4. (*Principal axis theorem*). *For every symmetric matrix $\boldsymbol{A}$ there exists an orthogonal matrix $\boldsymbol{U}$ which transforms $\boldsymbol{A}$ into a diagonal form $\boldsymbol{D} = \boldsymbol{U}^{\mathrm{T}}\boldsymbol{AU}$. The diagonal elements of $\boldsymbol{D}$ are the eigenvalues of $\boldsymbol{A}$.*

Proof: By Theorem 4–2, there exists for every symmetric matrix A of order n a system of n orthonormal eigenvectors $x_1, x_2, \ldots, x_n$. This system of orthonormal vectors will be used as a new basis. According to the general rules governing the transformation of the representation of an operator A in switching to another basis (see Sec. 1–1–2), we may use a matrix U which contains in its kth column the coordinates of the kth new basis vector referred to the old basis. If e_k is the kth unit vector, then $x_k = Ue_k$. The orthonormal property of the columns of U assures that $U^TU = I$, so that $U^T = U^{-1}$, and the matrix U is orthogonal. In the change of basis, the matrix A is subject to the similarity transformation

$$B = U^{-1}AU = U^TAU.$$

The eigenvectors of B are now the unit vectors e_k, while the eigenvalues λ_k remain unchanged. It follows that

$$Be_k = (U^TAU)e_k = U^TA(Ue_k) = U^TAx_k = U^T\lambda_k x_k = \lambda_k U^T x_k = \lambda_k e_k.$$

Since the equation $Be_k = \lambda_k e_k$ holds for every unit vector $e_1, e_2, \ldots, e_n$, matrix B must necessarily have a diagonal form, and its diagonal elements consist of the eigenvalues λ_k of A.

Note: Since the transformation matrix U of the principal axis theorem contains the normed eigenvector of A corresponding to the eigenvalue λ_k in its kth column, U is not unique in the case of multiple eigenvalues.

4-4 TRANSFORMATION TO DIAGONAL FORM; SIMULTANEOUS COMPUTATION OF ALL EIGENVALUES

The principal axis theorem is the basis of a class of techniques which produce a diagonal form from a sequence of orthogonal similarity transformations and thereby determine all the eigenvalues of a symmetric matrix simultaneously. These methods should be used when we want to know all the eigenvalues, and possibly their eigenvectors as well.

4-4-1 Elementary Orthogonal Two-Dimensional Rotations

For the similarity transformation of a matrix A to a given normal form, we will use transformation matrices which are as simple as possible, whose effects are easily visualized, and which can be used as building blocks of more intricate approaches. The simplest nontrivial orthogonal matrices have the form

$$U = \begin{bmatrix} 1 & & & & & & \\ & 1 & & & & & \\ & & \cos\phi & \cdots & \sin\phi & & \\ & & & 1 & & & \\ & & & & 1 & & \\ & & -\sin\phi & \cdots & \cos\phi & & \\ & & & & & 1 & \\ & & & & & & 1 \end{bmatrix} \begin{matrix} \\ \\ \leftarrow p \\ \\ \\ \leftarrow q \\ \\ \\ \end{matrix}, \quad U^{-1} = U^{\mathrm{T}}. \qquad (4\text{–}26)$$

Geometrically, this corresponds to a two-dimensional rotation of angle ϕ in the plane which is spanned by the coordinate directions p and q. The matrix U is characterized by a *rotation index pair* p and q, with $1 \leq p < q \leq n$, and a real *angle of rotation* ϕ. Its elements are defined as

$$\left.\begin{aligned} u_{ii} &= 1 & i &\neq p, q \\ u_{pp} &= \cos\phi, & u_{pq} &= \sin\phi \\ u_{qp} &= -\sin\phi, & u_{qq} &= \cos\phi \\ u_{ij} &= 0 \quad \text{otherwise} & & \end{aligned}\right\}. \qquad (4\text{–}27)$$

The effect of a similarity transformation of a matrix A, using a matrix U as in (4–26), will be examined in two steps.

First step: $A' = U^{\mathrm{T}}A$. Left multiplication of the matrix A by U^{T} causes a mutual linear combination of the pth and qth *rows* of A as a result of the particular form of U; the remaining elements are unaltered:

$$\left.\begin{aligned} a'_{pj} &= a_{pj}\cos\phi - a_{qj}\sin\phi \\ a'_{qj} &= a_{pj}\sin\phi + a_{qj}\cos\phi \\ a'_{ij} &= a_{ij} \quad \text{for} \quad i \neq p, q \end{aligned}\right\} \qquad j = 1, 2, \ldots, n. \qquad (4\text{–}28)$$

Second step: $A'' = A'U = U^{\mathrm{T}}AU$. Right multiplication of matrix A' by U causes a mutual linear combination of the pth and qth *columns* of A'; the other elements remain unaltered:

$$\left.\begin{aligned} a''_{ip} &= a'_{ip}\cos\phi - a'_{iq}\sin\phi \\ a''_{iq} &= a'_{ip}\sin\phi + a'_{iq}\cos\phi \\ a''_{ij} &= a'_{ij} \quad \text{for} \quad j \neq p, q \end{aligned}\right\} \qquad i = 1, 2, \ldots, n. \qquad (4\text{–}29)$$

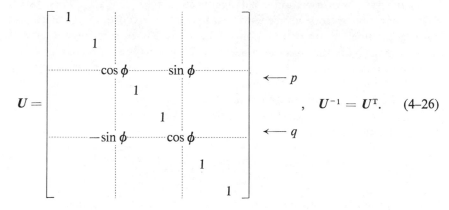

Through the similarity transformation $A'' = U^T A U$, only the elements of A in the pth and qth rows and columns are altered (see Fig. 15).

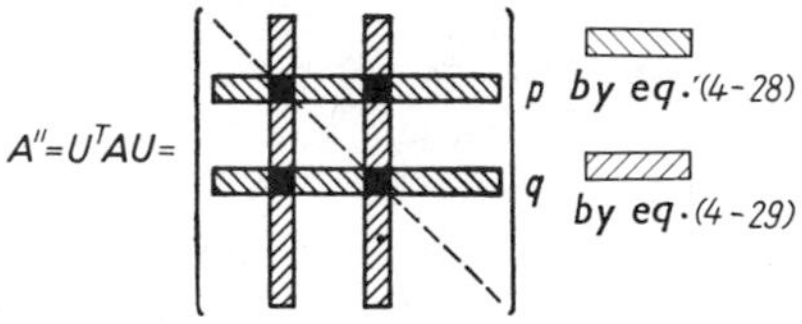

Figure 15. Effect of similarity transformation.

The elements at the intersections of the pth and qth rows and columns experience the most drastic changes, as they are altered by (4–28) and (4–29). For a *symmetric* matrix A, the transformation formulas for the elements at the intersections are given by

$$\left.\begin{aligned} a''_{pp} &= a_{pp}\cos^2\phi - 2a_{pq}\cos\phi\sin\phi + a_{qq}\sin^2\phi \\ a''_{qq} &= a_{pp}\sin^2\phi + 2a_{pq}\cos\phi\sin\phi + a_{qq}\cos^2\phi \\ a''_{pq} &= a''_{qp} = (a_{pp} - a_{qq})\cos\phi\sin\phi + a_{pq}(\cos^2\phi - \sin^2\phi) \end{aligned}\right\}. \tag{4–30}$$

The above technique of elementary orthogonal similarity transformations was first used by *Jacobi* in 1846 (Ref. 34) in conjunction with his method described in Sec. 4–4–2. In line with the geometrical interpretation, such a transformation is called a *Jacobi rotation* or, to emphasize the rotation index pair, a (p, q)-rotation. The nondiagonal intersection element a_{pq} is called a *pivot element.*

ALGOL Procedure for a Jacobi Rotation. In light of the retention of the symmetry property under orthogonal similarity transformations, it suffices in practice to carry out the transformation on the elements on and above the diagonal. Since the intent is to apply a series of Jacobi rotations to a matrix A, the transformed matrix A'' is ultimately relabeled A. The transformation is decomposed into four parts: (1) Calculation of the new intersection elements by (4–30). (2) Transformation of the elements of the pth and qth column lying above the pth row, by (4–29). (3) Treatment of the elements between intersection elements, making use of symmetry. (4) Row combinations right of the qth column, using equations (4–28). Because of the relabeling of A'' as A, auxiliary variables g and h are necessary for temporary storage of the original values of matrix elements to be reused.

Procedure parameters are identified as:

n	Order of matrix A
c, s	$c = \cos\phi$, $s = \sin\phi$
p, q	rotation index pair, $p < q$
a	elements of the matrix A before and after the (p, q)-rotation

ALGOL Procedure No. 8

```
procedure rotation (n, c, s, p, q, a);
        value n, c, s, p, q;
        real c, s;  integer n, p, q;  array a;
begin real g, h;  integer j;
   h: = c × c × a[p, p] − 2 × c × s × a[p, q] + s × s × a[q, q];
   g: = s × s × a[p, p] + 2 × c × s × a[p, q] + c × c × a[q, q];
   a[p, q]: = c × s × (a[p, p] − a[q,q]) + (c × c − s × s) × a[p, q];
   a[p, p]: = h;  a[q, q]: = g;
   for j: = 1 step 1 until p − 1 do
   begin
      h: = c × a[j, p] − s × a[j, q];
      a[j, q]: = s × a[j, p] + c × a[j, q];
      a[j, p]: = h
   end j;
   for j: = p + 1 step 1 until q − 1 do
   begin
      h: = c × a[p, j] − s × a[j, q];
      a[j, q]: = s × a[p, j] + c × a[j, q];
      a[p, j]: = h
   end j;
   for j: = q + 1 step 1 until n do
   begin
      h: = c × a[p, j] − s × a[q, j];
      a[q, j]: = s × a[p, j] + c × a[q, j];
      a[p, j]: = h
   end j
end rotation
```

4-4-2 The Classical Jacobi Method

In the Jacobi method (Ref. 34), the given symmetric matrix $\boldsymbol{A}$ is transformed by a series of steps into a diagonal form $\boldsymbol{D}$ through a sequence of Jacobi rotations. The number of rotations needed to achieve diagonalization is in general infinite, since the diagonal elements of $\boldsymbol{D}$ are the eigenvalues of $\boldsymbol{A}$. The eigenvalues being also zeros of the characteristic polynomial, they cannot be found through a finite rational process. In practice, diagonalization is terminated when the nondiagonal elements become sufficiently small.

In its classical approach, the successive rotations are selected such that the instantaneous nondiagonal element of largest magnitude in the current matrix is made to vanish through rotation. At the same time we should bear in mind that the zero element achieved thusly will usually be destroyed by later rotations. In order to avoid superfluous indices, let $\boldsymbol{A}$ be the current matrix $\boldsymbol{A}_k$ of the sequence of transformed matrices, and let a_{pq} be the largest-magnitude nondiagonal element *above* the diagonal, with $1 \leq p < q \leq n$. This element is to be made zero through a (p, q)-rotation which has a_{pq} at its nondiagonal intersection. This requirement leads by equations (4–30) to a trigonometric equation for the angle of rotation ϕ

$$(a_{pp} - a_{qq}) \sin \phi \cos \phi + a_{pq}(\cos^2 \phi - \sin^2 \phi) = 0,$$

or, after appropriate trigonometric substitution, to

$$\cot (2\phi) = \frac{a_{qq} - a_{pp}}{2a_{pq}} = \theta. \tag{4–31}$$

The equation for ϕ is formulated in terms of $\cot (2\phi)$, and to good advantage at that, since this value is finite with a_{pq} nonzero. The relation $\theta = \cot (2\phi)$ yields the values $c = \cos \phi$ and $s = \sin \phi$ through the trigonometric identities

$$\tan \phi = \frac{s}{c} = t, \quad \theta = \frac{1 - t^2}{2t}, \quad c^2 + s^2 = 1.$$

Then t is found as the solution to the quadratic equation $t^2 + 2\theta t - 1 = 0$ to be

$$t = -\theta \pm \sqrt{\theta^2 + 1} = \frac{1}{\theta \pm \sqrt{\theta^2 + 1}}. \tag{4–32}$$

Between the two possible solutions, we choose the one leading to no loss of significant figures in the denominator of (4–32); in other words, we choose

$$\begin{aligned} t &= \frac{1}{\theta + \operatorname{sign}(\theta) \cdot \sqrt{\theta^2 + 1}} && \text{for} \quad \theta \neq 0, \\ t &= 1 && \text{for} \quad \theta = 0. \end{aligned} \tag{4–33}$$

We end up with the results

$$c = \cos \phi = \frac{1}{\sqrt{1 + t^2}}, \quad s = \sin \phi = ct. \tag{4–34}$$

Since the product of the two solutions of t is -1 by the quadratic equation, we can easily show that the other solution of t leads to values of c and s in (4–34) which are simply interchanged. In the end, both solutions have the same effect of making a_{pq} vanish through an appropriate Jacobi rotation. By (4–33) it follows that $|t| \leq 1$, so that (4–34) leads to $c = \cos \phi \geq 1/\sqrt{2}$ and thus $|s| = |\sin \phi| \leq 1/\sqrt{2}$. Consequently, since c is greater than zero, the rotation angle ϕ is confined to the region $-\pi/4 < \phi \leq \pi/4$. This observation will be important in Sec. 4–4–3.

As a preliminary to the convergence proof of the classical Jacobi technique, we examine the change in the sum of the squares of the nondiagonal elements

$$S(A) = \sum_{i=1}^{n} \sum_{\substack{j=1 \\ j \neq i}}^{n} a_{ij}^2 \tag{4–35}$$

under a (p, q)-rotation. For this purpose, $S(A'')$ of the transformed matrix $A'' = U^{\mathrm{T}}AU$ is subdivided into the partial sums

$$S(A'') = \sum_{\substack{i=1 \\ i \neq p,q}}^{n} \sum_{\substack{j=1 \\ j \neq i \\ j \neq p,q}}^{n} a_{ij}''^2 + \sum_{\substack{i=1 \\ i \neq p,q}}^{n} (a_{ip}''^2 + a_{iq}''^2) + \sum_{\substack{j=1 \\ j \neq p,q}}^{n} (a_{pj}''^2 + a_{qj}''^2) + 2a_{pq}''^2. \tag{4–36}$$

Only the elements of the pth and qth rows and columns are altered; also by (4–28) and (4–29) we have the relations for the coupled elements

$$a_{ip}''^2 + a_{iq}''^2 = a_{ip}^2 + a_{iq}^2 \qquad i \neq p, q,$$
$$a_{pj}''^2 + a_{qj}''^2 = a_{pj}^2 + a_{qj}^2 \qquad j \neq p, q,$$

so that in general for a (p, q)-rotation, (4–36) yields

$$S(A'') = S(U^{\mathrm{T}}AU) = \{S(A) - 2a_{pq}^2\} + 2a_{pq}''^2. \tag{4–37}$$

Theorem 4-5. *In the classical Jacobi method, the sequence of values of $S(A_k)$ decreases monotonically to zero.*

Proof: Let $A = A_0$, and let A_k indicate the matrix resulting after k rotations. In the general kth step, $a_{pq}^{(k)} = 0$, so that (4–37) yields the relation

$$S(A_k) = S(A_{k-1}) - 2a_{pq}^{(k-1)^2} \qquad k = 1, 2, \ldots. \tag{4–38}$$

The sequence of values $S(A_k)$ thus decreases monotonically. Since the magnitude of $a_{pq}^{(k-1)}$ is the greatest among the nondiagonal elements of A_{k-1}, we have the inequality

$$S(A_{k-1}) \leq (n^2 - n)a_{pq}^{(k-1)^2}. \tag{4–39}$$

Therefore, Eq. (4–38) yields the inequality

$$S(A_k) = S(A_{k-1}) - 2a_{pq}^{(k-1)^2} \leq \left(1 - \frac{2}{n^2 - n}\right) S(A_{k-1}).$$

Through repeated application of this inequality, which is independent of

$a_{pq}^{(k-1)}$, we have

$$S(A_k) \leq \left(1 - \frac{2}{n^2 - n}\right)^k S(A_0). \tag{4–40}$$

Since $(1 - 2/(n^2 - n)) < 1$, $S(A_k)$ in (4–40) converges to zero, and after a sufficiently large number of steps A_k is arbitrarily close to a diagonal matrix. The classical Jacobi technique converges.

Therein, incidentally, lies a new constructive proof of the principal axis theorem.

The inequality (4–40) gives an estimate of the necessary Jacobi rotations in order to arrive at a value of the quotient $S(A_k)/S(A_0) < \epsilon$, where ϵ is any prescribed tolerance. We must have

$$\left(1 - \frac{2}{n^2 - n}\right)^k < \epsilon, \quad \epsilon < 1,$$

or, by going over to natural logarithms,

$$k \cdot \log\left(1 - \frac{2}{n^2 - n}\right) < \log \epsilon, \quad \text{or} \quad k \cdot \log\left(\frac{n^2 - n}{n^2 - n - 2}\right) > \log\left(\frac{1}{\epsilon}\right),$$

$$k > \frac{-\log(\epsilon)}{\log\left(\dfrac{n^2 - n}{n^2 - n - 2}\right)}. \tag{4–41}$$

This estimate is exact. Now let us try to estimate the relationship between the number of rotations k and the order n of matrix A with the assumption that $n \gg 1$. The denominator in (4–41) can after linearization (Ref. 64) be further approximated by means of

$$\log\left(\frac{n^2 - n}{n^2 - n - 2}\right) = \log\left(1 + \frac{2}{n^2 - n - 2}\right) \sim \frac{2}{n^2 - n - 2} \sim \frac{2}{n^2}, \quad n \gg 1.$$

In place of the inequality (4–41), then, we have the approximation

$$k \sim \frac{n^2}{2} \log\left(\frac{1}{\epsilon}\right). \tag{4–42}$$

The number of rotations is proportional to the square of the order n of the given matrix A and is dependent on the tolerance ϵ. Since one rotation requires about $4n$ multiplications for a symmetric matrix, neglecting two square root operations, the classical Jacobi method for the determination of the eigenvalues of a real symmetric matrix is an n^3-process, where the factor is significantly dependent on ϵ.

The estimate (4–42) is very coarse, since it applies for all symmetric matrices and is in general too pessimistic. Convergence is actually significantly better than what (4–42) predicts. By (4–40) the values $S(A_k)$ converge at least as well as a geometric sequence of the quotient $q = (1 - 2/(n^2 - n))$ toward zero. In the customary terminology, this method is termed *linearly convergent*. With the additional information that matrix A has n *distinct eigenvalues*, Ref. 29 shows that at latest when the largest magnitude nondiagonal element falls below a certain quantity, the classical Jacobi method converges *quadratically*. The greater the minimal distance of two successive eigenvalues on the real axis, the sooner the quadratic convergence takes over. This phenomenon greatly improves the rate of convergence. Lately the proof of quadratic convergence has been improved in several ways. Aside from the diminishing of the constant in the estimate of Ref. 29, Ref. 52 also shows that in the case of (no worse than) *double eigenvalues*, quadratic convergence applies. And as a result of a new convergence proof, Ref. 53 demonstrates further improvement in the results.

In practice, a criterion for the deviation of the diagonal elements $a_{ii}^{(k)}$ of matrix A_k from the eigenvalues of the given matrix A is indispensable to knowing the accuracy of the approximately derived eigenvalues. Thus we cite

Theorem 4-6. *Let λ_j $(j = 1, 2, \ldots, n)$ be the eigenvalues of a symmetric matrix A, numbered in the order $\lambda_1 \geq \lambda_2 \geq \lambda_3 \geq \cdots \geq \lambda_n$. Furthermore, let $d_j^{(k)}$ be the diagonal elements $a_{ii}^{(k)}$ of matrix A_k similar to A, similarly numbered $d_1^{(k)} \geq d_2^{(k)} \geq d_3^{(k)} \geq \cdots \geq d_n^{(k)}$. Then the bound*

$$|d_j^{(k)} - \lambda_j| \leq \sqrt{S(A_k)} \qquad j = 1, 2, \ldots, n; \quad k = 1, 2, \ldots \tag{4–43}$$

holds simultaneously for all eigenvalues.

The proof given in Ref. 29 for Theorem 4–6 uses the *maximum-minimum principle* (Ref. 11), which we will not go into further, to characterize the ith eigenvalue as a maximum or minimum problem.

Theorem 4–6 yields a *general absolute* bound on the accuracy with which the diagonal elements approximate the eigenvalues. If, for instance, the value of $S(A_k)$ has decreased to 10^{-8} after k Jacobi rotations, the diagonal elements $a_{ii}^{(k)}$ yield the eigenvalues in one particular permutation with an absolute deviation of, at most, 10^{-4}. In other words, the eigenvalues are correct to four decimals.

In the classical Jacobi method, the value of $S(A_k)$ is not immediately available. In principle, it could be computed recursively by Eq. (4–38) without much trouble. But in light of the continued loss of significant figures, this approach is not recommended. At each step, however, the largest-magnitude nondiagonal element $|a_{pq}^{(k)}| = \max_{i<j} |a_{ij}^{(k)}|$ is readily found, and from it

$S(A_k)$ can be estimated by

$$S(A_k) \leq (n^2 - n)a_{pq}^{(k)^2} < n^2 a_{pq}^{(k)^2}. \tag{4–44}$$

In order to avoid the time-consuming computation of $S(A_k)$ at every step, we can use in place of (4–43) the generally poorer bound

$$|d_j^{(k)} - \lambda_j| < n \cdot \max_{i<j} |a_{ij}^{(k)}| = n |a_{pq}^{(k)}|, \tag{4–45}$$

which can serve as a termination criterion.

Aside from the computation of the eigenvalues of a symmetric matrix $\boldsymbol{A}$, the Jacobi method can be utilized to determine the corresponding eigenvectors at the same time. The sequence of matrices $\boldsymbol{A}_k$ is computed recursively through $\boldsymbol{A}_k = \boldsymbol{U}_k^{\mathrm{T}} \boldsymbol{A}_{k-1} \boldsymbol{U}_k$ ($k = 1, 2, \ldots$). With $\boldsymbol{A}_0 = \boldsymbol{A}$, we derive $\boldsymbol{A}_k$ from $\boldsymbol{A}$ by

$$\boldsymbol{A}_k = \boldsymbol{U}_k^{\mathrm{T}} \boldsymbol{U}_{k-1}^{\mathrm{T}} \ldots \boldsymbol{U}_2^{\mathrm{T}} \boldsymbol{U}_1^{\mathrm{T}} \boldsymbol{A} \boldsymbol{U}_1 \boldsymbol{U}_2 \ldots \boldsymbol{U}_{k-1} \boldsymbol{U}_k = \boldsymbol{V}^{\mathrm{T}} \boldsymbol{A} \boldsymbol{V}, \tag{4–46}$$

with $\boldsymbol{V} = \boldsymbol{U}_1 \boldsymbol{U}_2 \ldots \boldsymbol{U}_{k-1} \boldsymbol{U}_k$. Under the assumption that $\boldsymbol{A}_k$ is sufficiently close to a diagonal matrix $\boldsymbol{D}$, we can apply an approximation $\boldsymbol{V}^{\mathrm{T}} \boldsymbol{A} \boldsymbol{V} \approx \boldsymbol{D}$. The matrix $\boldsymbol{V}$, a product of orthogonal matrices, is itself orthogonal. By the principal axis theorem, $\boldsymbol{D}$ contains the eigenvalues of $\boldsymbol{A}$ in an unspecified sequence, and $\boldsymbol{V}$ contains the eigenvectors of $\boldsymbol{A}$. The jth column of $\boldsymbol{V}$ is the eigenvector of the eigenvalue $\lambda = a_{jj}^{(k)}$. Matrix $\boldsymbol{V}$ is best found recursively by repeated multiplication of the individual transformation matrices $\boldsymbol{U}_k$, as given by

$$\boldsymbol{V}_0 = \boldsymbol{I}; \quad \boldsymbol{V}_k = \boldsymbol{V}_{k-1} \boldsymbol{U}_k \qquad k = 1, 2, \ldots. \tag{4–47}$$

The multiplication in (4–47) again corresponds just to a combination of pth and qth columns of $\boldsymbol{V}_{k-1}$. In contrast to the symmetric matrices $\boldsymbol{A}_k$, $\boldsymbol{V}_k$ must be evaluated as a complete matrix. Since the computation time for finding $\boldsymbol{V}_k$ from $\boldsymbol{V}_{k-1}$ corresponds to one Jacobi rotation, the time is doubled whenever we also want to find the eigenvectors.

The eigenvectors are not found with the same accuracy as the eigenvalues. The convergence of the matrix sequence $\boldsymbol{V}_k$ toward $\boldsymbol{U}$ obeys more complex laws than those governing convergence of the diagonal elements $a_{ij}^{(k)}$ to the eigenvalues. It is nonetheless important to note that the eigenvector approximations, at least within the limits of numerical accuracy, always form a system of orthonormal vectors.

ALGOL Procedure for the Computation of Eigenvalues by the Classical Jacobi Method. Only the elements on or above the diagonal of the symmetric matrix $\boldsymbol{A}$ of order n will be considered. The program calls on the procedure *rotation* defined in Sec. 4–4–1 as a global parameter. The procedure consists of four parts. By means of $n(n-1)/2$ comparisons, the largest-magnitude nondiagonal element is identified with the rotation index pair (p, q). For a termination criterion, (4–45) is used. There follows the computation of values $c = \cos\phi$ and $s = \sin\phi$ by (4–33) and (4–34), and finally the (p, q)-rotation. The eigenvalues appear as diagonal elements of matrix A. In general, they do not appear in order of size.

Procedure parameters are defined as:

n Order of matrix A

eps Allowable tolerance for eigenvalue accuracy, in absolute value

a Elements of matrix A. At the end, its diagonal elements are the eigenvalues.

ALGOL Procedure No. 9

```
procedure jacobi 1 (n, eps, a);
           value n, eps;
           integer n;  real eps;  array a;
begin integer p, q, i, j;  real max, theta, t, c, s;
search: max: = 0;
         for i: = 1 step 1 until n − 1 do
           for j: = i + 1 step 1 until n do
             if abs (a[i, j]) < max then
             begin max: = abs (a[i, j]);
                 p: = i;  q: = j
             end;
test:    if n × max < eps then goto out;
         theta: = 0.5 × (a[q, q] − a[p, p])/a[p, q];
         if theta = 0 then
           t: = 1
         else
           t: = 1/(theta + sign(theta) × sqrt(1 + theta ↑ 2));
         c: = 1/sqrt(1 + t ↑ 2);  s: = c × t;
         rotation (n, c, s, p, q, a);
         goto search;
out:
end jacobi 1
```

Example 4-2. Table 10 shows a 4×4 matrix undergoing the first five steps of the Jacobi method; it also shows the matrices A_{10} and A_{16} after 10 and 16 rotations, respectively. Below each matrix, the value of $S(A_k)$ computed for illustration purposes is given. Below the first six matrices, the rotation index pairs (p, q) and the values of $c = \cos\phi$ and $s = \sin\phi$ leading to the next

Table 10 Classical Jacobi Technique

$$A_0 = \begin{bmatrix} 20.000000 & \boxed{-7.000000} & 3.000000 & -2.000000 \\ -7.000000 & 5.000000 & 1.000000 & 4.000000 \\ 3.000000 & 1.000000 & 3.000000 & 1.000000 \\ -2.000000 & 4.000000 & 1.000000 & 2.000000 \end{bmatrix}$$

$S(A_0) = 1.600000 \cdot 10^2$

$p = 1, \quad q = 2, \quad c = 0.930337, \quad s = 0.366705$

$$A_1 = \begin{bmatrix} 22.759142 & 0 & 2.424307 & \boxed{-3.327494} \\ 0 & 2.240858 & 2.030452 & 2.987940 \\ 2.424307 & 2.030452 & 3.000000 & 1.000000 \\ -3.327494 & 2.987940 & 1.000000 & 2.000000 \end{bmatrix}$$

$S(A_1) = 6.200000 \cdot 10^1$

$p = 1, \quad q = 4, \quad c = 0.987994, \quad s = 0.154494$

$$A_2 = \begin{bmatrix} 23.279466 & -0.461618 & 2.240707 & 0 \\ -0.461618 & 2.240858 & 2.030452 & \boxed{2.952066} \\ 2.240707 & 2.030452 & 3.000000 & 1.362534 \\ 0 & 2.952066 & 1.362534 & 1.479676 \end{bmatrix}$$

$S(A_2) = 3.985557 \cdot 10^1$

$p = 2, \quad q = 4, \quad c = 0.750954, \quad s = -0.660354$

$$A_3 = \begin{bmatrix} 23.279466 & -0.346654 & 2.240707 & 0.304831 \\ -0.346654 & 4.836766 & \boxed{2.424531} & 0 \\ 2.240707 & 2.424531 & 3.000000 & -0.317616 \\ 0.304831 & 0 & -0.317616 & -1.116232 \end{bmatrix}$$

$S(A_3) = 2.242618 \cdot 10^1$

$p = 2, \quad q = 3, \quad c = 0.822869, \quad s = -0.568231$

$$A_4 = \begin{bmatrix} 23.279466 & 0.987988 & \boxed{2.040788} & 0.304831 \\ 0.987988 & 6.511022 & 0 & -0.180479 \\ 2.040788 & 0 & 1.325743 & -0.261356 \\ 0.304831 & -0.180479 & -0.261356 & -1.116232 \end{bmatrix}$$

$S(A_4) = 1.066948 \cdot 10^1$

$p = 1, \quad q = 3, \quad c = 0.995779, \quad s = -0.091780$

$$A_5 = \begin{bmatrix} 23.467563 & \boxed{0.983818} & 0 & 0.279557 \\ 0.983818 & 6.511022 & -0.090677 & -0.180479 \\ 0 & -0.090677 & 1.137646 & -0.288230 \\ 0.279557 & -0.180479 & -0.288230 & -1.116232 \end{bmatrix}$$

$S(A_5) = 2.339846$

$p = 1, \quad q = 2, \quad c = 0.998332, \quad s = -0.057730$

$$A_{10} = \begin{bmatrix} 23.527319 & -0.001741 & -0.038765 & -0.000060 \\ -0.001741 & 6.460515 & 0 & 0.000017 \\ -0.038765 & 0 & 1.173115 & -0.001349 \\ -0.000060 & 0.000017 & -0.001349 & -1.160949 \end{bmatrix}$$

$S(A_{10}) = 3.015230 \cdot 10^{-3}$

$$A_{16} = \begin{bmatrix} 23.527386 & 0 & 0 & 0 \\ 0 & 6.460515 & 0 & 0 \\ 0 & 0 & 1.173049 & 0 \\ 0 & 0 & 0 & -1.160950 \end{bmatrix}$$

$S(A_{16}) = 2.268476 \cdot 10^{-15}$

matrix are listed, and the element a_{pq} which is to vanish is boxed. In A_{16}, the biggest nondiagonal element is smaller than 0.25×10^{-6}, thus the diagonal elements yield the eigenvalues correct to the number of places shown. Since $S(A_{16}) \cong 2.2 \times 10^{-15}$, the maximum remaining error in each diagonal element is actually no more than 5×10^{-8} by (4–43). Just by chance, the eigenvalues appear in decreasing order. The quadratic convergence begins early, since the eigenvalues are widely separated. Elements smaller than 10^{-7} in magnitude are shown as zero.

4-4-3 Cyclic Jacobi Methods

The search procedure of the classical Jacobi method for finding the largest-magnitude nondiagonal element requires $N = n(n-1)/2$ operations. Comparison with the approximately $4n$ multiplications of an actual rotation makes this procedure uneconomical on a computer. The situation is analogous to that of relaxation by hand computation (see Sec. 2–2–1). Just as in that case, each of the N elements above the diagonal is made zero exactly once in a *cycle* of $N = n(n-1)/2$ rotations by a modification of the classical version known as the *cyclic* Jacobi method. And this occurs within each cycle in the same sequence. Two special approaches surface quite naturally, depending on whether we handle the nondiagonal elements by *rows*, or by *columns*.

In the *special cyclic Jacobi method* using the row approach, the rotation index pairs (p_k, q_k) for one cycle are chosen in the following order:

$$\begin{array}{rl} (1, 2), (1, 3), (1, 4), \ldots, (1, n), & \\ (2, 3), (2, 4), \ldots, (2, n), & \\ (3, 4), \ldots, (3, n), & \\ \cdots \quad \cdot\ \cdot & \\ & (n-1, n). \end{array} \tag{4–48}$$

The rotation angle ϕ and the values of $c_k = \cos \phi_k$ and $s_k = \sin \phi_k$ are chosen for the individual rotations, using (4–31) through (4–34) in such a way that the angles are confined to the interval $-\pi/4 < \phi_k \leq \pi/4$. If the element $a_{p_k q_k} = 0$, no rotation is made.

Theorem 4-7. *The special cyclic Jacobi method using the row approach and choosing the rotation angle ϕ_k in accord with* (4–31) *through* (4–34) *is convergent. The values $S(A_k)$ make up a monotonic, nonincreasing sequence.*

Proof of Theorem 4–7 is not as simple as that of the classical counterpart, and it will only be sketched here. Reference 18 takes off from the observation that the sequence $S(A_k)$ is monotonically nonincreasing and bounded from below. Thus it has a lower limit. From the relation $S(A_{k+1}) = S(A_k) - 2\,(a_{pq}^{(k)})^2$ and the existence of the bound follows the result that the current pivot element $a_{pq}^{(k)}$ must necessarily converge to zero. Examination of the history of the other nondiagonal elements can show that they also become arbitrarily small.

Theorem 4–7 assures convergence of the cyclic Jacobi method. Going even further, Ref. 29 demonstrates analogous to the classical technique that for a symmetric matrix A with n different eigenvalues, *quadratic convergence* takes place as soon as the largest-magnitude nondiagonal element falls below a threshold value. By Theorem 4–7, this condition must occur, and the further that the eigenvalues are separated from one another, the sooner it occurs. More precise ramifications of this proof are given in Refs. 52, 73, and 53.

The convergence and the rate of convergence of the special cyclic Jacobi method using the column approach follow from that of the row approach since the two methods are equivalent in the sense that, starting with the same matrix and the same pivot element, the same matrix is produced after a full cycle of $N = n(n-1)/2$ rotations. This is shown in Ref. 53 with the systematic examination of how the matrix elements are coupled by the rotations and in what sequence they are altered.

A *general cyclic Jacobi method* is the name of a technique in which the pivot elements are the N elements above the diagonal, taken in an entirely arbitrary but cyclic order. Convergence of the general cyclic Jacobi method has not yet been proved in general terms. Reference 27 points out the principal difficulties in such a proof. It is especially fortunate that both special cyclic Jacobi methods converge.

ALGOL Procedure for Computation of Eigenvalues by the Special Cyclic Jacobi Method. In the given symmetric matrix A of order n, only the elements on and above the diagonal are utilized. For a criterion of termination, (4–43) will be used in that the value $S(A)$ is calculated before each cycle. The calculation time here in comparison with the one of $N = n(n-1)/2$ rotations is not as significant. The subroutine *rotation* defined in Sec. 4–4–1 is used as a global parameter. Compare also to the procedure given in Ref. 49.

The procedure parameters are defined as:

- *n* Order of matrix A
- *eps* Tolerance in absolute value for the accuracy of each eigenvalue
- *a* Elements of matrix A. Its diagonal elements at completion are the eigenvalues, in no particular sequence.

ALGOL Procedure No. 10

```
procedure jacobi 2 (n, eps, a);
          value n, eps;
          integer n;  real eps;  array a;
begin integer p, q, i, j;  real ss, theta, t, c, s;
sweep: ss: = 0;
        for i: = 1 step 1 until n − 1 do
            for j: = i + 1 step 1 until n do
                ss: = ss + a[i, j] ↑ 2;
test:   if 2 × ss < eps ↑ 2 then goto out;
        for p: = 1 step 1 until n − 1 do
            for q: = p + 1 step 1 until n do
            begin
                if a[p, q] ≠ 0 then
rotatepq:       begin
                    theta: = 0.5 × (a[q, q] − a[p, p])/a[p, q];
                    if theta = 0 then
                        t: = 1
                    else
                        t: = 1/theta + sign(theta) × sqrt(1 + theta ↑ 2));
                    c: = 1/sqrt(1 + t ↑ 2);  s = c × t;
                    rotation (n, c, s, p, q, a)
                end rotatepq
            end q;
        goto sweep;
out:
end jacobi 2
```

Example 4-3. The same matrix as used in Example 4-2 will be attacked with the special cyclic Jacobi method. Table 11 shows the matrices after complete cycles. After the fourth cycle we have $S(A_4) \cong 1.54 \times 10^{-30}$, thus a very high accuracy of eigenvalues is already assured. The quadratic convergence of the diminishing sequences $S(A_k)$ is quite evident. Nondiagonal elements are equated to zero as soon as their magnitudes are below 10^{-7}.

PROBLEMS

4-5. Find the matrix similar to A,

$$A = \begin{bmatrix} 5 & -2 & 1 & -1 \\ -2 & 10 & -3 & 4 \\ 1 & -3 & 20 & 9 \\ -1 & 4 & 9 & 30 \end{bmatrix},$$

Table 11 CYCLIC JACOBI METHOD

$$A_0 = \begin{bmatrix} 20.000000 & -7.000000 & 3.000000 & -2.000000 \\ -7.000000 & 5.000000 & 1.000000 & 4.000000 \\ 3.000000 & 1.000000 & 3.000000 & 1.000000 \\ -2.000000 & 4.000000 & 1.000000 & 2.000000 \end{bmatrix}$$

$S(A_0) = 1.600000 \cdot 10^{2}$

$$A_1 = \begin{bmatrix} 23.523089 & -0.009053 & -0.238471 & 0.151640 \\ -0.009053 & -0.437554 & -1.397689 & 0.931475 \\ -0.238471 & -1.397689 & 6.174371 & 0 \\ 0.151640 & 0.931475 & 0 & 0.740095 \end{bmatrix}$$

$S(A_1) = 5.802252$

$$A_2 = \begin{bmatrix} 23.527359 & 0.022392 & -0.005358 & 0.010605 \\ 0.022392 & -1.160102 & 0.079297 & 0.002477 \\ -0.005358 & 0.079297 & 6.459691 & 0 \\ 0.010605 & 0.002477 & 0 & 1.173051 \end{bmatrix}$$

$S(A_2) = 1.387334 \cdot 10^{-2}$

$$A_3 = \begin{bmatrix} 23.527386 & -0.000023 & 0 & 0 \\ -0.000023 & -1.160950 & 0 & 0 \\ 0 & 0 & 6.460515 & 0 \\ 0 & 0 & 0 & 1.173049 \end{bmatrix}$$

$S(A_3) = 1.094265 \cdot 10^{-9}$

$$A_4 = \begin{bmatrix} 23.527386 & 0 & 0 & 0 \\ 0 & -1.160950 & 0 & 0 \\ 0 & 0 & 6.460515 & 0 \\ 0 & 0 & 0 & 1.173049 \end{bmatrix}$$

$S(A_4) = 1.542515 \cdot 10^{-30}$

resulting from a (2, 4) rotation of A through an angle $\phi = \pi/2$. What is the effect of this transformation?

4-6. Transform the matrices

$$A_1 = \begin{bmatrix} 3 & 1 \\ 1 & 5 \end{bmatrix} \quad \text{and} \quad A_2 = \begin{bmatrix} 3 & 2 \\ 2 & 3 \end{bmatrix}$$

to diagonal form by the Jacobi method. Find the transformation matrices, eigenvalues, and eigenvectors.

4-7. In order to abbreviate the search process of the classical Jacobi method, it may be altered as follows. Set up the sum s_i of the squares of the elements of each line, omitting the diagonal element.

$$s_i = \sum_{\substack{j=1 \\ j \neq i}}^{n} a_{ij}^2 \qquad i = 1, 2, \ldots, n.$$

Clearly, $S(A) = \sum_{i=1}^{n} s_i$. In which straightforward fashion are the row sums s_i transformed in a (p, q)-rotation? In order to find the pivot element of the rotation, identify the largest-magnitude nondiagonal element a_{pq} $(= a_{qp}!)$ of the line p with the largest value s_p. For this, only about $2n$ comparisons are needed. In other words, avoid the actual evaluation of the largest-magnitude nondiagonal element at every step. Prove the convergence of this modified Jacobi method, and then set up a computer program for it.

4-8. How accurately do the diagonal elements of

$$A = \begin{bmatrix} 5 & 10^{-4} & 2\cdot10^{-4} & 2\cdot10^{-4} \\ 10^{-4} & -9 & 10^{-5} & -10^{-5} \\ 2\cdot10^{-4} & 10^{-5} & 10 & 10^{-6} \\ 2\cdot10^{-4} & 10^{-5} & 10^{-6} & 8 \end{bmatrix}$$

approximate the eigenvalues? What are the estimates, using the Gerschgorin circle theorem (p. 150)?

4-5 TRANSFORMATION TO TRIDIAGONAL FORM: STURM SEQUENCES; COMPUTATION OF INDIVIDUAL EIGENVALUES

This section describes methods whereby a given symmetric matrix A is transformed into a *tridiagonal* form by an orthogonal similarity transformation. Subsequently, individual eigenvalues, or even eigenvalues lying within a given interval, may be found. These methods are useful when certain eigenvalues, not necessarily the largest or the smallest, are to be evaluated.

4-5-1 Givens' Method

In contrast to the Jacobi technique, the method of *Givens* (Ref. 23) makes possible the reduction of the given symmetric matrix A to a symmetric, tridiagonal matrix $\boldsymbol{J}$ (see below) through a *finite* number of suitable rotations:

$$\boldsymbol{J} = \begin{bmatrix} \alpha_1 & \beta_1 & & & & \\ \beta_1 & \alpha_2 & \beta_2 & & & \\ & \beta_2 & \alpha_3 & \beta_3 & & \\ & & \cdot & \cdot & \cdot & \\ & & & \beta_{n-2} & \alpha_{n-1} & \beta_{n-1} \\ & & & & \beta_{n-1} & \alpha_n \end{bmatrix}. \tag{4–49}$$

Matrix $\boldsymbol{J}$ contains nonzero elements only in the diagonal and in the two immediately adjacent diagonals. The elements of A lying outside these three nonzero diagonals of $\boldsymbol{J}$ are eliminated in such a way that, once an element is brought to zero through rotation, it stays there. In contrast to the Jacobi method, the rotation index pairs are selected such that the element brought to zero does not lie on the nondiagonal intersection of altered rows and columns. For instance, in the elimination of the elements a_{ik} by rows, with $k \geq i + 2$, the rotations given respectively in Table 12 accomplish the desired transformation. There are numerous variants giving the same result.

Table 12 ROTATION BY GIVENS' METHOD

Elimination of	*Rotation Index Pair*
a_{13} a_{14} a_{15} $\cdots$ a_{1n}	(2, 3) (2, 4) (2, 5) $\cdots$ (2, n)
a_{24} a_{25} $\cdots$ a_{2n}	(3, 4) (3, 5) $\cdots$ (3, n)
a_{35} $\cdots$ a_{3n}	(4, 5) $\cdots$ (4, n)
$\cdots$	$\cdots$ $\cdot$
$n_{n-2,n}$	($n-1$, n)

For the elimination of the current element a_{jk} $(k \geq j + 2)$ a rotation $(j + 1, k)$ is employed. Thus the element a_{jk} is influenced only by the column operation, and the vanishing of the transformed element a'_{jk} by (4–29) leads to the equation

$$a'_{jk} = a_{j,j+1} \sin \phi + a_{j,k} \cos \phi = 0. \tag{4–50}$$

In conjunction with the identity $\sin^2 \phi + \cos^2 \phi = 1$, Eq. (4–50) yields

$$c = \cos \phi = \frac{a_{j,j+1}}{\sqrt{a_{j,j+1}^2 + a_{j,k}^2}}, \quad s = \sin \phi = -\frac{a_{j,k}}{\sqrt{a_{j,j+1}^2 + a_{j,k}^2}}. \tag{4–51}$$

It remains to verify that the rotation sequence of Table 12 actually achieves the transformation of A to the form J. We must merely show that the zero elements already produced are not changed by later rotations. For the first row, this is obvious, since a $(2, k)$ rotation affects only the second and kth columns and rows, but never the entire first row. In the first row, only element a_{12} is changed and a_{1k} $(k = 3, 4, \ldots, n)$ is made to vanish. From symmetry considerations, an analogous result holds for the first column. Assume further that the first r rows and columns have been brought to the desired form. For $r = 2$, we have the situation of Fig. 16. Elimination of the elements of the $(r + 1)$th row requires $(r + 2, k)$-rotations. Figure 16 shows the case $k = 5$. In the first r rows, only zero elements are combined with each other by the column operations, thus nothing actually changes there. Since both rotation indices are greater than $(r + 1)$, only rows below the $(r + 1)$th are affected.

$$\begin{bmatrix} \times & \times & 0 & 0 & 0 & 0 \\ \times & \times & \times & 0 & 0 & 0 \\ 0 & \times & \times & \times & \otimes & \times \\ 0 & 0 & \times & \times & \times & \times \\ 0 & 0 & \otimes & \times & \times & \times \\ 0 & 0 & \times & \times & \times & \times \end{bmatrix}$$

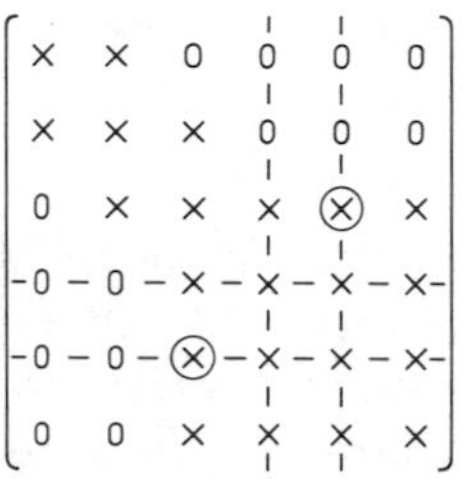

Figure 16. Givens' method.

In all, $(n - 1)(n - 2)/2$ rotations accomplish the desired reduction. The whole computational procedure exhibits great simplicity and unity. The technique goes entirely automatically, requiring no special provisions for exceptions. A rotation is omitted only if the element in question a_{jk} is already equal to zero. The determination of the values c and s by (4–51) offers no problems, even when $a_{j,j+1} = 0$. In that case, $c = 0$ and $s = -1$; that is, the rotation corresponds essentially to a permutation.

ALGOL Procedure for Givens' Method. For a given symmetric matrix A of order n, the procedure uses Givens' method to yield the values α_i and β_i of the reduced matrix J of (4–49) as elements of the transformed matrix A by $\alpha_i = a_{ii}$ $(i = 1, 2, \ldots, n)$ and $\beta_i = a_{i,i+1}$ $(i = 1, 2, \ldots, n - 1)$. Utilizing symmetry, the program involves only the elements a_{ik} on and above the diagonal. The procedure *rotation* defined in Sec. 4–4–1 is used as a global parameter. The resultant program is not optimal, as we do not take advantage of the zero elements that have already been created.

The procedure parameters are defined as:

n Order of matrix A

a Elements of matrix A. At the end, $a_{ii} = \alpha_i$, $a_{i,i+1} = \beta_i$.

ALGOL Procedure No. 11

```
procedure givens (n, a);
          value n;  integer n;  array a;
       begin integer j, k;  real w, c, s;
```

```
for j: = 1 step 1 until n - 2 do
    for k: = j + 2 step 1 until n do
    begin
        if a[j, k] ≠ 0 then
        begin
            w: = sqrt(a[j, j + 1] ↑ 2 + a[j, k] ↑ 2);
            c: = a[j, j + 1]/w;   s: = -a[j, k]/w;
            rotation (n, c, s, j + 1, k, a)
        end
    end k
end givens
```

As shown, the method of Givens produces the transformation of a full matrix $\boldsymbol{A}$ to a tridiagonal form $\boldsymbol{J}$ as defined in (4–49). In the special case of a *band matrix* with given band width m, naturally the given sequence of Jacobi rotations can transform this matrix into tridiagonal form. Yet we would not be taking advantage of the band structure, since the rotations would quickly fill up many initially zero elements of the lower rows. References 47 and 56 show that through an appropriate sequence of rotations we can take advantage of the band structure and retain it by systematic rotations which eliminate the nonzero elements outside the band.

For band matrices whose band width is small compared to the order of the matrix, the resultant computation is less time consuming than the customary Givens' method. In addition it requires less storage space in memory, since with the special indexing of Sec. 1–4–5, the band structure can be fully exploited to maximum advantage.

4-5-2 Householder's Method

Instead of achieving the tridiagonal form by simple rotations, element by element, *Householder* (Ref. 32) suggested another transformation method whereby at every step an entire row or column is brought into the desired configuration. He examined transformation matrices $\boldsymbol{U}$ of the form

$$\boldsymbol{U} = \boldsymbol{I} - 2\boldsymbol{w}\boldsymbol{w}^{\mathrm{T}}. \tag{4–52}$$

Here $\boldsymbol{w}$ indicates a *normalized* vector with n components, and $\boldsymbol{w}\boldsymbol{w}^{\mathrm{T}}$, a *dyadic product* of a column vector and a transposed row vector, represents an $n \times n$ matrix. The matrix $\boldsymbol{U}$ (4–52) is evidently symmetric, and also *involutary* in that its square is the identity matrix. It turns out that

$$\begin{aligned}\boldsymbol{U}\boldsymbol{U} &= (\boldsymbol{I} - 2\boldsymbol{w}\boldsymbol{w}^{\mathrm{T}})(\boldsymbol{I} - 2\boldsymbol{w}\boldsymbol{w}^{\mathrm{T}}) \\ &= \boldsymbol{I} - 2\boldsymbol{w}\boldsymbol{w}^{\mathrm{T}} - 2\boldsymbol{w}\boldsymbol{w}^{\mathrm{T}} + 4\boldsymbol{w}\boldsymbol{w}^{\mathrm{T}}\boldsymbol{w}\boldsymbol{w}^{\mathrm{T}} \\ &= \boldsymbol{I},\end{aligned}$$

since through the normalization of $\boldsymbol{w}$, $\boldsymbol{w}^T\boldsymbol{w} = (\boldsymbol{w}, \boldsymbol{w}) = 1$. Matrix $\boldsymbol{U}$ (4–52) is thus symmetric and also *orthogonal.*

In order to make the elements $a_{13}, a_{14}, \ldots, a_{1n}$ and their symmetric counterparts vanish simultaneously in the first step of the Householder transformation, the vector $\boldsymbol{w} = (0, w_2, w_3, \ldots, w_n)^T$ is set such that its first component vanishes. As a result of the normalization condition $(\boldsymbol{w}, \boldsymbol{w}) = 1$, there are $n - 2$ quantities available for making the $n - 2$ elements a'_{1k} $(k = 3, 4, \ldots, n)$ vanish after a similarity transformation $\boldsymbol{U}^T\boldsymbol{A}\boldsymbol{U} = \boldsymbol{A}'$. In the case $n = 5$, the transformation matrix $\boldsymbol{U}$ has the following form:

$$\boldsymbol{U} = \boldsymbol{U}^T = \boldsymbol{U}^{-1} = \begin{bmatrix} 1 & 0 & 0 & 0 & 0 \\ 0 & 1 - 2w_2^2 & -2w_2w_3 & -2w_2w_4 & -2w_2w_5 \\ 0 & -2w_2w_3 & 1 - 2w_3^2 & -2w_3w_4 & -2w_3w_5 \\ 0 & -2w_2w_4 & -2w_3w_4 & 1 - 2w_4^2 & -2w_4w_5 \\ 0 & -2w_2w_5 & -2w_3w_5 & -2w_4w_5 & 1 - 2w_5^2 \end{bmatrix}. \tag{4–53}$$

Let us examine first the transformation of elements of the first row. In the formation of $\boldsymbol{U}^T\boldsymbol{A}$, the first row remains unaltered because of the particular form of $\boldsymbol{U}$. Thus

$$\left.\begin{aligned} a'_{11} &= a_{11} \\ a'_{12} &= (1 - 2w_2^2)a_{12} - 2w_2w_3a_{13} - 2w_2w_4a_{14} - \cdots - 2w_2w_na_{1n} \\ a'_{13} &= -2w_2w_3a_{12} + (1 - 2w_3^2)a_{13} - 2w_3w_4a_{14} - \cdots - 2w_3w_na_{1n} \\ &\cdot \qquad \cdot \qquad \cdot \qquad \cdot \qquad \cdots \qquad \cdot \\ a'_{1n} &= -2w_2w_na_{12} - 2w_3w_na_{13} - 2w_4w_na_{14} - \cdots + (1 - 2w_n^2)a_{1n} \end{aligned}\right\}. \tag{4–54}$$

A trivial computation shows that the sum s^2 of the squares of nondiagonal elements of the first row is invariant:

$$a'^2_{12} + a'^2_{13} + \cdots + a'^2_{1n} = a^2_{12} + a^2_{13} + \cdots + a^2_{1n} = s^2, \quad s \geq 0. \tag{4–55}$$

For brevity, we define the quantity

$$h = w_2a_{12} + w_3a_{13} + \cdots + w_na_{1n}. \tag{4–56}$$

Then the formulas (4–54) for the altered elements of the first line are

$$a'_{1k} = a_{1k} - 2w_kh \quad k = 2, 3, \ldots, n. \tag{4–57}$$

The requirement $a'_{1k} = 0$ $(k = 3, 4, \ldots, n)$ and the invariance of the sum of the squares (4–55) yields, in conjunction with (4–57), the equation for $a_{12}'^2$

$$a_{12}'^2 = (a_{12} - 2w_2h)^2 = \sum_{k=2}^{n} a_{1k}^2 = s^2,$$

or

$$a_{12} - 2hw_2 = \pm s. \tag{4–58}$$

The requirement $a_{1k}' = 0\,(k = 3, 4, \ldots, n)$ leads from (4–57) to the equations

$$a_{1k} - 2hw_k = 0 \qquad k = 3, 4, \ldots, n. \tag{4–59}$$

Multiply (4–58) by w_2 and each of the equations (4–59) by the corresponding w_k, and add the resultant equations. If we invoke the normalization condition for the vector $\boldsymbol{w}$ and the definition of h of Eq. (4–56), we are left with

$$-h = \pm sw_2, \quad \text{or} \quad h = \mp sw_2. \tag{4–60}$$

The signs correspond inversely to those of Eq. (4–58). Equations (4–58) and (4–60) yield first the value of w_2, then (4–60) gives that of h and, finally, by (4–59), we have the values of $w_3, w_4, \ldots, w_n$.

$$w_2 = \sqrt{\frac{1}{2}\left(1 \mp \frac{a_{12}}{s}\right)}, \quad h = \mp sw_2, \quad w_k = \frac{a_{1k}}{2h} \qquad k = 3, 4, \ldots, n. \tag{4–61}$$

The sign still open for selection is taken such that there is no loss of accuracy in the radicand; in other words, the sign corresponds to that of a_{12}. The transformation matrix $\boldsymbol{U}$ (4–52) has thus been derived through the components of the vector $\boldsymbol{w}$. There remains only carrying out the similarity transformation of matrix $\boldsymbol{A}$, using $\boldsymbol{U}$. It turns out that the matrix products can be found without finding $\boldsymbol{U}$ explicitly, element by element. By definition, the elements u_{ik} of $\boldsymbol{U}$ are, by (4–52),

$$u_{ik} = \delta_{ik} - 2w_iw_k \qquad i, k = 1, 2, \ldots, n.$$[1]

The elements b_{ij} of the product matrix $\boldsymbol{B} = \boldsymbol{U}^{\mathrm{T}}\boldsymbol{A} = \boldsymbol{U}\boldsymbol{A}$ are

$$b_{ij} = \sum_{k=1}^{n} u_{ik}a_{kj} = \sum_{k=1}^{n} (\delta_{ik} - 2w_iw_k)a_{kj} = a_{ij} - 2w_i \sum_{k=1}^{n} w_ka_{kj}.$$

Noting that $w_1 = 0$, we define the j-dependent quantities

$$p_j = \sum_{k=1}^{n} w_ka_{kj} = \sum_{k=2}^{n} w_ka_{kj} \qquad j = 1, 2, \ldots, n. \tag{4–62}$$

[1]The *Kronecker delta* symbol δ_{ik} is used here, with the value $\delta_{ik} = 1$ when $i = k$, and $\delta_{ik} = 0$ when $i \neq k$.

In particular, $p_1 = h$ and is thus already known. In these terms, the elements b_{ij} are

$$b_{ij} = a_{ij} - 2w_i p_j \qquad i, j = 1, 2, \ldots, n.$$

The elements a'_{ij} of matrix $\boldsymbol{A}' = \boldsymbol{U}^{\mathrm{T}}\boldsymbol{A}\boldsymbol{U} = \boldsymbol{B}\boldsymbol{U}$ are then

$$\begin{aligned} a'_{ij} &= \sum_{k=1}^{n} b_{ik} u_{kj} = \sum_{k=1}^{n} (a_{ik} - 2w_i p_k)(\delta_{kj} - 2w_k w_j) \\ &= a_{ij} - 2w_j \sum_{k=1}^{n} a_{ik} w_k - 2w_i p_j + 4w_i w_j \sum_{k=1}^{n} p_k w_k. \end{aligned}$$

By symmetry and equations (4–62),

$$\sum_{k=1}^{n} a_{ik} w_k = \sum_{k=1}^{n} a_{ki} w_k = p_i.$$

Furthermore, we will define a quantity independent of i and j,

$$g = \sum_{k=1}^{n} p_k w_k = \sum_{k=2}^{n} p_k w_k.$$

Then the transformed quantities a'_{ij} may be written

$$\begin{aligned} a'_{ij} &= a_{ij} - 2w_j p_i - 2w_i p_j + 4g w_i w_j \\ &= a_{ij} - 2w_i(p_j - g w_j) - 2w_j(p_i - g w_i). \end{aligned}$$

For calculations, it is useful to introduce the quantities

$$q_i = 2(p_i - g w_i) \qquad i = 1, 2, \ldots, n \tag{4–63}$$

so that the elements a_{ij} of the transformed matrix may be computed by a completely symmetric formulation (4–64):

$$a'_{ij} = a_{ij} - w_i q_j - q_i w_j \qquad i, j = 1, 2, \ldots, n. \tag{4–64}$$

After the first step is completed, the transformed matrix $\boldsymbol{A}'$ exhibits the desired form in the first row and column. The technique is used analogously on the matrix to the right and below having $(n - 1)$ rows. In the second step, a vector $\boldsymbol{w}$ is chosen with its first two components zero. The corresponding transformation matrix $\boldsymbol{U}$ thus retains the first row and column. After $n - 2$ transformation steps, we are left with the tridiagonal form $\boldsymbol{J}$.

ALGOL Precedure Using Householder's Method. The symmetry of the given matrix $\boldsymbol{A}$ of order n is exploited by using only elements on or above the diagonal. The program produces the elements of the tridiagonal matrix J of (4–49) with $\alpha_i = a_{ii}$ $(i = 1, 2, \ldots, n)$ and $\beta_i = a_{i,i+1}$ $(i = 1, 2, \ldots, n-1)$.

Procedure parameters are defined as:

n Order of matrix A

a Elements of matrix A. At the end, $a_{ii} = \alpha_i$, $a_{i,i+1} = \beta_i$.

ALGOL Procedure No. 12

```
procedure householder (n, a);
          value n;  integer n;  array a;
begin integer i, j, k, t;  real sigma, s, h, g;
   array w, p, q[1 : n];
   for t: = 1 step 1 until n - 2 do
   begin sigma: = 0;
      for k: = t + 1 step 1 until n do
         sigma: = sigma + a[t, k] ↑ 2;
      if sigma = 0 then goto zero;
      if a[t, t + 1] > 0 then
        s: = sqrt (sigma)
      else
        s: = -sqrt (sigma);
      w[t]: = 0;  w[t + 1]: = sqrt(0.5 × (1 + a[t, t + 1]/s));
      h: = s × w[t + 1];
      for k: = t + 2 step 1 until n do
         w[k]: = a[t, k]/(2 × h);
      p(t): = h;
      for j: = t + 1 step 1 until n do
      begin p(j): = 0;
         for k: = t + 1 step 1 until j do
            p(j): = p[j] + w[k] × a[k, j];
         for k: = j + 1 step 1 until n do
            p[j]: = p[j] + w[k] × a[j, k]
      end j:
      g: = 0;
      for k: = t + 1 step 1 until n do
         g: = g + p[k] × w[k];
      for i: = t step 1 until n do
         q[i]: = 2 × (p[i] - g × w[i]);
      for i: = t step 1 until n do
         for j: = i step 1 until n do
            a[i, j]: = a[i, j] - w[i] × q[j] - w[j] × q[i];
zero:  end
end householder
```

Although from the computational point of view the methods of Givens and Householder are totally different, it is interesting that the two methods yield essentially the same tridiagonal matrix in which, at most, the signs of the nondiagonal elements differ. The reason for this noteworthy result is that in both methods, the product of all transformation matrices is an orthogonal matrix $\boldsymbol{U}$ having the unit vector in the first column (and thus also in the first row). Starting from this fact, simple computations reveal that $\boldsymbol{U}$ and the tridiagonal matrix $\boldsymbol{J}$ are uniquely determined save for certain signs (see Refs. 19, 75).

Despite its apparent complications, Householder's method has two advantages over that of Givens. In Givens' method, *round-off errors* in the elements appear at each individual rotation. If in Householder's method we carry out the scalar multiplications on a computer with more significant places, without rounding off until the final solution is reached, the round-off error will remain much smaller, and the tridiagonal matrix $\boldsymbol{J}$ has greater numerical accuracy (see Ref. 75).

In addition, the computational effort for the two methods is significantly different. A summation of the substantial operations in the reduction of symmetric matrices of large order n reveals asymptotically the number $4n^3/3$ for Givens' method and only $2n^3/3$ for Householder's. Besides, Givens' method requires for every element rotated to zero the computation of a square root, in all $(n - 1)(n - 2)/2$ of them, while Householder's method needs only two of them per step, or $2(n - 2)$ in all. *The Householder transformation is thus doubly preferable, both on grounds of computational economy and numerical accuracy.*

As described here, Householder's method does not take advantage of the band form of a symmetric matrix. But just as in the case of Givens, however, a modification exists, allowing exploitation of the special matrix structure. Reference 47 shows how the nonzero elements produced outside the band during the first step of the Householder transformation and lying in a triangular domain can again be eliminated by additional transformations of a similar nature. Basically, these elements upsetting the band structure are shoved to the lower left until they cross the boundary of the band matrix and again vanish. Subsequently, we take up the transformation of the second row and column, and the procedure is carried out analogously to the first step. In this fashion, the band structure can be maintained throughout the computation and can be evaluated with a minimum of memory storage space through the special indexing of Sec. 1–4–5.

4-5-3 The Sturm Sequence

After the reduction methods of Givens and Householder, we are left with the problem of finding the eigenvalues of tridiagonal symmetric matrices. The first technique is based on the concept of a *Sturm sequence*.

Definition 4-1

A Sturm sequence is a sequence of real functions $f_n(x), f_{n-1}(x), \ldots, f_1(x), f_0(x)$ *in the real variable* x, *fulfilling the following conditions on a closed interval* $[a, b]$ *on the* x *axis:*

1. *Functions* $f_i(x)$ *are continuous for* $i = n, n-1, \ldots, 1, 0$.
2. *For* $a \leq x \leq b$, $f_0(x)$ *does not change sign.*
3. *No two successive functions of the sequence have a common zero. In other words, if* $\bar{x}$ *is a zero on* $[a, b]$ *for a function* $f_i(x)$ $(i = n-1, n-2, \ldots, 2, 1)$, *then* $f_{i-1}(\bar{x}) \neq 0$ and $f_{i+1}(\bar{x}) \neq 0$.
4. *For a zero* $\bar{x}$ *of a function* $f_i(x)$ $(i = n-1, n-2, \ldots, 2, 1)$ *on* $[a, b]$, *the adjacent functions are of opposite sign, to wit:*

$$f_i(\bar{x}) = 0: \qquad \operatorname{sign} f_{i+1}(\bar{x}) = -\operatorname{sign} f_{i-1}(\bar{x}). \tag{4–65}$$

5. *For a zero* $\bar{x}$ *of the function* $f_n(x)$ *on* $[a, b]$, *for* $h > 0$ *sufficiently small, the following relations hold:*

$$\operatorname{sign}\frac{f_n(\bar{x}-h)}{f_{n-1}(\bar{x}-h)} = -1, \quad \operatorname{sign}\frac{f_n(\bar{x}+h)}{f_{n-1}(\bar{x}+h)} = +1. \tag{4–66}$$

Condition (4–66) *means that at every zero of* $f_n(x)$, *the sign of* $f_{n-1}(x)$ *corresponds with that of the derivative of* $f_n(x)$, *if the latter exists.*

Theorem 4-8. *Let* $f_n(x), f_{n-1}(x), \ldots, f_1(x), f_0(x)$ *be a Sturm sequence, and let* $V(x)$ *indicate the number of sign changes within the sequence* $f_n(x), f_{n-1}(x), \ldots, f_1(x), f_0(x)$ *for a given* x *in the interval* $[a, b]$. *The number of zeros* m *of* $f_n(x)$ *in the interval* $[a, b]$ *is then* $m = V(a) - V(b)$.

Proof: Consider the number of sign changes in the sequence $f_n(x), f_{n-1}(x), \ldots, f_1(x), f_0(x)$ as a function of x as x grows monotonically from a to b. Because of the continuity of functions in the Sturm sequence, the value of $V(x)$ can change only if one or more functions pass through a zero. Since by definition $f_0(x)$ cannot change sign, two cases must be considered:

(a) Zero crossing of one of the "inner" functions $f_i(x)$ $(i = n-1, n-2, \ldots, 2, 1)$. Because of properties 1, 3, and 4, only the sign combinations shown in Table 13 are possible for $h > 0$ sufficiently small.

Table 13 A ZERO OF AN "INNER" FUNCTION

$x =$	$f_{i+1}(x)$	$f_i(x)$	$f_{i-1}(x)$	$x =$	$f_{i+1}(x)$	$f_i(x)$	$f_{i-1}(x)$
$\bar{x}-h$	+	±	−	$\bar{x}-h$	−	±	+
$\bar{x}$	+	0	−	$\bar{x}$	−	0	+
$\bar{x}+h$	+	∓	−	$\bar{x}+h$	−	∓	+

In any event, the number of sign changes remains unaltered. This is also true in the case of a false zero crossing of $f_i(x)$ if $f_i(x)$ has a multiple zero of even order. Every zero of an inner function leaves the number of sign changes invariant.

Table 14 A ZERO OF THE FIRST FUNCTION

$x =$	$f_n(x)$	$f_{n-1}(x)$	$x =$	$f_n(x)$	$f_{n-1}(x)$
$\bar{x} - h$	$+$	$-$	$\bar{x} - h$	$-$	$+$
$\bar{x}$	0	$-$	$\bar{x}$	0	$+$
$\bar{x} + h$	$-$	$-$	$\bar{x} + h$	$+$	$+$

(b) A zero of $f_n(x)$. Because of properties 1, 3, and 5, only the sign combinations depicted in Table 14 are possible, for $h > 0$ sufficiently small.

In both cases, one sign change is lost with increasing x. A multiple zero of $f_n(x)$ having even order is inconsistent with properties 1, 3, and 5; one of odd order has the same effect as a simple zero. Every simply tabulated zero of $f_n(x)$ reduces the number of sign changes by one.

In sum, then, m zeros of $f_n(x)$ in the interval $[a, b]$ diminish the sign changes $V(x)$ of the Sturm chain by m, and thus m is the difference of $V(a)$ and $V(b)$.

Example 4-4. The four polynomials

$$f_3(x) = x^3 - 3x^2 + 1, \quad f_2(x) = x^2 - 3x + 1, \quad f_1(x) = x - 1,$$

and

$$f_0(x) = 1$$

form a Sturm sequence. The conditions 1 and 2 are clearly fulfilled for every finite interval. In a graphic plot, the other three conditions are readily seen to be fulfilled. Table 15 shows the sign and the number of sign changes $V(x)$ for integer values of x.

Table 15 SIGN OF THE STURM SEQUENCE

x	$f_3(x)$	$f_2(x)$	$f_1(x)$	$f_0(x)$	$V(x)$
-2	$-$	$+$	$-$	$+$	3
-1	$-$	$+$	$-$	$+$	3
0	$+$	$+$	$-$	$+$	2
$+1$	$-$	$-$	(0)	$+$	1
$+2$	$-$	$-$	$+$	$+$	1
$+3$	$+$	$+$	$+$	$+$	0

For $x = 1$, $f_1(x) = 0$. In principle, the sign of zero is immaterial since,

interval. Determination of the number of sign changes for both bounds of the interval yields the index values of the eigenvalues within, by Theorem 4–12.

ALGOL Procedure for the Method of Continuous Interval Bisection. This procedure finds the kth eigenvalue λ_k (numbered in diminishing sequence) of a symmetric, tridiagonal, irreducible matrix by a number of divisions which is given in advance. The initial values are given by Gerschgorin's theorem. The squares of the nondiagonal elements, which must be repeatedly used, are computed in advance. The nonexistent element b_n is set equal to zero for simplicity. In addition, in order to avoid an indexed parameter for $f_i(\lambda)$, the nonindexed parameters p, q, and r are used respectively for $f_{i-2}(\lambda)$, $f_{i-1}(\lambda)$, and $f_i(\lambda)$, and these must correspondingly be redefined. The sign changes are effected with the help of function sign (x). If the signs of q and r match, sign (q) − sign $(r) = 0$, and if they differ, $|\text{sign}(q) - \text{sign}(r)| = 2$; thus the sign changes usually count double. But the result is correct even if one of the inner polynomials $f_i(\lambda)$ $(i = 1, 2, \ldots, n-1)$ vanishes, since then sign $(0) = 0$, but the absolute difference of two successive steps is each equal to unity. If it should also happen that $f_n(\lambda) = 0$, then the number of sign changes by this manner of counting is odd. If λ is the eigenvalue being sought, it is immaterial whether the pertinent interval midpoint is taken as the next upper or lower bound.

The procedure parameters are defined as:

- n Order of the tridiagonal matrix
- a Elements a_i of the diagonals $(i = 1, 2, \ldots, n)$
- b Elements b_i of the offdiagonals $(i = 1, 2, \ldots, n)$; $b_n = 0$
- t Number of interval bisections to be carried out
- k Index of the eigenvalue λ_k to be found
- eig Value of λ_k as solution

ALGOL Procedure No. 13

```
procedure bisection (n, a, b, t, k, eig);
          value n, t, k;
          integer n, t, k;  real eig;  array a, b;
begin integer i, l, v;  real lambda, min, max, p, q, r;
   array b2[1 : n];
   min: = a[1] - abs(b[1]);  max: = a[1] + abs(b[1]);
   b2[1]: = b[1] ↑ 2;
   for i: = 2 step 1 until n do
   begin b2[i]: = b[i] ↑ 2;
      r: = abs(b[i - 1]) + abs(b[1]);
      if a[i] - r < min then min: = a[i] - r;
      if a[i] + r > max then max: = a[i] + r
```

```
    end i;
    for l: = 1 step 1 until t do
    begin lambda: = (min + max)/2;
        p: = 1;  q: = lambda - a[1];
        v: = abs(sign (p) - sign(q));
        for i: = 2 step 1 until n do
        begin
            r: = (lambda - a[i]) × q -b2[i - 1] × p;
            v: = v + abs(sign(q) - sign(r));
            p: = q; q: = r
        end i;
        v: = entier(v/2);
        if v ≥ k then min: = lambda
                  else max: = lambda
    end l;
    eig: = (min + max)/2
end bisection
```

Example 4-5. A tridiagonal matrix of order $n = 10$ has the diagonal element values $a_i = 2$ and the immediately adjacent offdiagonal element values $b_i = 1$, so that the recursion polynomials are

$$f_0(\lambda) = 1, \quad f_1(\lambda) = \lambda - 2, \quad f_k(\lambda) = (\lambda - 2)f_{k-1}(\lambda) - f_{k-2}(\lambda) \quad k \geq 2.$$

Since the matrix is irreducible, symmetric, and diagonal dominant in the weak sense, having also positive diagonal elements, it is positive definite. Gerschgorin's theorem yields the bounds $a = 0$ and $b = 4$ for the eigenvalues. Twenty interval bisections evaluate the eigenvalues with an absolute accuracy of $(b - a)\cdot 2^{-21} = 4\cdot 1^{-21} = 2^{-19} \cong 1.9\cdot 10^{-6}$. The sixth decimal place is thus correct to within two units. Table 16 shows the evaluation of the smallest and fifth eigenvalues.

The values "min" and "max" are the interval bounds for that eigenvalue at the start of the lth step; λ indicates the interval midpoint, and $V(\lambda)$ the number of sign changes of the Sturm sequence for λ.

4-5-5 The Eigenvectors of Tridiagonal Matrices

If we want to find the eigenvectors corresponding to the evaluated eigenvalues of a given symmetric matrix A, and if we have reduced the matrix A to a tridiagonal form J of Eq. (4–49) by the method of Givens or Householder, we can proceed to find the eigenvectors of J. The eigenvectors of A are then found by a finite number of matrix multiplications appropriate to the individual steps of the transformation utilized.

The eigenvectors x of a symmetric tridiagonal matrix J whose eigenvalues λ_k have been found are the solutions of the following homogeneous

Table 16 METHOD OF REPEATED INTERVAL BISECTION

$k = 10$: *Smallest Eigenvalue* λ_{10}					$k = 5$: *Eigenvalue* λ_5				
l	min	max	λ	$V(\lambda)$	l	min	max	λ	$V(\lambda)$
1	0	4.000000	2.000000	5	1	0	4.000000	2.000000	5
2	0	2.000000	1.000000	7	2	2.000000	4.000000	3.000000	3
3	0	1.000000	0.500000	8	3	2.000000	3.000000	2.500000	4
4	0	0.500000	0.250000	9	4	2.000000	2.500000	2.250000	5
5	0	0.250000	0.125000	9	5	2.250000	2.500000	2.375000	4
6	0	0.125000	0.062500	10	6	2.250000	2.375000	2.312500	4
7	0.062500	0.125000	0.093750	9	7	2.250000	2.312500	2.281250	5
8	0.062500	0.093750	0.078125	10	8	2.281250	2.312500	2.296875	4
9	0.078125	0.093750	0.085938	9	9	2.281250	2.296875	2.289063	4
10	0.078125	0.085938	0.082031	9	10	2.281250	2.289063	2.285156	4
11	0.078125	0.082031	0.080078	10	11	2.281250	2.285156	2.283203	5
12	0.080078	0.082031	0.081055	9	12	2.283203	2.285156	2.284180	5
13	0.080078	0.081055	0.080566	10	13	2.284180	2.285156	2.284668	4
14	0.080566	0.081055	0.080811	10	14	2.284180	2.284668	2.284424	5
15	0.080811	0.081055	0.080933	10	15	2.284424	2.284668	2.284546	5
16	0.080933	0.081055	0.080994	10	16	2.284546	2.284668	2.284607	5
17	0.080994	0.081055	0.081024	9	17	2.284607	2.284668	2.284637	4
18	0.080994	0.081024	0.081009	10	18	2.284607	2.284637	2.284622	5
19	0.081009	0.081024	0.081017	9	19	2.284622	2.284637	2.284630	4
20	0.081009	0.081017	0.081013	10	20	2.284622	2.284630	2.284626	5
$\lambda_{10} = 0.081015$					$\lambda_5 = 2.284628$				

system of equations having a nontrivial solution:

$$\left.\begin{aligned}(a_1-\lambda_k)x_1+b_1x_2 \qquad\qquad\qquad\qquad &= 0\\ b_1x_1+(a_2-\lambda_k)x_2+b_2x_3 \qquad\qquad &= 0\\ b_2x_2+(a_3-\lambda_k)x_3+b_3x_4 &= 0\\ \cdot\quad\cdot\quad\cdot\quad\cdot \qquad &\\ b_{n-1}x_{n-1}+(a_n-\lambda_k)x_n &= 0\end{aligned}\right\}. \tag{4–72}$$

With $x_1 = 1$, and proper use of the recursion formulas for the recursion polynomials $f_i(\lambda)$ (4–69) for the remaining components x_j of the kth eigenvector, the system (4–72) yields the explicit formula

$$x_1 = 1, \quad x_j = f_{j-1}(\lambda_k)\Big/\left(\prod_{i=1}^{j-1} b_i\right) \qquad (j = 2, 3, \ldots, n). \tag{4–73}$$

Therein the problem of solution of the eigenvectors appears to be trivially solved. The explicit formulas (4–73), however, turn out to be highly *unstable* numerically, to the point that in many cases they produce totally unsatisfactory eigenvectors. Even when the recursion polynomials $f_i(\lambda)$ in the form of a Sturm sequence locate the eigenvalues in a stable manner, the stability is entirely lost in the explicit formulas (see Ref. 75). The recommended numerical computation of eigenvectors of a tridiagonal matrix $\boldsymbol{J}$ with eigenvalues approximately known is based on the *inverse iteration of Wielandt** (see Ref. 72; refer also to Sec. 4–7–3). Starting with an arbitrary normed vector $\boldsymbol{x}^{(0)}$ and an approximate value $\bar{\lambda}$ of an eigenvalue λ_j, we are to compute the sequence of vectors $\boldsymbol{x}^{(k)}$ by the formula

$$(\boldsymbol{J} - \bar{\lambda}\boldsymbol{I})\boldsymbol{x}^{(k)} = \boldsymbol{x}^{(k-1)} \qquad (k = 1, 2, \ldots). \tag{4–74}$$

Note first that both symmetric matrices $\boldsymbol{J}$ and $\boldsymbol{B} = (\boldsymbol{J} - \bar{\lambda}\boldsymbol{I})$ possess the same complete system of orthonormal eigenvectors $\boldsymbol{y}_1, \boldsymbol{y}_2, \ldots, \boldsymbol{y}_n$, and that the eigenvalues are λ_k and $\lambda_k - \bar{\lambda}$, respectively, so that we have

$$\boldsymbol{J}\boldsymbol{y}_k = \lambda_k \boldsymbol{y}_k, \quad \boldsymbol{B}\boldsymbol{y}_k = (\boldsymbol{J} - \bar{\lambda}\boldsymbol{I})\boldsymbol{y}_k = (\lambda_k - \bar{\lambda})\boldsymbol{y}_k.$$

The given initial vector $\boldsymbol{x}^{(0)}$ and the vectors $\boldsymbol{x}^{(k)}$ found by iteration from (4–74) are expanded in terms of the eigenvectors $\boldsymbol{y}_i$,

$$\boldsymbol{x}^{(0)} = \sum_{i=1}^{n} c_i^{(0)}\boldsymbol{y}_i, \quad \boldsymbol{x}^{(k)} = \sum_{i=1}^{n} c_i^{(k)}\boldsymbol{y}_i. \tag{4–75}$$

*In German, "gebrochene Vektoriteration." (Translator's note.)

The relation between the expansion coefficients $c_i^{(0)}$ and $c_i^{(k)}$ is found by substituting (4–75) into (4–74).

$$(J - \bar{\lambda}I)\left(\sum_{i=1}^{n} c_i^{(k)} y_i\right) = \sum_{i=1}^{n} c_i^{(k)}(J - \bar{\lambda}I)y_i = \sum_{i=1}^{n} c_i^{(k)}(\lambda_i - \bar{\lambda})y_i$$
$$= \sum_{i=1}^{n} c_i^{(k-1)} y_i.$$

The linear independence of the vectors y_i leads to

$$c_i^{(k)} = \frac{c_i^{(k-1)}}{\lambda_i - \bar{\lambda}} \qquad i = 1, 2, \ldots, n; \quad k = 1, 2, \ldots, \tag{4–76}$$

so that for the kth iterated vector $x^{(k)}$ we have the form

$$x^{(k)} = \sum_{i=1}^{n} c_i^{(0)} \frac{1}{(\lambda_i - \bar{\lambda})^k} y_i \qquad k = 1, 2, \ldots. \tag{4–77}$$

We take it for granted that $\bar{\lambda}$ is a good approximation of λ_1 such that $\lambda = \lambda_1 + \epsilon$ the relation $|\lambda - \lambda_i| \gg \epsilon$ $(i = 2, 3, \ldots, n)$. Under these circumstances, $x^{(k)}$, by (4–77), equals

$$x^{(k)} = \frac{c_1^{(0)}}{\epsilon^k} y_1 + \sum_{i=2}^{n} c_i^{(0)} \frac{1}{(\lambda_i - \bar{\lambda})^k} y_i = \frac{1}{\epsilon^k}\left[c_1^{(0)} y_1 + \sum_{i=2}^{n} c_i^{(0)} \left\{\frac{\epsilon}{(\lambda_i - \bar{\lambda})}\right\}^k y_i\right]. \tag{4–78}$$

If $c_1^{(0)} \neq 0$, with the given assumptions, Eq. (4–78) shows that $x^{(k)}$ will quickly become proportional to the eigenvector y_1 with increasing k. The further the minimal distance from λ to one of the other eigenvalues λ_i compared to ϵ, the better the convergence.

Example 4-6. Even in unfavorable situations, the convergence is still satisfactory. Suppose that the eigenvalue λ_1 has been found by the method of repeated interval bisection with an absolute accuracy of $\epsilon = 10^{-10}$. It is unfortunately close to its neighbors, with $\min_{i=2,\ldots,n} |\lambda_1 - \lambda_i| = 10^{-5}$. In addition, $c_1^{(0)} = 10^{-5}$ or, in other words, given the starting vector $x^{(0)}$ normalized by $\sum_{i=1}^{n} c_i^{(0)^2} = 1$, the first eigenvector y_1 is only weakly represented in $x^{(0)}$. After just three iterations in accord with (4–78), we have

$$x^{(3)} = 10^{30}\left[10^{-5} y_1 + \sum_{i=2}^{n} c_i^{(0)} \left\{\frac{10^{-10}}{\lambda_i - \bar{\lambda}}\right\}^3 y_i\right] = 10^{25} y_1 + r$$

with

$$r = \sum_{i=2}^{n} c_i^{(0)} \frac{1}{(\lambda_i - \bar{\lambda})^3} y_i.$$

Because of the orthonormal property of the eigenvectors y_1 and the normalization of $x^{(0)}$, the Euclidean norm of r is

$$\|r\| = \left\{\sum_{i=2}^{n}\left[c_i^{(0)} \frac{1}{(\lambda_i - \bar{\lambda})^3}\right]^2\right\}^{1/2} \leq 10^{15}\left\{\sum_{i=2}^{n} c_i^{(0)2}\right\}^{1/2} \leq 10^{15}.$$

The norm of r compared to the first part is at least 10^{10} times smaller. The normed vector $x^{(3)}$ provides the eigenvector accurate to 10 decimals.

The practical execution of the inverse iteration method raises several questions. Each iteration step requires the solution of a system of equations (4–74) for $x^{(k)}$ with a given right-hand side $x^{(k-1)}$. The system is symmetric, and the coefficient matrix is tridiagonal but usually no longer positive definite. Cholesky's method thus cannot be employed. Besides, the system is nearly singular, exhibiting a dreadfully bad condition. This central difficulty can be largely circumvented by applying an *unsymmetric triangular decomposition* to the tridiagonal matrix $(J - \bar{\lambda} I)$ such that we avoid dividing by small (and thus inaccurate) quantities. The system can thus be solved with great accuracy (see Refs. 74, 75).

Choice of the initial vector $x^{(0)}$ must also be made with care to avoid having only a pathologically small component of the appropriate eigenvector. The unit vector e_1 often represents a very poor initial vector, since often it is virtually orthogonal to an eigenvector. References 74 and 75 show how the initial vector $x^{(0)}$ can be computed by using $\bar{\lambda}$ in half an iteration step, thereby guaranteeing that the desired eigenvector will not be a negligibly small component of $x^{(0)}$.

PROBLEMS

4-9. Transform the matrix

$$A = \begin{bmatrix} 10 & 4 & 3 \\ 4 & 15 & 10 \\ 3 & 10 & 20 \end{bmatrix}$$

into tridiagonal form by the methods of Givens and Householder. Give the transformation matrices U and the tridiagonal matrices. Verify that, in this case, the resultant matrices show essential agreement.

4-10. Show that any tridiagonal matrix of order n having the form

$$\boldsymbol{J} = \begin{bmatrix} a_1 & b_1 & & & & \\ c_1 & a_2 & b_2 & & & \\ & c_2 & a_3 & b_3 & & \\ & & \cdot & \cdot & \cdot & \\ & & & c_{n-2} & a_{n-1} & b_{n-1} \\ & & & & c_{n-1} & a_n \end{bmatrix}$$

and possessing the property $b_i \cdot c_i > 0$ $(i = 1, 2, \ldots, n-1)$ has real eigenvalues. (*Hint:* $\boldsymbol{J}$ is similar to a symmetric tridiagonal matrix, and can thus be symmetricized.)

4-11. Lattice interpolation produces tridiagonal matrices of the form

$$\boldsymbol{J} = \begin{bmatrix} 2 & 1 & & & & \\ 1 & 4 & 1 & & & \\ & 1 & 4 & 1 & & \\ & & \cdot & \cdot & \cdot & \\ & & & 1 & 4 & 1 \\ & & & & 1 & 2 \end{bmatrix}$$

having a high order n. What are the upper and lower bounds on the eigenvalues, given by Gerschgorin's theorem? Correspondingly, what is the largest condition number of matrix $\boldsymbol{J}$, independent of n? For the case $n = 4$, carry out five steps in the bounding of the smallest eigenvalue, using the bisection method.

4-12. If all the eigenvalues of a tridiagonal matrix are to be computed by the method of bisection, the total number of interval bisections can be reduced, as follows. The information on the location of the eigenvalues being determined is continually processed, taking into account the number of changes in sign, so that as starting values for the eigenvalues in a computation in progress, narrower bounds may be utilized. Establish a computer program, taking advantage of the accumulating information on the bounds of the eigenvalue and improving constantly the bounds. The interval bisection is to be halted when the absolute values of the bound intervals are sufficiently close to each other.

4-13. For tridiagonal matrices of very high order, the values of the recursion polynomials $f_k(\lambda)$ can become very large or very small so that, in a computer, overflow or underflow could result. By way of demonstration, examine the behavior of the recursion polynomials for a matrix $\boldsymbol{J}$ in Problem 4-11 of order $n = 50$ for $\lambda = 1$ and $\lambda = 6$. What modification of the method can eradicate this difficulty?

4-6 LR TRANSFORMATION AND QD ALGORITHM: CALCULATION OF THE SMALLEST EIGENVALUES

This section describes a technique which produces a sequence of similar matrices yielding the *smallest eigenvalues in ascending order*. A minor modification can also yield the largest eigenvalues in descending order. The method is useful if we seek only *a few* eigenvalues of the matrix at *one end of the spectrum*. This formulation corresponds to many vibration problems in which only the fundamental frequency and a few higher frequencies are of interest. The technique is especially handy for band matrices, where this property can be amply exploited.

4-6-1 The LR Transformation

The fundamental notion (see Ref. 44) for the solution of eigenvalues of a given general square matrix $\boldsymbol{A} = \boldsymbol{A}_1$ lies in a decomposition of $\boldsymbol{A}_1$ into a product of two triangular matrices $\boldsymbol{L}_1$ and $\boldsymbol{R}_1$, as in

$$\boldsymbol{A}_1 = \boldsymbol{L}_1\boldsymbol{R}_1, \tag{4–79}$$

in order to multiply the two triangular matrices in inverse order and form a new matrix $\boldsymbol{A}_2$, as in

$$\boldsymbol{A}_2 = \boldsymbol{R}_1\boldsymbol{L}_1. \tag{4–80}$$

There $\boldsymbol{L}_1$ is a *lower triangular* matrix and $\boldsymbol{R}_1$ an *upper triangular* matrix, as given by

$$\boldsymbol{L}_1 = \begin{bmatrix} 1 & & & & & \\ l_{21} & 1 & & & 0 & \\ l_{31} & l_{32} & 1 & & & \\ \cdot & \cdot & \cdot & \cdot & & \\ \cdot & \cdot & \cdot & \cdot & \cdot & \\ l_{n1} & l_{n2} & l_{n3} & \cdot & \cdot & 1 \end{bmatrix}, \quad \boldsymbol{R}_1 = \begin{bmatrix} r_{11} & r_{12} & r_{13} & \cdots & r_{1n} \\ & r_{22} & r_{23} & \cdots & r_{2n} \\ & & r_{33} & \cdots & r_{2n} \\ & 0 & & \cdot & \cdot \\ & & & \cdot & \cdot \\ & & & & r_{nn} \end{bmatrix}. \tag{4–81}$$

The diagonal elements of the lower triangular matrix $\boldsymbol{L}_1$ are normalized to unity so as to render the decomposition unique. The triangular decomposition (4–79) of an arbitrary (neither symmetric nor positive definite) matrix $\boldsymbol{A}$ is more general than that of Cholesky, which is for symmetric and positive definite matrices. It is not always feasible however; in particular, no principal minor[1] of A may vanish (Ref. 76). Under this assumption, the nontrivial

[1] Refer to the footnote of Sec. 4–5–4.

elements r_{ik} and l_{ik} may be explicitly evaluated, row by row:

$$\begin{aligned} r_{1k} &= a_{1k} && k = 1, 2, \ldots, n \\ l_{ik} &= \left(a_{ik} - \sum_{j=1}^{k-1} l_{ij} r_{jk}\right) \Big/ r_{kk} && \left.\begin{aligned} &k = 1, 2, \ldots, i-1 \\ &k = i, i+1, \ldots, n \end{aligned}\right\} \quad i(= 2, 3, \ldots, n). \\ r_{ik} &= a_{ik} - \sum_{j=1}^{i-1} l_{ij} r_{jk} && \end{aligned} \tag{4–82}$$

When $k = 1$, the sums in (4–82) are to be considered empty.

Theorem 4-13. *Given that the decomposition* (4–79) $A_1 = L_1 R_1$ *exists, then the matrix* $A_2 = R_1 L_1$ *is similar to* A_1.

Proof: If decomposition (4–79) is feasible, L_1 is nonsingular as a result of normalization of the diagonal elements, and thus its inverse exists. Equation (4–79) yields $R_1 = L_1^{-1} A_1$, and thus by (4–80) also $A_2 = L_1^{-1} A_1 L_1$, which proves the similarity sought. The eigenvalues of A_1 and A_2 are therefore identical.

The transition from A_1 to A_2 by (4–79) and (4–80) is called an *LR step*, as A_1 is decomposed into a product of a lower triangular matrix L and an upper triangular matrix R. The *LR* step can be repeated, using A_2, and an infinite sequence of similar matrices A_k can be generated by the definition

$$A_k = L_k R_k, \quad A_{k+1} = R_k L_k \qquad (k = 1, 2, \ldots). \tag{4–83}$$

Equations (4–83) yield the recursive result that the products of the triangular matrices

$$\Lambda_k = L_1 L_2 \ldots L_k, \quad P_k = R_k R_{k-1} \ldots R_1 \tag{4–84}$$

are the transformation matrices converting A_1 to A_{k+1}.

$$A_{k+1} = \Lambda_k^{-1} A_1 \Lambda_k = P_k A_1 P_k^{-1}. \tag{4–85}$$

Because of the group property of triangular matrices, the product matrices Λ_k and P_k are themselves lower and upper triangular matrices respectively. Since $L_i R_i = R_{i-1} L_{i-1}$, the product $\Lambda_k P_k$ reduces after repeated application to the form

$$\Lambda_k P_k = (L_1 R_1)^k = A_1^k.$$

The product matrices Λ_k and P_k represent the triangular decomposition of the kth power of the initial matrix $A_1 = A$.

Under appropriate conditions, the sequence of similar matrices A_k (4–83) converges toward an upper triangular matrix (Refs. 44, 76).

Theorem 4-14. *If the sequence of matrices $\mathbf{\Lambda}_k$ converges as $k \longrightarrow \infty$, then the sequence of matrices A_k converges toward an upper triangular matrix.*

Proof: The known existence of the limit value $\lim_{k\to\infty} \mathbf{\Lambda}_k = \mathbf{\Lambda}_\infty$ leads to

$$\lim_{k\to\infty} \boldsymbol{L}_k = \lim_{k\to\infty} \mathbf{\Lambda}_{k-1}^{-1}\mathbf{\Lambda}_k = \boldsymbol{I}.$$

Furthermore, by (4–85) and (4–83),

$$\boldsymbol{A}_k = \mathbf{\Lambda}_{k-1}^{-1}\boldsymbol{A}_1\mathbf{\Lambda}_{k-1} = \boldsymbol{L}_k\boldsymbol{R}_k,$$

and thus

$$\boldsymbol{R}_k = \mathbf{\Lambda}_k^{-1}\boldsymbol{A}_1\mathbf{\Lambda}_{k-1}.$$

Therefore, there must also exist the limit value

$$\boldsymbol{L}_\infty = \lim_{k\to\infty} \boldsymbol{R}_k = \lim_{k\to\infty} \mathbf{\Lambda}_k^{-1}\boldsymbol{A}_1\mathbf{\Lambda}_{k-1}.$$

Since $\lim_{k\to\infty} \boldsymbol{L}_k = \boldsymbol{I}$, $\boldsymbol{A}_k$ must also have a limit value, as

$$\boldsymbol{A}_\infty = \lim_{k\to\infty} \boldsymbol{A}_k = \lim_{k\to\infty} \boldsymbol{L}_k\boldsymbol{R}_k = \boldsymbol{R}_\infty.$$

In the convergent case, the eigenvalues of the given matrix A turn up as diagonal elements of the upper triangular matrix $\boldsymbol{A}_\infty = \boldsymbol{R}_\infty$.

A statement on the convergence criteria for the LR transformation is given in

Theorem 4-15. *If matrix A has only real and simple eigenvalues of unequal magnitude, and if no principal minor of the transformation matrices $\boldsymbol{U}$ and $\boldsymbol{U}^{-1}$ vanishes, where $\boldsymbol{U}$ is a matrix transforming $\boldsymbol{A}$ to diagonal form by $\boldsymbol{A} = \boldsymbol{U}\boldsymbol{D}\boldsymbol{U}^{-1}$, then there exists the limit value $\mathbf{\Lambda}_\infty = \lim_{k\to\infty} \mathbf{\Lambda}_k$, and $\boldsymbol{A}_\infty = \lim_{k\to\infty} \boldsymbol{A}_k$ is an upper triangular matrix in which the eigenvalues of A are given in the diagonal elements in order of diminishing magnitude.*

For the proof, the reader is referred to Refs. 44 and 76.

The LR transformation for a general matrix A manifests the numerical disadvantage that the decomposition (4–79) cannot always be carried out in the form described, or at least that because of the small magnitude of a principal minor, large round-off errors are encountered. This difficulty can, in principle, be eliminated through a triangular decomposition with an interchange of rows (Refs. 75, 76).

In the ensuing, we will confine ourselves to the calculation of eigenvalues of *symmetric, positive definite* matrices by the method of LR transformations, since the method proves to be stable for that case. The limitation to positive definite matrices is in no way restrictive, as we will presently see.

A *spectral transformation* consists of making simple alterations of a given matrix A in order to alter the spectrum in such a manner as to simplify the calculation of eigenvalues. For instance, the transition from A to $(A + cI)$ involves a translation of the whole spectrum of A by an amount c. In the ensuing, such an alteration will be termed a *coordinate translation.* Another example of a spectral transformation is given by the conversion from A to A^{-1}, which is equivalent to replacing the eigenvalues of A by their reciprocal values.

Addition of cI to matrix A causes a movement of the spectrum of A by a distance c. An opportune choice of coordinate translation with $c > 0$ can always guarantee that all the eigenvalues of $(A + cI)$ are positive, which is tantamount to saying that $(A + cI)$ is positive definite. Subsequent to evaluation of the positive eigenvalues of $(A + cI)$, the eigenvalues of A are found by subtracting c.

In a symmetric, positive definite matrix A undergoing an LR transformation, the lower triangular matrices L_k are replaced by the transpose of the upper triangular matrices R_k, so that (4–79) is replaced by the *Cholesky decomposition.* The computational criterion (4–83) is replaced by

$$A_k = R_k^T R_k, \quad A_{k+1} = R_k R_k^T, \qquad k = 1, 2, \ldots. \tag{4–86}$$

The symmetry of matrix A_{k+1} is obvious and the similarity of matrices A_{k+1} and A_k is unchanged; thus the LR steps can always be carried out in this variant. The modification (4–86) is called *LR-Cholesky*, and we have

Theorem 4-16. *For a symmetric, positive definite matrix* $\mathbf{A}_1$, *the LR-Cholesky method produces a sequence of symmetric, similar matrices by the rule* (4–86).

4-6-2 Convergence Proof of the LR-Cholesky Method

An important property of the LR-Cholesky method is that it always converges.

Theorem 4-17. *For a symmetric, positive definite matrix* $\mathbf{A}_1$, *the LR-Cholesky method* (4–86) *converges. The sequence of similar matrices* A_k *converges toward a diagonal matrix* A_∞ *whose diagonal elements equal the eigenvalues of* A_1.

Proof: Define $A_k = (a_{ij}^{(k)})$ and $R_k = (r_{ij}^{(k)})$. By (4–86), the matrices A_k and A_{k+1} obey the relations

$$a_{ij}^{(k)} = \sum_{p=1}^{i} r_{pi}^{(k)} r_{pj}^{(k)}, \quad a_{ij}^{(k+1)} = \sum_{p=j}^{n} r_{ip}^{(k)} r_{jp}^{(k)} \qquad i \leq j. \tag{4–87}$$

Furthermore, let $s_m^{(k)}$ be a *partial trace* of iterated matrices A_k, made up of the first m diagonal elements of A_k:

$$s_m^{(k)} = a_{11}^{(k)} + a_{22}^{(k)} + \cdots + a_{mm}^{(k)} = \sum_{i=1}^{m} a_{ii}^{(k)} \qquad m = 1, 2, \ldots, n. \tag{4–88}$$

Examine now the change in the partial trace for fixed m in going from A_k to A_{k+1}. For the diagonal elements of A_k and A_{k+1} we have, by (4–87),

$$a_{ii}^{(k)} = \sum_{p=1}^{i} (r_{pi}^{(k)})^2, \qquad a_{ii}^{(k+1)} = \sum_{p=i}^{n} (r_{ip}^{(k)})^2.$$

The mth partial traces are thus

$$s_m^{(k)} = \sum_{i=1}^{m} a_{ii}^{(k)} = \sum_{i=1}^{m} \sum_{p=1}^{i} (r_{pi}^{(k)})^2,$$

$$s_m^{(k+1)} = \sum_{i=1}^{m} a_{ii}^{(k+1)} = \sum_{i=1}^{m} \sum_{p=i}^{n} (r_{ip}^{(k)})^2.$$

The partial traces $s_m^{(k)}$ and $s_m^{(k+1)}$ of two successive matrices A_k and A_{k+1} appear as expressions of elements of the same matrix R_k. The quantity $s_m^{(k)}$ is the sum of the squares of the elements of R_k in the first m *columns* (see Fig. 17a), while $s_m^{(k+1)}$ is the sum of the squares of the elements of R_k in the first m *rows* (see Fig. 17b).

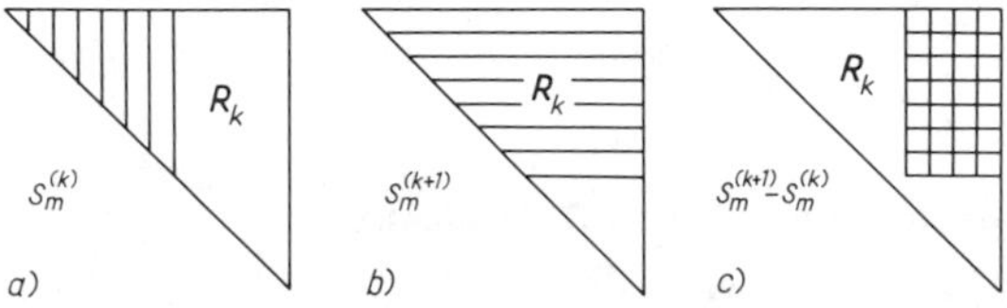

Figure 17. Diagrams of partial traces.

The difference of the partial traces is made up of the squares of the elements of R_k in the first m rows and the last $(n - m)$ columns (see Fig. 17c):

$$s_m^{(k+1)} - s_m^{(k)} = \sum_{i=1}^{m} \sum_{p=m+1}^{n} (r_{ip}^{(k)})^2. \tag{4–89}$$

Equation (4–89) leaves the result for the partial traces $s_m^{(k)}$ that they form a monotonic, nondecreasing sequence with increasing k, since the difference above is nonnegative. Thus, for fixed m,

$$s_m^{(1)} \le s_m^{(2)} \le \cdots \le s_m^{(k)} \le s_m^{(k+1)} \le \cdots \qquad m = 1, 2, \ldots, n. \qquad (4\text{-}90)$$

The given positive definiteness of matrix $\boldsymbol{A}_1$ is carried over to matrices $\boldsymbol{A}_k$ by Theorem 4–16. Thus all diagonal elements obey $a_{ii}^{(k)} > 0$ ($i = 1, 2, \ldots, n$; $k = 1, 2, \ldots$). In addition, the sum of all the eigenvalues $s_n^{(k)}$, called the *trace*, is an *invariant* of the LR-Cholesky method, and it has a certain finite value. As a result of these two observations, every partial trace $s_m^{(k)}$, which is a partial sum of the trace itself with several positive ingredients missing, is bounded from above. For any fixed m, the monotonic nondecreasing sequence of values $s_m^{(k)}$ converges for $k \longrightarrow \infty$, and thus we must have

$$\lim_{k\to\infty} (s_m^{(k+1)} - s_m^{(k)}) = \lim_{k\to\infty} \left\{\sum_{i=1}^{m} \sum_{p=m+1}^{n} (r_{ip}^{(k)})^2\right\} = 0 \qquad m = 1, 2, \ldots, n. \qquad (4\text{–}91)$$

Each individual element $r_{ip}^{(k)}$ of matrix $\boldsymbol{R}_k$ shown in the rectangular cross-hatched region of Fig. 17c converges to zero for fixed m. Invoking this result successively for $m = 1, 2, \ldots, n-1$, then for $m = 1$ the nondiagonal elements of the first row of $\boldsymbol{R}_k$ must converge to zero, and for $m = 2$ also those of the second row, and so forth. Finally,

$$\lim_{k\to\infty} r_{ij}^{(k)} = 0 \quad \text{for} \quad 1 \le i < j \le n. \qquad (4\text{–}92)$$

The upper triangular matrix $\boldsymbol{R}_k$ converges to a diagonal matrix. The same applies to the matrices $\boldsymbol{A}_k$, since symmetry considerations of (4–87) and (4–92) lead to

$$\lim_{k\to\infty} a_{ij}^{(k+1)} = \lim_{k\to\infty} a_{ji}^{(k+1)} = \lim_{k\to\infty} \sum_{p=1}^{n} r_{ip}^{(k)} r_{jp}^{(k)} = 0 \qquad 1 \le i < j \le n.$$

The proof of the convergence of the matrix sequence $\boldsymbol{A}_k$ (4–86) to a diagonal matrix $\boldsymbol{A}_\infty$ for symmetric positive definite matrices is thereby completed. Similarity of the matrices assures that the diagonal elements of the limit matrix $\boldsymbol{A}_\infty$ are identical to the eigenvalues of the given matrix $\boldsymbol{A}_1$.

Theorem 4–17 guarantees the convergence of the LR-Cholesky method, and the eigenvalues turn up in *arbitrary sequence* on the diagonal of $\boldsymbol{A}_\infty$. In the form of the LR transformation described above, we find all the eigenvalues together, analogous to the Jacobi method.

In the following modification, the situation is different.

4-6-3 Convergence Rate, Coordinate Translation

From the fundamental convergence proof, no general statements are forthcoming regarding the manner of convergence of the nondiagonal elements $a_{ij}^{(k)}$ of A_k to zero. With one additional assumption, however, the convergence rate of the elements $a_{ij}^{(k)}$ can be examined more precisely.

If the limit matrix A_∞ of the matrix sequence A_k (4–86) contains the eigenvalues *in diminishing sequence* on its diagonal, it is called the *conventional case*, and it is the one usually encountered. But there is an obvious exception in the reducible symmetric matrices, which contain nonzero elements only in square submatrices along the diagonal. This form is retained during the LR-Cholesky procedure, and the eigenvalues of the various submatrices cannot all be "shuffled" into proper order, since the necessary coupling is missing. But this exception can be summarily dismissed, since the evaluation of the eigenvalues of a reducible matrix decomposes into the subordinate problems of finding eigenvalues of the submatrices. In addition, there is the rare theoretical case of unsequenced eigenvalues where one of the principal minors of the orthogonal transformation matrix U which transforms A to the diagonal form D by $U^T A U = D$ vanishes altogether (see the original work in Ref. 44, or Refs. 75, 76). This exception, however, is broken down by the inevitable round-off errors of numerical computations, so that even here, after admittedly a great many iterations, the conventional case predominates.

Theorem 4-18. *In the conventional case, if the eigenvalues λ_i of matrix $\mathbf{A}_1$ all differ from one another, then for k sufficiently large we have the asymptotic result*

$$a_{ij}^{(k+1)} \sim a_{ij}^{(k)} \sqrt{\frac{a_{jj}^{(k)}}{a_{ii}^{(k)}}} \sim a_{ij}^{(k)} \cdot \frac{r_{jj}^{(k)}}{r_{ii}^{(k)}} \sim a_{ij}^{(k)} \sqrt{\frac{\lambda_j}{\lambda_i}} \qquad 1 \leq i < j \leq n. \tag{4–93}$$

The nondiagonal elements $a_{ij}^{(k)}$ $(i < j)$ manifest a linear convergence rate in that they converge to zero as a geometric sequence with the quotient $q_{ij} = \sqrt{(\lambda_j/\lambda_i)}$.

Proof: By Theorem 4–17, for any arbitrary small $\epsilon < 0$, there exists a quantity $k(\epsilon)$ such that

$$|r_{ij}^{(k)}| \leq \epsilon \quad \text{for} \quad 1 \leq i < j \leq n \quad \text{and} \quad k > K(\epsilon).$$

This result can be expressed in the alternate fashion[1]

$$r_{ij}^{(k)} = O(\epsilon) \quad k \longrightarrow \infty, \qquad 1 \leq i < j \leq n. \tag{4–94}$$

[1] Here we use the order symbol $O(\epsilon^m)$, by which is meant a function $f(\epsilon)$ such that the quotient $f(\epsilon)/\epsilon^m$ has a finite (and possibly vanishing) value for $\epsilon \longrightarrow 0$. Then $f(\epsilon)$ is said to be of order ϵ^m, and we write $f(\epsilon) = O(\epsilon^m)$. The following corrolaries apply as well: (a) If $f(\epsilon) = O(\epsilon^m)$ and $g(\epsilon) = O(\epsilon^n)$, then $f(\epsilon) \cdot g(\epsilon) = O(\epsilon^m) \cdot O(\epsilon^n) = O(\epsilon^{m+n})$. (b) With $f(\epsilon)$ and $g(\epsilon)$ thus defined, and $m \leq n$, then $f(\epsilon) + g(\epsilon) = O(\epsilon^m) + O(\epsilon^n) = O(\epsilon^m)$.

Because of the positive definiteness of A_1, all the diagonal elements of A_k are greater than zero, and since they converge to the positive eigenvalues of A_1 by Theorem 4–17, we can make the asymptotic statement

$$a_{jj}^{(k)} = O(1) \quad k \longrightarrow \infty, \qquad j = 1, 2, \ldots, n. \tag{4–95}$$

The Cholesky decomposition $A_k = R_k^T R_k$ expressed for diagonal elements

$$a_{jj}^{(k)} = \sum_{i=1}^{j} (r_{ij}^{(k)})^2 = (r_{jj}^{(k)})^2 + \sum_{i=1}^{j-1} (r_{ij}^{(k)})^2$$

leads via (4–94) and (4–95) to

$$r_{jj}^{(k)} = O(1) \quad k \longrightarrow \infty, \qquad j = 1, 2, \ldots, n. \tag{4–96}$$

For the nondiagonal elements $a_{ij}^{(k)}$ and $a_{ij}^{(k+1)}$ of identical indices in two successive matrices A_k and A_{k+1}, it follows from their representation (4–87) for $k \longrightarrow \infty$ that

$$a_{ij}^{(k)} = \sum_{p=1}^{i} r_{pi}^{(k)} r_{pj}^{(k)} = r_{ii}^{(k)} r_{ij}^{(k)} + \sum_{p=1}^{i-1} r_{pi}^{(k)} r_{pj}^{(k)} = r_{ii}^{(k)} r_{ij}^{(k)} + O(\epsilon^2) \qquad i < j,$$

$$a_{ij}^{(k+1)} = \sum_{p=j}^{n} r_{ip}^{(k)} r_{jp}^{(k)} = r_{jj}^{(k)} r_{ij}^{(k)} + \sum_{p=j+1}^{n} r_{ip}^{(k)} r_{jp}^{(k)} = r_{jj}^{(k)} r_{ij}^{(k)} + O(\epsilon^2) \qquad i < j.$$

The first two terms in these representations of $a_{ij}^{(k)}$ and $a_{ij}^{(k+1)}$ are asymptotically $O(\epsilon)$. If terms of order ϵ^2 are neglected in comparison to those of order ϵ, these two equations yield the asymptotic result

$$a_{ij}^{(k+1)} \sim r_{jj}^{(k)} \frac{a_{ij}^{(k)}}{r_{ii}^{(k)}} = a_{ij}^{(k)} \frac{r_{jj}^{(k)}}{r_{ii}^{(k)}} \qquad 1 \leq i < j \leq n.$$

Because of the descending sequence of eigenvalues in A_∞, the quotients $r_{jj}^{(k)}/r_{ii}^{(k)}$ are less than unity for k sufficiently large, since numerator and denominator approach $\sqrt{\lambda_j}$ and $\sqrt{\lambda_i}$, respectively, and $\sqrt{\lambda_j} < \sqrt{\lambda_i}$ for $i < j$. Since $r_{jj}^{(k)} \approx a_{jj}^{(k)}$, the three equivalent asymptotic statements (4–93) are substantiated, and so, too, the linear convergence of the nondiagonal elements to zero.

The statement about convergence rate holds water only if the eigenvalues appear in sequence. Conversely, we cannot assume that because of the convergence of the LR-Cholesky technique the eigenvalues are certain to turn up in proper sequence. In the theoretical exception of unsequenced eigenvalues, another law governing asymptotic decrease of nondiagonal elements takes effect (see Refs. 44, 75, 76).

In computations, the transformation of the exceptional case to the conventional one can be artificially accelerated by replacing nondiagonal elements which are either zero or smaller in magnitude than a certain limit by small quantities, such that these perturbations introduced into the matrix have no effect on the required accuracy of eigenvalues (see Ref. 75).

Example 4-7. Consider matrix **A** of Example 4-2, which is not positive definite. Adding 2 to each diagonal element of A produces a positive definite matrix $A_1 = (A + 2I)$, and the LR-Cholesky method can be applied. Table 17 shows the first three LR steps with the upper triangular matrices. The

Table 17 LR-CHOLESKY, THREE ITERATION STEPS

$$A_1 = \begin{bmatrix} 22.000000 & -7.000000 & 3.000000 & -2.000000 \\ -7.000000 & 7.000000 & 1.000000 & 4.000000 \\ 3.000000 & 1.000000 & 5.000000 & 1.000000 \\ -2.000000 & 4.000000 & 1.000000 & 4.000000 \end{bmatrix}$$

$$R_1^T = \begin{bmatrix} 4.690416 & & & \\ -1.492405 & 2.184657 & & \\ 0.639602 & 0.894669 & 1.946915 & \\ -0.426401 & 1.539663 & -0.053809 & 1.201967 \end{bmatrix}$$

$$A_2 = \begin{bmatrix} 24.818182 & -3.344676 & 1.268195 & -0.512520 \\ -3.344676 & 7.943723 & 1.658996 & 1.850624 \\ 1.268195 & 1.658996 & 3.793372 & -0.064677 \\ -0.512520 & 1.850624 & -0.064677 & 1.444724 \end{bmatrix}$$

$$R_2^T = \begin{bmatrix} 4.981785 & & & \\ -0.671381 & 2.737329 & & \\ 0.254566 & 0.668501 & 1.811539 & \\ -0.102879 & 0.650836 & -0.261420 & 0.970676 \end{bmatrix}$$

$$A_3 = \begin{bmatrix} 25.344322 & -1.734570 & 0.488052 & -0.099862 \\ -1.734570 & 8.363452 & 1.040874 & 0.631751 \\ 0.488052 & 1.040874 & 3.350014 & -0.253754 \\ -0.099862 & 0.631751 & -0.253754 & 0.942212 \end{bmatrix}$$

$$R_3^T = \begin{bmatrix} 5.034314 & & & \\ -0.344549 & 2.871365 & & \\ 0.096945 & 0.374134 & 1.789033 & \\ -0.019836 & 0.217637 & -0.186277 & 0.927229 \end{bmatrix}$$

$$A_4 = \begin{bmatrix} 25.472828 & -0.957374 & 0.177133 & -0.018393 \\ -0.957374 & 8.432080 & 0.628798 & 0.201800 \\ 0.177133 & 0.628798 & 3.235338 & -0.172722 \\ -0.018393 & 0.201800 & -0.172722 & 0.859753 \end{bmatrix}$$

nondiagonal elements of A_4 are perceptibly smaller than those of A_1, and the diagonal elements converge toward the properly sequenced eigenvalues. To illustrate the convergence rate, Table 18 depicts the twelfth LR step.

Table 18 LR-Cholesky, Twelfth Iteration

$$A_{12} = \begin{bmatrix} 25.527379 & -0.011249 & 0.000043 & -0.000000 \\ -0.011249 & 8.460493 & 0.012435 & 0.000020 \\ 0.000043 & 0.012435 & 3.173078 & -0.000962 \\ -0.000000 & 0.000020 & -0.000962 & 0.839051 \end{bmatrix}$$

$$R_{12}^{T} = \begin{bmatrix} 5.052463 & & & \\ -0.002226 & 2.908692 & & \\ 0.000008 & 0.004275 & 1.781308 & \\ -0.000000 & 0.000007 & -0.000540 & 0.915997 \end{bmatrix}$$

$$A_{13} = \begin{bmatrix} 25.527384 & -0.006476 & 0.000015 & -0.000000 \\ -0.006476 & 8.460506 & 0.007616 & 0.000006 \\ 0.000015 & 0.007616 & 3.173060 & -0.000495 \\ -0.000000 & 0.000006 & -0.000495 & 0.839050 \end{bmatrix}$$

In the latter, (4–93) can be verified. The elements $a_{i,i+1}^{(k)}$ $(i = 1, 2, \ldots, n-1)$ of the longest offdiagonal converge to zero in the most leisurely fashion, since for them the convergence quotients are greatest and are closest to unity.

The *linear convergence* of nondiagonal elements to zero, which is the conventional case of Theorem 4–18, can render the LR-Cholesky method unusable in the form described, at least for the case of neighboring eigenvalues where the linear convergence is too slow and, consequently, where the matrix diagonalization takes too many steps. An improvement in convergence is possible in the conventional case through the *coordinate translation* of the spectrum which diminishes the crucial quotients $q_{ij} = (\lambda_j/\lambda_i)^{1/2}$ $(i < j)$. If the given matrix A_1 is replaced by $A_1' = A_1 - yI$, all the eigenvalues $\lambda_i'(A_1') = \lambda_i(A_1) - y$ are diminished by y. In order for the Cholesky decomposition and the LR-Cholesky procedure to be at all usable for A_1', we must have $0 \leq y < \lambda_n(A)$. To be sure, not all the convergence quotients q_{ij} can be arbitrarily reduced through a suitable choice of y, but they can, at least in the case of the quantities $q_{in} = (\lambda_n/\lambda_i)^{1/2}$, in that for A_1' they acquire the values $q_{in}' = [(\lambda_n - y)/(\lambda_i - y)]^{1/2} < q_{in}$ $(i = 1, 2, \ldots, n-1)$. The closer the value of y to the smallest eigenvalue λ_n, the smaller the values q_{in}', and the faster the convergence to zero of the nondiagonal elements of the last row and column in the sequence of matrices A_k'. The LR-Cholesky method (4–86) for the initial matrix $A_1' = A_1 - yI$ produces the matrix sequence

A'_k which, in the conventional case for suitable choice of y, converges rapidly to a matrix of the form

$$A'_m = \left[\begin{array}{cccc:c} a'^{(m)}_{11} & a'^{(m)}_{12} & \cdots & a'^{(m)}_{1,n-1} & 0 \\ \cdot & \cdot & & \cdot & \cdot \\ \cdot & \cdot & & \cdot & \cdot \\ \cdot & \cdot & & \cdot & \cdot \\ a'^{(m)}_{n-1,1} & a'^{(m)}_{n-1,2} & \cdots & a'^{(m)}_{n-1,n-1} & 0 \\ \hdashline 0 & 0 & \cdots & 0 & \lambda'_n \end{array}\right]. \tag{4–97}$$

Here λ'_n indicates the smallest eigenvalue of the matrix $A'_1 = A_1 - yI$, thus the smallest eigenvalue of A_1 is $\lambda_n = \lambda'_n + y$. Thus the *smallest eigenvalue* λ_n is the first to be computed. Since matrix A'_m (4–97) uncouples, the LR-Cholesky method can be used on the $(n - 1)$-row principal minor, whose eigenvalues $\lambda'_1, \lambda'_2, \ldots, \lambda'_{n-1}$ are the remaining eigenvalues of A'_1. Another well-chosen coordinate translation will quickly and analogously yield the smallest eigenvalue λ'_{n-1} of the principal minor of A'_m, which is the *second smallest eigenvalue* λ_{n-1} of A_1. When thus modified, the LR-Cholesky method no longer produces the eigenvalues simultaneously, but rather individually in ascending sequence.

Example 4-8. The quantity $y = 0.8$ is subtracted from each of the diagonal elements of matrix A_1 in Example 4–7 before the first Cholesky decomposition, and the LR transformation is begun with the matrix $A'_1 = A_1 - 0.8I$. Table 19 shows the first five LR steps with the matrices $A'_1, A'_2, \ldots, A'_6$. The rapid convergence of elements $a'^{(k)}_{in}$ and $a'^{(k)}_{ni}$ to zero is evident. The largest convergence quotient in this case is

$$q_{3,4} = \sqrt{\frac{\lambda_4(A_1) - 0.8}{\lambda_3(A_1) - 0.8}} = \sqrt{\frac{0.03905}{2.373}} = 0.1285.$$

Asymptotically, the element $a'^{(k)}_{34}$ is diminished in each LR step by about a factor of eight, and other elements of the last row and column even more.

The coordinate translation described is done not just before the first LR step, but at each subsequent step as well, in order to improve the essentially linear convergence. In continuing the method, the last eigenvalue turns up ever more clearly in the last diagonal element of the instantaneous matrix; thus the possible coordinate translations assuring retention of the positive definiteness can be picked with greater accuracy. Naturally, the *total coordinate translation* z_k representing the sum of individual translations must be logged in this modification. The computational instruction (4–86) is replaced by the following algorithm:

Start:

$$\boxed{A_1 = A, \quad z_1 = 0} \tag{4–98}$$

The kth LR-Cholesky Step ($k = 1, 2, \ldots$):

$$\boxed{\begin{array}{ll} \text{(a)} & \text{Choice of } y_k < \lambda_n(A_k) \\ \text{(b)} & \text{Cholesky-Decomposition } A_k - y_k I = R_k^{\mathrm{T}} R_k \\ \text{(c)} & A_{k+1} = R_k R_k^{\mathrm{T}}, \quad z_{k+1} = z_k + y_k \end{array}} \tag{4–99}$$

Termination criterion: If at every step y_k is taken as large as possible but still smaller than the smallest eigenvalue of A_k, in general, the last diagonal element $a_{nn}^{(k)}$ of A_k converges to zero and the quantity z_k converges to the smallest eigenvalue λ_n of A. As soon as $a_{nn}^{(k)}$ is small enough, $z_k + a_{nn}^{(k)}$ yields the smallest eigenvalue of A with an accuracy of magnitude $a_{nn}^{(k)}$. Thus the current values of z_k and $z_k + a_{nn}^{(k)}$ continually provide lower and upper bounds on the smallest eigenvalue.

The remaining eigenvalues are successively found by dropping off the last row and column of A_k and by continuing the algorithm (4–99) with the reduced submatrix, remembering to replace (4–98) by the current value of z_k.

The LR-Cholesky method with coordinate translation presents one difficulty, namely, that of selecting an effective value of y_k, since in general nothing is known about the magnitude of the smallest eigenvalue. For a positive definite matrix, zero is a lower bound for the eigenvalues, and by Theorem 3–2 the smallest diagonal element of A_k is an upper bound for the smallest eigenvalue. Thus an interval $[0, \min_{i=1,2,\ldots,n} a_{ii}^{(k)}]$ is established within which y_k must lie. The decomposition of $A_k - y_k I$ is attempted with a fraction of the smallest diagonal element of A_k by using the computer to determine a possible value of y_k. An unsuccessful selection always produces a better upper bound for the smallest eigenvalue. Reference 45 goes so far as to show that in certain cases the information obtained in an unsuccessful choice can be used to yield a very good *lower* bound for the smallest eigenvalue, resulting in a significantly better convergence rate.

As already mentioned, the LR-Cholesky method is also usable with symmetric matrices that are not positive definite if a suitable coordinate translation $(A_1 - z_1 I)$ with $z_1 < 0$ is chosen. The initial value z_1 should not be taken too small; otherwise, significant decimal places can be lost in the calculation $\lambda_i(A) = \lambda_i(A_1 - z_1 I) + z_1$.

Using the computer, a value of z_1 as large as possible can be found through trial and error until z_1 approaches the magnitude of λ_n.

Table 19 LR-Cholesky Method

$$A'_1 = \begin{bmatrix} 21.200000 & -7.000000 & 3.000000 & -2.000000 \\ -7.000000 & 6.200000 & 1.000000 & 4.000000 \\ 3.000000 & 1.000000 & 4.200000 & 1.000000 \\ -2.000000 & 4.000000 & 1.000000 & 3.200000 \end{bmatrix}$$

$$A'_2 = \begin{bmatrix} 24.124528 & -3.075924 & 1.193350 & -0.120724 \\ -3.075924 & 7.775714 & 1.240896 & 0.470684 \\ 1.193350 & 1.240896 & 2.822513 & -0.071394 \\ -0.120724 & 0.470684 & -0.071394 & 0.077244 \end{bmatrix}$$

$$A'_3 = \begin{bmatrix} 24.576349 & -1.581241 & 0.386559 & -0.004879 \\ -1.581241 & 7.674430 & 0.794669 & 0.033262 \\ 0.386559 & 0.794669 & 2.509813 & -0.018996 \\ -0.004879 & 0.033262 & -0.018996 & 0.039408 \end{bmatrix}$$

$$A'_4 = \begin{bmatrix} 24.684167 & -0.854527 & 0.121191 & -0.000195 \\ -0.854527 & 7.661529 & 0.462642 & 0.002366 \\ 0.121191 & 0.462642 & 2.415249 & -0.002859 \\ -0.000195 & 0.002366 & -0.002859 & 0.039055 \end{bmatrix}$$

$$A'_5 = \begin{bmatrix} 24.714345 & -0.471032 & 0.037680 & -0.000008 \\ -0.471032 & 7.660503 & 0.261029 & 0.000169 \\ 0.037680 & 0.261029 & 2.386102 & -0.000384 \\ -0.000008 & 0.000169 & -0.000384 & 0.039050 \end{bmatrix}$$

$$A'_6 = \begin{bmatrix} 24.723379 & -0.261372 & 0.011686 & -0.000000 \\ -0.261372 & 7.660480 & 0.145892 & 0.000012 \\ 0.011686 & 0.145892 & 2.377090 & -0.000050 \\ -0.000000 & 0.000012 & -0.000050 & 0.039050 \end{bmatrix}$$

Use of a *spectral transformation* also permits employing the LR-Cholesky method to find the *largest eigenvalues* in descending order. For this purpose, let c be a quantity that is larger than the largest eigenvalue of the given symmetric matrix $\boldsymbol{A}$. The matrix $\boldsymbol{B} = c\boldsymbol{I} - \boldsymbol{A}$ is likewise symmetric and positive definite, and its eigenvalues $\lambda_i(\boldsymbol{B})$ are related to those of $\boldsymbol{A}$ by $\lambda_i(\boldsymbol{B}) = c - \lambda_i(\boldsymbol{A})$. Since $c > \lambda_1(\boldsymbol{A})$, the largest eigenvalues of $\boldsymbol{A}$ are hereby transformed to the smallest (positive) eigenvalues of $\boldsymbol{B}$. The LR-Cholesky method produces the smallest eigenvalues of $\boldsymbol{B}$ in ascending sequence, thus also the largest eigenvalues of $\boldsymbol{A}$. Here, too, c should be chosen as the smallest possible quantity larger than the largest eigenvalue. We may use, for instance, trial decompositions, starting with a value equal to the largest diagonal element and successively increasing it so as to minimize the loss of significant places in the formulation $\lambda_i(A) = c - \lambda_i(B)$.

Note: Recently another transformation rather analogous to the LR transformation was developed in which the given matrix A_1 is decomposed by Theorem 3–4 into the product of an orthogonal matrix Q_1 and an upper triangular matrix R_1 as in $A_1 = Q_1R_1$, in order to create a matrix $A_2 = R_1Q_1$ similar to A_1. By a continuation of this procedure we carry out the *QR transformation* (Ref. 20), which was originally thought up to handle unsymmetric matrices, but which is applicable to symmetric matrices as well. The QR transformation also exhibits the property of retaining the symmetry of given matrix A_1 in the sequence of matrices A_k, and numerically it is very stable. For further details, see the original publication (Ref. 20) or Refs. 75, 76 and 77.

4-6-4 Symmetric Definite Band Matrices

In principle, the LR-Cholesky method is usable for any symmetric, positive definite matrices. However, the computation time becomes large when it is used for full matrices having few or no zero elements. The computation time is appreciably reduced for symmetric definite band matrices. We note

Theorem 4-19. *The band structure of a symmetric definite matrix is retained through the LR-Cholesky method.*

Proof: Let $A = (a_{ik})$ be a symmetric definite band matrix with band width m, so that

$$a_{ik} = 0 \quad \text{for all} \quad i, k \quad \text{with} \quad |i - k| > m.$$

The Cholesky decomposition $A = R^{\mathrm{T}}R$ creates an upper triangular matrix $R = (r_{ik})$ of band form, in accord with Theorem 1–13:

$$r_{ik} = 0 \quad \text{for all} \quad i, k \quad \text{with} \quad k - i > m. \tag{4–100}$$

The elements a'_{ik} of the matrix $A' = RR^{\mathrm{T}}$ on and above the diagonals are given by

$$a'_{ik} = \sum_{j=k}^{n} r_{ij}r_{kj} \qquad i \leq k. \tag{4–101}$$

For Eq. (4–101), Eq. (4–100) tells us that all elements r_{ij} vanish as soon as $j > i + m$. Because of the relation $k \geq j$, where k is the summation index, and because of symmetry considerations, we have the assertion

$$a'_{ik} = 0 \quad \text{for all} \quad i, k \quad \text{with} \quad |i - k| > m.$$

The methods of Jacobi and Givens-Householder destroy the band form of a matrix and thus offer no particular advantage here. If the band width m is small compared to the order n of the matrix, the LR-Cholesky method is superior to the others as regards the computational effort. For the other methods, computation time for finding eigenvalues is proportional to n^3, while for the LR transformation, it is m^2n.

Symmetric definite band matrices having a small band width m compared to order n appear in the discretization of differential equations. In eigenvalue problems of this class, only a few eigenvalues at the lower end of the spectrum are usually of interest, precisely where the LR-Cholesky method is particularly applicable.

ALGOL Procedure for the LR-Cholesky Method Applied to Band Matrices. For a positive definite band matrix A of order n and band width m, the optimally economical storage scheme of Sec. 1–4–5 is employed, and the corresponding procedure *choleskyband* is used as a global variable. The coordinate translations are carried out according to a crude scheme which is certainly not optimal. At each step, a fraction ϕ of the last diagonal element is employed as the next trial translation y. If the Cholesky decomposition is successful with this trial value, the LR step is completed, the total coordinate translation logged, and the value of ϕ increased. If, however, the decomposition is unsuccessful with this positive value of y, it is carried out without any coordinate translation at all, $y = 0$, and the factor ϕ is decreased for safer future values of y. As the coordinate translation is already entered in the diagonal elements of A before the calling of the subroutine *choleskyband*, it must be retracted in case of an unsuccessful decomposition. As soon as the last diagonal element $a_{n0}^{(k)}$ falls below a certain given tolerance ϵ, the value $z + a_{n0}^{(k)}$ is inserted as the eigenvalue in place of the appropriate diagonal element, and simultaneously the order n is reduced by one. This continues until the specified number n_{eig} of eigenvalues has been evaluated. Perturbations of magnitude $\epsilon \cdot 10^{-3}$ replacing the vanishing elements assures that the conventional case of properly sequenced eigenvalues occurs. The given matrix A is modified successively in the program such that after ending the program, the last n_{eig} diagonal elements contain the smallest eigenvalues, and the resultant principal minor of order $n - n_{\text{eig}}$ contains the remaining eigenvalues of the original matrix A, since the total coordinate translation is again retracted.

The procedure parameters are defined as:

n Order of the given matrix A. Note that parameter n is not stored, but rather changed, during the process. The resultant n is the order of the principal minor.

m Band width of matrix *A*.

a Elements *a*[*i*, *k*] of the band matrix using the special indexing of Sec. 1–4–5. The elements *a*[*i*, *k*] with $n < i + k \leq n + m$ are initially set equal to zero. The meaning of *a*[*i*, *k*] after program termination is as explained above.

eps Absolute tolerance for the accuracy of eigenvalues.

neig Number of smallest eigenvalues desired ($neig \leq n$).

ALGOL Procedure No. 14

```
procedure lrcholband (n, m, a, eps, neig) ;
            value m, eps, neig ;
            integer n, m, neig ;  real eps ;  array a ;
begin integer i, j, k, keig, min ;  real y, z, phi ;
            array r[1 : n, 0 : m] ;
            z : = 0 :  keig : = 0 ;  phi = 0.5 ;
iter :      y : = phi × a[n, 0] ;
            for k : = 1 step 1 until n do a[k, 0] : = a[k, 0] − y ;
rep :       choleskyband (a, n, m, r, fail) ;
success :   z : = z + y ;
            if y ≠ 0 then phi : = (1 + phi) × 0.5 ;
            goto recomb ;
fail :      phi : = 0.5 × phi ;
            for k : = 1 step 1 until n do a[k, 0] : = a[k, 0] + y ;
            y : = 0 ;  goto rep ;
recomb :    for k : = 1 step 1 until n do
            begin if m < n − k then min : = m else min : = n − k ;
               for j : = 0 step 1 until min do
               begin a[k, j] : = 0 ;
                 for i : = j step 1 until min do
                     a[k, j] : = a[k, j] + r[k, i] × r[k + j, i − j] ;
perturb :        if a[k, j] = 0 then a[k, j] : = 10-3 × eps
               end j
            end k ;
            if a[n, 0] > eps then goto iter ;
eigen :     begin keig : = keig + 1 ;
               a[n, 0] : = a[n, 0] + z ;
               for j : = 1 step 1 until m do
                 if j < n then a[n − j, j] : = 0 ;
               n : = n − 1 ;  phi : = 0.9 ;
               if keig < neig then
               begin if n = 1 then goto eigen ;
                     goto iter
               end
            end eigen ;
            for k : = 1 step 1 until n do a[k, 0] : = a[k, 0] + z
end lrcholband
```

A procedure utilizing all available numerical and theoretical fine points is given in Ref. 48.

Example 4-9. Finding the smallest eigenvalues of a homogeneous vibrating beam leads after discretization into 12 intervals to the problem of finding the smallest eigenvalues of the following symmetric definite band matrix of order $n = 11$ and the commonly encountered band width $m = 2$:

$$A = \begin{bmatrix} 5 & -4 & 1 & & & & & & & & \\ -4 & 6 & -4 & 1 & & & & & & & \\ 1 & -4 & 6 & -4 & 1 & & & & & & \\ & 1 & -4 & 6 & -4 & 1 & & & & & \\ & & 1 & -4 & 6 & -4 & 1 & & & & \\ & & & 1 & -4 & 6 & -4 & 1 & & & \\ & & & & 1 & -4 & 6 & -4 & 1 & & \\ & & & & & 1 & -4 & 6 & -4 & 1 & \\ & & & & & & 1 & -4 & 6 & -4 & 1 \\ & & & & & & & 1 & -4 & 6 & -4 \\ & & & & & & & & 1 & -4 & 5 \end{bmatrix}.$$

Nine LR steps with the ALGOL procedure *lrcholband* produces the smallest eigenvalue with an absolute accuracy of $\epsilon = 10^{-8}$. Table 20 depicts the values $a_{n0}^{(k)}$ of the last diagonal element, the total displacements z_k, and the values $z_k + a_{n0}^{(k)}$, which are upper bounds for the smallest eigenvalue, plus the factors ϕ controlling the next displacement, and the actual displacements y_k. In the first two LR steps, the program's trial displacements y_k were too large; thus the attempted decomposition broke down both times, and both steps had to be repeated with $y = 0$. Subsequent trial values of y_k were all successful,

Table 20 LR-Cholesky Method for a Band Matrix

k	$a_{n0}^{(k)}$	z_k	$z_k + a_{n0}^{(k)}$	ϕ	y_k
1	5.000 000	0	5.000 000 000	0.5000	0
2	$2.845\,850 \cdot 10^{-1}$	0	0.284 584 979	0.2500	0
3	$6.672\,845 \cdot 10^{-3}$	0	0.006 672 845	0.1250	$8.34106 \cdot 10^{-4}$
4	$3.872\,768 \cdot 10^{-3}$	0.000 834 106	0.004 706 873	0.5625	$2.17843 \cdot 10^{-3}$
5	$1.633\,022 \cdot 10^{-3}$	0.003 012 537	0.004 645 559	0.7813	$1.27580 \cdot 10^{-3}$
6	$3.558\,686 \cdot 10^{-4}$	0.004 288 336	0.004 644 204	0.8906	$3.16945 \cdot 10^{-4}$
7	$3.891\,606 \cdot 10^{-5}$	0.004 605 281	0.004 644 197	0.9453	$3.67878 \cdot 10^{-5}$
8	$2.128\,218 \cdot 10^{-6}$	0.004 642 069	0.004 644 197	0.9727	$2.07003 \cdot 10^{-6}$
9	$5.819\,346 \cdot 10^{-8}$	0.004 644 139	0.004 644 197	0.9863	$5.73979 \cdot 10^{-8}$
10	$7.956\,138 \cdot 10^{-10}$	0.004 644 196	0.004 644 197	—	—

as luck would have it. The pairs of values z_k and $z_k + a_{n0}^{(k)}$ yield better bounds with increasing k for the smallest eigenvalue $\lambda_{11} = 0.004644197$.

For the remaining eigenvalues the procedure *lrcholband* takes only five to seven steps each, since the reduced matrices provide good approximations for the smallest eigenvalues through the last diagonal elements.

4-6-5 The QD Algorithm

For a symmetric definite tridiagonal matrix, the LR transformation leads to a particularly simple calculational procedure which, in turn, leads to another method for eigenvalue computation. For convenience, the irreducible symmetric tridiagonal matrix $\boldsymbol{J}$ of (4–49) is altered by a similarity transformation into the unsymmetric form

$$\boldsymbol{A}_1 = \begin{bmatrix} \alpha_1 & 1 & & & & \\ \beta_1^2 & \alpha_2 & 1 & & & \\ & \beta_2^2 & \alpha_3 & 1 & & \\ & & \cdot & \cdot & \cdot & \\ & & & \cdot & \cdot & \cdot \\ & & & \cdot & \cdot & \cdot \\ & & & \beta_{n-2}^2 & \alpha_{n-1} & 1 \\ & & & & \beta_{n-1}^2 & \alpha_n \end{bmatrix}. \tag{4–102}$$

For the transformation, with $k = 2, 3, \ldots, n$, the kth column was divided by a suitable value and the kth row was simultaneously multiplied by the same number. The matrix $\boldsymbol{A}_1$ (4–102) serves as an initial matrix for the LR transformation in its original form of Sec. 4–6–1. Since the band form is retained even in an unsymmetrical decomposition, the matrices $\boldsymbol{L}_1$ and $\boldsymbol{R}_1$ of $\boldsymbol{A}_1 = \boldsymbol{L}_1\boldsymbol{R}_1$ are given by

$$\boldsymbol{L}_1 = \begin{bmatrix} 1 & & & & & \\ e_1^{(1)} & 1 & & & & \\ & e_2^{(1)} & 1 & & & \\ & & e_3^{(1)} & 1 & & \\ & & & \cdot & \cdot & \\ & & & & e_{n-1}^{(1)} & 1 \end{bmatrix}, \quad \boldsymbol{R}_1 = \begin{bmatrix} q_1^{(1)} & 1 & & & & \\ & q_2^{(1)} & 1 & & & \\ & & q_3^{(1)} & 1 & & \\ & & & \cdot & \cdot & \\ & & & & q_{n-1}^{(1)} & 1 \\ & & & & & q_n^{(1)} \end{bmatrix}. \tag{4–103}$$

For the values $q_k^{(1)}$ and $e_k^{(1)}$, we have from (4–102) and (4–103)

$$\left.\begin{aligned} &q_1^{(1)} = \alpha_1, \\ &e_k^{(1)} = \beta_k^2/q_k^{(1)}, \qquad q_{k+1}^{(1)} = \alpha_{k+1} - e_k^{(1)} \qquad k = 1, 2, \ldots, n-1. \end{aligned}\right\} \tag{4–104}$$

The multiplication $R_1L_1 = A_2$ produces a matrix A_2 analogous to (4–102), having diagonal elements identified by $\alpha_k^{(2)}$ and elements in the diagonal immediately below given by $\beta_k^{(2)^2}$. These are defined as

$$\left.\begin{aligned} \alpha_k^{(2)} &= q_k^{(1)} + e_k^{(1)}, \quad \beta_k^{(2)^2} = q_{k+1}^{(1)}\cdot e_k^{(1)} \qquad k = 1, 2, \ldots, n-1, \\ \alpha_n^{(2)} &= q_n^{(1)}. \end{aligned}\right\} \tag{4–105}$$

The decomposition of matrix $A_2 = L_2R_2$ into corresponding matrices L_2 and R_2 yields for their elements, analogous to (4–104),

$$\left.\begin{aligned} q_1^{(2)} &= \alpha_1^{(2)}, \\ e_k^{(2)} &= \beta_k^{(2)^2}/q_k^{(2)}, \quad q_{k+1}^{(2)} = \alpha_{k+1}^{(2)} - e_k^{(2)} \qquad k = 1, 2, \ldots, n-1. \end{aligned}\right\} \tag{4–106}$$

In this computation scheme the matrix A_2 is eliminated and the direct switch-over from $q_k^{(1)}$ and $e_k^{(1)}$ to $q_k^{(2)}$ and $e_k^{(2)}$ can be made. Equations (4–105) and (4–106) lead to the relations

$$\left.\begin{aligned} q_1^{(2)} &= q_1^{(1)} + e_1^{(1)}, \\ e_k^{(2)} &= \frac{q_{k+1}^{(1)}}{q_k^{(2)}} e_k^{(1)}, \quad q_{k+1}^{(2)} = q_{k+1}^{(1)} + e_{k+1}^{(1)} - e_k^{(2)}, \qquad k = 1, 2, \ldots, n-1. \end{aligned}\right\} \tag{4–107}$$

The formula (4–107) for $q_n^{(2)}$ remains valid if $e_n^{(1)}$ is defined as $e_n^{(1)} = 0$. The continuation of the first computational step defined through (4–107) for $q_k^{(2)}$ and $e_k^{(2)}$ produces sequences of $q_k^{(s)}$ and $e_k^{(s)}$, which are arranged in the so-called *quotient difference scheme* of (4–108). The q and e values of the same upper index are seen to appear alternately in one skew line:

$$\begin{array}{ccccccccc}
q_1^{(1)} & & & & & & & & \\
 & e_1^{(1)} & & & & & & & \\
q_1^{(2)} & & q_2^{(1)} & & & & & & \\
 & e_1^{(2)} & & e_2^{(1)} & & & & & \\
q_1^{(3)} & & q_2^{(2)} & & q_3^{(1)} & \cdot & & & \\
 & e_1^{(3)} & & e_2^{(2)} & & \cdot & q_{n-1}^{(1)} & & \\
\cdot & & q_2^{(3)} & & q_3^{(2)} & \cdot & & e_{n-1}^{(1)} & \\
 & \cdot & & e_2^{(3)} & & \cdot & q_{n-1}^{(2)} & & q_n^{(1)} \\
\cdot & & \cdot & & q_3^{(3)} & \cdot & & e_{n-1}^{(2)} & \\
 & \cdot & & \cdot & & \cdot & q_{n-1}^{(3)} & & q_n^{(2)} \\
 & & \cdot & & \cdot & \cdot & & e_{n-1}^{(3)} & \\
 & & & \cdot & & \cdot & \cdot & & q_n^{(3)} \\
 & & & & \cdot & \cdot & & \cdot & \\
 & & & & & & \cdot & & \cdot \\
 & & & & & & & \cdot & \\
 & & & & & & & & \cdot
\end{array} \tag{4–108}$$

As soon as one skew line has been found, the next one can be computed element by element from left to right by (4–107). Thus there are always four elements of the chart linked together, positioned such as to form the

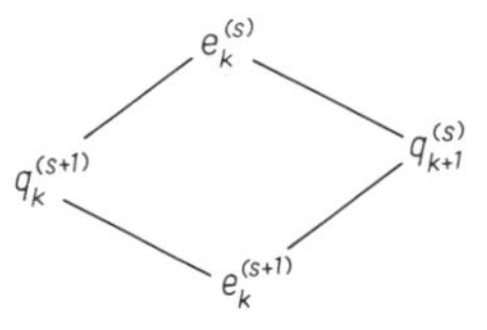

Figure 18. First rhombus rule.

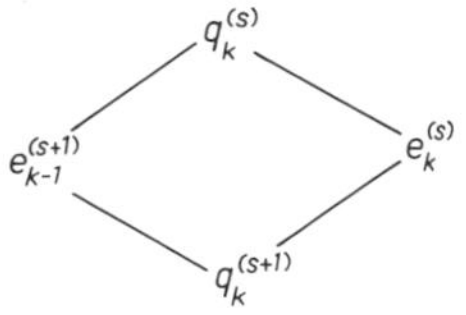

Figure 19. Second rhombus rule.

corners of a rhombus. The element to be computed is at the bottom corner. Figures 18 and 19 show typical rhombi for computing e and q values, respectively.

According to (4–107), the *product* (for Fig. 18) or *sum* (for Fig. 19) of the elements in the new line equal those in the old line. These are the *rhombus rules* of the *quotient difference algorithm* (QD algorithm) (see Refs. 42, 43, 28). The *first rhombus rule* (Fig. 18) says that $e_k^{(s+1)}$ is retained, in that the product of the upper and right elements is divided by the element at left. For the computation of $q_k^{(s+1)}$, the *second rhombus rule* (Fig. 19) is used in that the left element is subtracted from the sum of the top and right-hand elements. The two rhombus rules are dual to one another, as the multiplicative and additive operations match one another if the q's and e's are interchanged.

We can expect that the QD algorithm, being a special case of the LR transformation, also converges, and that the q values of the upper triangular matrices R_s converge to the eigenvalues of $\boldsymbol{A}$ and $\boldsymbol{J}$, and that the e values go to zero. Since the QD algorithm is based on asymmetric decomposition, Theorem 4–17 cannot be invoked. On the contrary, convergence must be proven anew. In preparation thereof, we demonstrate one facet of the QD technique.

Theorem 4-20. *The initial values $q_k^{(1)}$ and $e_k^{(1)}$ of the first skew line of the QD technique for a symmetric definite, irreducible tridiagonal matrix $\boldsymbol{J}$ (or else $\boldsymbol{A}_1$ of* (4–102)) *are greater than zero.*

Proof: In essence this is fulfilled for $q_1^{(1)} = \alpha_1$ and $e_1^{(1)} = \beta_1^2/q_1^{(1)}$ through the given positive definiteness and the irreducibility of $\boldsymbol{J}$ with $\alpha_1 > 0$ and β_1 nonzero. Since $\beta_k^2 > 0$ $(k = 1, 2, \ldots, n-1)$, then $e_k^{(1)} > 0$ by (4–104) if we establish that $q_k^{(1)} > 0$. In general, $q_k^{(1)}$ $(k = 1, 2, 3, \ldots, n)$ is the quotient of the *principal minor* having k and $(k-1)$ rows, and this will now be proven via complete induction.

Induction premise: Let h_i be the principal minor of order i $(i = 1, 2, \ldots, n)$ of the matrix $\boldsymbol{J}$, and let

$$q_i^{(1)} = h_i/h_{i-1} \qquad i = 2, 3, \ldots, k-1.$$

Induction assertion: that

$$q_k^{(1)} = h_k/h_{k-1}.$$

Induction proof: By (4–104), by the induction premise, and finally by the tridiagonal form of $\boldsymbol{J}$, we have, in turn,

$$\begin{aligned} q_k^{(1)} &= \alpha_k - e_{k-1}^{(1)} = \alpha_k - \beta_{k-1}^2/q_{k-1}^{(1)} = \alpha_k - \beta_{k-1}^2 h_{k-2}/h_{k-1} \\ &= (\alpha_k h_{k-1} - \beta_{k-1}^2 h_{k-2})/h_{k-1} = h_k/h_{k-1}. \end{aligned}$$

Induction check: For $q_2^{(1)}$, we have

$$q_2^{(1)} = \alpha_2 - e_1^{(1)} = \alpha_2 - \beta_1^2/\alpha_1 = (\alpha_1\alpha_2 - \beta_1^2)\alpha_1 = h_2/h_1.$$

The principal minors of a positive definite matrix are positive; thus $q_k^{(1)} > 0$ $(k = 1, 2, \ldots, n)$. Q.E.D.

The property of the first diagonal's having only positive elements is sustained as an *invariant* in the diagonals that follow, since they can be visualized as a decomposition of a corresponding positive definite matrix similar to $\boldsymbol{J}$. Thus we speak of a *positive QD technique* (Ref. 46) in which it is significant that the new q values remain positive, even though by the second rhombus rule they are formed by an addition and a subtraction. Similarly, the e values cannot theoretically vanish, since by the first rhombus rule they originate through multiplication by a positive quotient. A positive QD technique has the numerical characteristic of avoiding division by zero and thus obviating the need for handling such a case.

After these observations, we now turn to the convergence question.

Theorem 4-21. *The QD algorithm for a symmetric definite, irreducible tridiagonal matrix $\boldsymbol{J}$, or $\boldsymbol{A}_1$ of* (4–102), *is convergent. In addition, the limit values $\lim_{s\to\infty} q_k^{(s)} = \lambda_k$ are the eigenvalues of the matrix $\boldsymbol{J}$ (or $\boldsymbol{A}_1$) arranged in order of magnitude: $\lambda_1 > \lambda_2 > \cdots > \lambda_n$.*

Proof: For the sequence of q values in the last column of the QD technique, we have by the second rhombus rule

$$q_n^{(s+1)} = q_n^{(s)} - e_{n-1}^{(s+1)}.$$

Since the QD technique is positive, it follows that $e_{n-1}^{(s+1)} > 0$ for all $s \geq 0$, and consequently $q_n^{(s+1)}$ is a monotonically decreasing sequence in the strict sense. It is nonetheless bounded from below by zero. The sequence is convergent and thus it must follow that

$$\lim_{s\to\infty} e_{n-1}^{(s+1)} = 0. \tag{4–109}$$

For the sequence of values in the next to last q column, we have from the second rhombus rule

$$q_{n-1}^{(s+1)} = q_{n-1}^{(s)} + e_{n-1}^{(s)} - e_{n-2}^{(s+1)}.$$

This sequence need no longer be monotonic. In order to show that the values $e_{n-2}^{(s+1)}$ form a sequence tending to zero, let us examine the sum $q_{n-1}^{(s+1)} + q_n^{(s)}$ of two q values in the same echelon of the QD technique. For this sum,

$$\begin{aligned} q_{n-1}^{(s+1)} + q_n^{(s)} &= q_{n-1}^{(s)} + e_{n-1}^{(s)} - e_{n-2}^{(s+1)} + q_n^{(s-1)} + e_{n-1}^{(s)} \\ &= q_{n-1}^{(s)} + q_n^{(s-1)} - e_{n-2}^{(s+1)}. \end{aligned}$$

The positiveness of the QD technique means that the sum $q_{n-1}^{(s+1)} + q_n^{(s)}$ is a monotonically decreasing sequence in the strict sense for increasing s, and this sequence is bounded from below by zero. Thus the result

$$\lim_{s\to\infty} e_{n-2}^{(s+1)} = 0. \tag{4–110}$$

The analogous application of this argument to sums of q values shows that all e columns converge to zero and thus the q columns converge to the eigenvalues of matrices $\boldsymbol{J}$ and $\boldsymbol{A}_1$. The limit values of the q columns, however, appear in order of magnitude. By Theorem 4–10, matrix $\boldsymbol{J}$ has positive, unequal eigenvalues under the given assumptions. All quotients $q_{k+1}^{(s)}/q_k^{(s+1)}$ in the limiting case must therefore differ from unity. Since $e_k^{(s)}$ is greater than zero, and since $e_k^{(s)}$ converges to zero if k is fixed and s increases, this is possible in light of the first rhombus rule only if

$$\lim_{s\to\infty} \frac{q_{k+1}^{(s)}}{q_k^{(s+1)}} < 1 \quad \text{for} \quad k = 1, 2, \ldots, n-1.$$

Example 4-10. The eigenvalues of the tridiagonal matrix

$$\boldsymbol{J} = \boldsymbol{A}_1 = \begin{bmatrix} 2 & 1 & & \\ 1 & 2 & 1 & \\ & 1 & 2 & 1 \\ & & 1 & 2 \end{bmatrix}$$

are $\lambda_k = 2 \pm \sqrt{0.5(3 \pm \sqrt{5})}$; thus $\lambda_1 = 3.61803$, $\lambda_2 = 2.61803$, $\lambda_3 = 1.38197$, $\lambda_4 = 0.38197$. The first eight lines of the corresponding QD technique are found in Table 21.

The convergence of the e values to zero, or of the q values to the eigenvalues, is *linear* in that for every k the values $e_k^{(s)}$ converge to zero like a geometric sequence with the quotients λ_{k+1}/λ_k. By appropriate *coordinate*

Table 21 QD Technique

$q_1^{(s)}$	$e_1^{(s)}$	$q_2^{(s)}$	$e_2^{(s)}$	$q_3^{(s)}$	$e_3^{(s)}$	$q_4^{(s)}$
2.00000						
	0.50000					
2.50000		1.50000				
	0.30000		0.66667			
2.80000		1.86667		1.33333		
	0.20000		0.47619		0.75000	
3.00000		2.14286		1.60714		1.25000
	0.14286		0.35714		0.58333	
3.14286		2.35714		1.83333		0.66667
	0.10714		0.27778		0.21212	
3.25000		2.52778		1.76767		0.45455
	0.08333		0.19425		0.05455	
3.33333		2.63870		1.62797		0.40000
	0.06597		0.11984		0.01340	
3.39930		2.69257		1.52153		0.38660
	0.05225		0.06772		0.00340	
↓		2.70804		1.45721		0.38320
		↓	0.03644		0.00089	
				1.42166		0.38231
				↓	0.00024	
						0.38207
						↓
3.61803		2.61803		1.38197		0.38197

translations, the convergence rate is improved through the diminishing of the convergence quotients of $e_{n-1}^{(s)}$. Note that the numerically advantageous property of the positive QD technique is retained. The translation may not be greater than the smallest eigenvalue. Too large a translation manifests itself in a negative q value.

The modification of the rhombus rules through a coordinate translation of y_s in the sth QD step affects only the second rhombus rule. After formation of $\boldsymbol{R}_s\boldsymbol{L}_s = \boldsymbol{A}_{s+1}$, the translation of y_s is taken into account by replacing the elements $\alpha_k^{(s+1)}$ with $\alpha_k^{(s+1)} - y_s$. In the subsequent decomposition of $(\boldsymbol{A}_{s+1} - y_s\boldsymbol{I}) = \boldsymbol{L}_{s+1}\boldsymbol{R}_{s+1}$, only the formulas for computation of $q_k^{(s+1)}$ are affected. We end up with the modified *QD algorithm with coordinate translations.*

The value of z_s represents the total coordinate translation. If in every step y_s is chosen as large as possible but still smaller than the smallest eigenvalue of matrix $\boldsymbol{A}_{s+1}$, the last q element converges to zero, and z_{s+1} converges to the smallest eigenvalue of $\boldsymbol{A}_1$ and $\boldsymbol{J}$. Accordingly, $e_{n-1}^{(s)}$ generally goes rapidly to zero. As soon as $e_{n-1}^{(s)}$ is small enough, z_s represents the smallest eigenvalue. For the continuation of the method, the last two columns of the QD technique are omitted, so that once again the smallest eigenvalues turn up in

Start:

$$\boxed{\begin{array}{lll} q_1^{(1)} = \alpha_1, & z_1 = 0 & \\ e_k^{(1)} = \beta_k^2/q_k^{(1)}, & q_{k+1}^{(1)} = \alpha_{k+1} - e_k^{(1)}, & k = 1, 2, \ldots, n-1 \end{array}} \qquad (4\text{–}111)$$

The sth QD step ($s = 1, 2, \ldots$):

$$\boxed{\begin{array}{ll} \text{(a)} & \text{Choice of } y_s < \lambda_n(A_{s+1}) \\ \text{(b)} & q_1^{(s+1)} = q_1^{(s)} + e_1^{(s)} - y_s \\ & e_k^{(s+1)} = \dfrac{q_{k+1}^{(s)}}{q_k^{(s+1)}} e_k^{(s)}, \quad q_{k+1}^{(s+1)} = q_{k+1}^{(s)} + e_{k+1}^{(s)} - e_k^{(s+1)} - y_s \\ & \qquad\qquad k = 1, 2, \ldots, n-1 \\ \text{(c)} & z_{s+1} = z_s + y_s \end{array}} \qquad (4\text{–}112)$$

increasing order. Analogous to the LR-Cholesky method, the permissible translation y_s for the sth QD step can be determined by a trial decomposition and appropriate control by means of a computer.

Example 4-11. The sped-up convergence through coordinate translation is demonstrated for the matrix of Example 4–10. Starting with the same initial diagonal, the translations are set equal to $y_s = \phi \cdot q_n^{(s)}$, where $q_n^{(s)}$ indicates the current last q element of the possibly reduced QD technique, and the factor ϕ, identical for two successive QD steps, takes on the particular values 1/4, 1/2, 3/4, 7/8, 15/16, ... Translations thus selected work for the given example. The computation is given in Table 22. Note the rapid convergence of $e_3^{(s)}$ to zero; after five steps, $e_3^{(6)} \approx 2 \cdot 10^{-7}$. In the corresponding matrix A_7 we have $\beta_3^{(7)^2} \approx 2 \cdot 10^{-7}$, so that it, too, reduces. The smallest eigenvalue is thus $\lambda_4 = z_6 + q_4^{(6)} = 0.38196$, which is correct within one unit in the last decimal. Table 22 also shows the continuation with the reduced QD technique.

The QD algorithm for finding eigenvalues of a symmetric definite, irreducible tridiagonal matrix is a *simple and numerically stable procedure* (see Ref. 46). The computation time for one QD step is virtually the same as for one step of the method of repeated interval bisection (see Sec. 4-5-4), but in the former case, fewer steps are usually needed to find eigenvalues than in the latter case. *If only a few of the smallest eigenvalues are sought, the QD algorithm is preferable.*

4-6-6 Applications of the QD Algorithm

The QD algorithm has numerous applications aside from computing eigenvalues of tridiagonal matrices. The sections that follow show how the

Table 22 QD ALGORITHM WITH COORDINATE TRANSLATION

Translations		$q_1^{(s)}$	$e_1^{(s)}$	$q_2^{(s)}$	$e_2^{(s)}$	$q_3^{(s)}$	$e_3^{(s)}$	$q_4^{(s)}$
$z_1 = 0$		2.00000						
	$y_1 = 0.31250$		0.50000					
$z_2 = 0.31250$		2.18750		1.50000				
	$y_2 = 0.03620$		0.34286		0.66667			
$z_3 = 0.34870$		2.49416		1.51131		1.33333		
	$y_3 = 0.01761$		0.20775		0.58816		0.75000	
$z_4 = 0.36631$		2.68430		1.85552		1.18267		1.25000
	$y_4 = 0.00784$		0.14361		0.37488		0.79270	
$z_5 = 0.37415$		2.82007		2.06918		1.56429		0.14480
	$y_5 = 0.00586$		0.10537		0.28341		0.07338	
$z_6 = 0.38001$		2.91958		2.23938		1.33665		0.03522
			0.08082		0.16916		0.00193	
				2.32186		1.16158		0.01568
					0.08463		0.00003	
	$y_6 = 0.80334$					1.07112		0.00781
$z_7 = 1.18335$		2.19706					0.00000	
	$y_7 = 0.18204$		0.08541					0.00195
$z_8 = 1.36539$		2.10043		1.51774				
	$y_8 = 0.01460$		0.06172		0.05973			
$z_9 = 1.37999$		2.14755		1.33371		0.20805		$\underline{\lambda_4 = 0.38196}$
	$y_9 = 0.00185$		0.03833		0.00932			
$z_{10} = 1.38184$		2.18403		1.29010		0.01669		
			0.02264		0.00012			
				1.26573		0.00197		
					0.00000			
	$y_{10} = 1.18662$					0.00012		
$z_{11} = 2.56846$		1.02005						
	$y_{11} = 0.04943$		0.02809					
$z_{12} = 2.61789$		0.99871		0.05102		$\underline{\lambda_3 = 1.38196}$		
	$y_{12} = 0.00014$		0.00144					
$z_{13} = 2.61803$		1.00001		0.00015				
			0.00000					
				0.00001				
		$\underline{\lambda_1 = 3.61804}$		$\underline{\lambda_2 = 2.61804}$				

initial values in the first diagonal are obtained in these applications. The reader is also referred to the resume of Ref. 30.

(a) *Finding the zeros of a polynomial.* The *Euclidean algorithm* for finding the greatest common divisor of two polynomials $p_n(x)$ and $p_{n-1}(x)$ of true degree n and $n - 1$ and with the highest coefficients equal to unity, produces a recursive sequence of polynomials of descending degree

$$\left.\begin{aligned} p_n(x) &= (x - \alpha_n)p_{n-1}(x) - \beta_{n-1}p_{n-2}(x) \\ p_{n-1}(x) &= (x - \alpha_{n-1})p_{n-2}(x) - \beta_{n-2}p_{n-3}(x) \\ &\cdots\cdots\cdots\cdots \\ p_2(x) &= (x - \alpha_2)p_1(x) - \beta_1 p_0(x) \\ p_1(x) &= (x - \alpha_1)p_0(x) \end{aligned}\right\}. \tag{4–113}$$

The values β_k are to be found such that the remaining polynomial $p_{k-1}(x)$ acquires the highest coefficient of unity. If the given polynomials $p_n(x)$ and $p_{n-1}(x)$ have no common divisor, the algorithm is terminated with the polynomial $p_0(x) = 1$, except for some exceptional cases where one is left with a polynomial of smaller degree. Because of the recursion formulas (4–113), $p_n(x)$ can be interpreted as a characteristic polynomial of the tridiagonal matrix

$$A = \begin{bmatrix} \alpha_1 & 1 & & & \\ \beta_1 & \alpha_2 & 1 & & \\ & \beta_2 & \alpha_3 & 1 & \\ & & \cdot & \cdot & \cdot \\ & & & \beta_{n-1} & \alpha_n \end{bmatrix}.$$

The zeros of $p_n(x)$ are thus identical to the eigenvalues of A, which can be computed through the QD algorithm. The first diagonal of the QD technique is computed from the coefficients α_k and β_k of the Euclidean algorithm by

$$\left.\begin{aligned} &q_1^{(1)} = \alpha_1, \\ &e_k^{(1)} = \beta_k/q_k^{(1)}, \quad q_{k+1}^{(1)} = \alpha_k - e_k^{(1)}, \qquad k = 1, 2, \ldots, n-1. \end{aligned}\right\} \tag{4–114}$$

If $p_n(x)$ has simple zeros, the derivative of $p_n(x)$ can be used for $p_{n-1}(x)$, normalized such that the highest coefficient is unity. If in addition the *zeros* of $p_n(x)$ should be *real* and *positive*, the resultant QD technique is *positive*.[1]

[1]This opportune property is achieved by forming the first diagonal with the Euclidean algorithm. The simpler introduction to the QD technique of Ref. 64 lacks this property; even if all the zeros are positive, both positive and negative values appear in the QD technique.

Example 4-12. For the polynomial

$$p_4(x) = x^4 - 10x^3 + 35x^2 - 50x + 24 = (x-1)(x-2)(x-3)(x-4)$$

with real, positive zeros, the Euclidean algorithm with the normalized derivative,

$$p_3(x) = 0.25p_4'(x) = x^3 - 7.5x^2 + 17.5x - 12.5,$$

yields successively

$$\begin{aligned}
p_4(x) &= x^4 - 10x^3 + 35x^2 - 50x + 24 \\
p_3(x) &= \qquad x^3 - 7.5x^2 + 17.5x - 12.5, \quad \alpha_4 = 2.5, \quad \beta_3 = 1.25 \\
p_2(x) &= \qquad\qquad x^2 - 5x + 5.8 \quad\;\; , \quad \alpha_3 = 2.5, \quad \beta_2 = 0.8 \\
p_1(x) &= \qquad\qquad\quad x - 2.5 \quad\;\; , \quad \alpha_2 = 2.5, \quad \beta_1 = 0.45 \\
p_0(x) &= \qquad\qquad\qquad 1 \quad\;\; , \quad \alpha_1 = 2.5.
\end{aligned}$$

The corresponding QD technique (without translation) is given in Table 23.

In this case, the QD algorithm converges very slowly for the first two

Table 23 COMPUTATION OF ZEROS OF A POLYNOMIAL

$q_1^{(s)}$	$e_1^{(s)}$	$q_2^{(s)}$	$e_2^{(s)}$	$q_3^{(s)}$	$e_3^{(s)}$	$q_4^{(s)}$
2.50000						
	0.18000					
2.68000		2.32000				
	0.15582		0.34483			
2.83582		2.50901		2.15517		
	0.13786		0.29620		0.58000	
2.97368		2.66735		2.43897		1.92000
	0.12366		0.27084		0.45659	
3.09734		2.81453		2.62472		1.46341
	0.11237		0.25257		0.25457	
3.20971		2.95473		2.62672		1.20884
	0.10344		0.22453		0.11716	
3.31315		3.07582		2.51935		1.09168
↓	0.09603		0.18391		0.05077	
		3.16370		2.38621		1.04091
		↓	0.13871		0.02215	
				2.26965		1.01876
				↓	0.00994	
						1.00882
						↓
4.00000		3.00000		2.00000		1.00000

q columns because the q values are almost equal. Nevertheless, the smallest zero of $p_4(x)$ emerges quite clearly.

(b) *The method of conjugate gradients and the determination of eigenvalues of operator A.* According to the computation specifications (2–96), the method of conjugate gradients produces two vector sequences,

$$\boldsymbol{r}^{(k)} = \boldsymbol{r}^{(k-1)} + q_k(A\boldsymbol{p}^{(k)}), \qquad \boldsymbol{p}^{(k)} = -\boldsymbol{r}^{(k-1)} + e_{k-1}\boldsymbol{p}^{(k-1)},$$

with values $e_{k-1} > 0$ and $q_k > 0$. For the residual vector $\boldsymbol{r}^{(k)}$ there is a three-term recursion formula

$$\begin{aligned}\boldsymbol{r}^{(k)} &= \boldsymbol{r}^{(k-1)} + q_k(A\boldsymbol{p}^{(k)}) = \boldsymbol{r}^{(k-1)} + q_k A(-\boldsymbol{r}^{(k-1)} + e_{k-1}\boldsymbol{p}^{(k-1)})\\ &= \boldsymbol{r}^{(k-1)} - q_k A\boldsymbol{r}^{(k-1)} + e_{k-1}q_k A\boldsymbol{p}^{(k-1)}\\ &= \boldsymbol{r}^{(k-1)} - q_k A\boldsymbol{r}^{(k-1)} + e_{k-1}q_k\frac{1}{q_{k-1}}(\boldsymbol{r}^{(k-1)} - \boldsymbol{r}^{(k-2)}).\end{aligned}$$

When solved for $A\boldsymbol{r}^{(k-1)}$ this yields a relation between three successive residual vectors

$$A\boldsymbol{r}^{(k-1)} = -\frac{e_{k-1}}{q_{k-1}}\boldsymbol{r}^{(k-2)} + \left(\frac{1}{q_k} + \frac{e_{k-1}}{q_{k-1}}\right)\boldsymbol{r}^{(k-1)} - \frac{1}{q_k}\boldsymbol{r}^{(k)}. \tag{4–115}$$

The relation (4–115) is valid for $k = 1$ if we set $e_0 = 0$. By Theorem 2–12, the residual vectors $\boldsymbol{r}^{(0)}, \boldsymbol{r}^{(1)}, \ldots, \boldsymbol{r}^{(n-1)}$ constitute a complete orthogonal system in the event that the method is accidentally successful beforehand. The residual vectors are to be normalized into vectors

$$\boldsymbol{u}^{(k)} = \frac{\boldsymbol{r}^{(k-1)}}{\rho_{k-1}}, \qquad \rho_{k-1} = (\boldsymbol{r}^{(k-1)}, \boldsymbol{r}^{(k-1)})^{1/2} \qquad k = 1, 2, \ldots, n, \tag{4–116}$$

which form a complete orthonormal system. For them, (4–115) becomes

$$\begin{aligned}A\boldsymbol{u}^{(k)} = \frac{1}{\rho_{k-1}}\Bigg[&-\frac{e_{k-1}}{q_{k-1}}\rho_{k-2}\boldsymbol{u}^{(k-1)} + \left(\frac{1}{q_k} + \frac{e_{k-1}}{q_{k-1}}\right)\rho_{k-1}\boldsymbol{u}^{(k)}\\ &-\frac{1}{q_k}\rho_k\boldsymbol{u}^{(k+1)}\Bigg] \qquad k = 1, 2, \ldots, n.\end{aligned} \tag{4–117}$$

For $k = n$, $\boldsymbol{u}^{(n+1)} = 0$ corresponds to the fact that $\boldsymbol{r}^{(n)}$ vanishes. The representation of operator A by means of the basis of orthonormal vectors $\boldsymbol{u}^{(k)}$ is given by $\boldsymbol{B} = \boldsymbol{U}^{\mathrm{T}}A\boldsymbol{U}$, where $\boldsymbol{U}$ is the orthogonal matrix whose kth column contains the vector $\boldsymbol{u}^{(k)}$. The elements b_{ik} of $\boldsymbol{B}$ can be computed from

$$b_{ik} = (\boldsymbol{u}^{(i)}, A\boldsymbol{u}^{(k)}).$$

Because of (4–117) and the orthonormal property of the vectors $\boldsymbol{u}^{(i)}$, matrix $\boldsymbol{B}$ must be tridiagonal. Noting that by (2–90)

$$e_k = \left(\frac{\rho_k}{\rho_{k-1}}\right)^2,$$

we see that the nontrivial elements of $\boldsymbol{B}$, $\alpha_k = b_{kk}$ and $\beta_k = b_{k,k+1} = b_{k+1,k}$ become

$$\left.\begin{aligned}
\alpha_k &= \frac{1}{q_k} + \frac{e_{k-1}}{q_{k-1}} \qquad && k = 1, 2, \ldots, n \qquad e_0 = 0, \\
\beta_k &= \frac{e_k}{q_k}\frac{\rho_{k-1}}{\rho_k} && k = 1, 2, \ldots, n-1, \\
\beta_k^2 &= \frac{e_k^2}{q_k^2}\frac{\rho_{k-1}^2}{\rho_k^2} = \frac{e_k}{q_k^2} && k = 1, 2, \ldots, n-1.
\end{aligned}\right\} \tag{4–118}$$

The connection of given matrix $\boldsymbol{A}$ to a similar tridiagonal matrix $\boldsymbol{B}$ is thereby established for the method of conjugate gradients.[1] Its elements are found from the q and e values of the conjugate gradients method by (4–118). The eigenvalues of $\boldsymbol{A}$ can be computed from $\boldsymbol{B}$ by using the QD algorithm, since the initial values $q_k^{(1)}$ and $e_k^{(1)}$ of the first diagonal of the QD technique can be found through a trivial calculation involving quantities q_k and e_k of the method of conjugate gradients.

$$\left.\begin{aligned}
q_1^{(1)} &= \frac{1}{q_1}, \\
e_k^{(1)} &= \frac{e_k}{q_k}, \quad q_{k+1}^{(1)} = \frac{1}{q_{k+1}} \qquad k = 1, 2, \ldots, n-1.
\end{aligned}\right\} \tag{4–119}$$

The resultant QD technique is positive.

PROBLEMS

4-14. Carry out at least two steps of the LR-Cholesky technique for the matrix

$$\boldsymbol{A} = \begin{bmatrix} 4 & -2 & 6 \\ -2 & 17 & 1 \\ 6 & 1 & 19 \end{bmatrix}.$$

4-15. The smallest eigenvalue of matrix $\boldsymbol{A}$ in Problem 4-14 is $\lambda_3(\boldsymbol{A}) = 1.5762$. What happens if the coordinates are displaced by $y = 1.55$ and $y = 1.60$,

[1]The conjugate gradients method thus theoretically does the same as the methods of Givens and Householder, but because of numerical instability it is not recommended for eigenvalue calculations (see Ref. 13).

respectively, prior to the first decomposition? For the first case, carry out two or three LR-Cholesky steps and compare with the corresponding results of Problem 4-14. Note particularly element $a_{33}^{(k)}$. Slide-rule accuracy will suffice. What is the convergence quotient with which element $a_{23}^{(k)}$ converges to zero, given $\lambda_2(A) \cong 17.30$?

4-16. To find the smallest eigenvalue of

$$J = \begin{bmatrix} 2 & 1 & & \\ 1 & 4 & 1 & \\ & 1 & 4 & 1 \\ & & 1 & 2 \end{bmatrix},$$

carry out several QD steps:

(a) without coordinate translation;

(b) with a single initial coordinate translation, using the lower bound $y = 1.3125$ found in Problem 4-11;

(c) with a trial coordinate translation based on $q_4^{(s)}$, using a scheme of your own.

What is the convergence rate in the three cases?

4-17. Using the QD algorithm, find the zeros of the polynomial $p_4(x) = x^4 - 12x^3 + 49x^2 - 80x + 45$.

4-18. Utilizing the values of q_k and e_k found in Problem 2-9 for the solution of the linear system of equations of Problem 2-2, find the smallest eigenvalue of the coefficient matrix A of Problem 4-16. How good is the agreement with the value found in Problem 4-16 if the method of conjugate gradients is used with a small (say, 5 or 6) number of significant figures?

4-7 VECTOR ITERATION: LARGEST AND SMALLEST EIGENVALUES

In this section iterative methods produce a result which yields the largest or smallest eigenvalue and the corresponding eigenvector through an appropriate sequence of iterations on an arbitrary initial vector. A modification of the iteration method can produce through the simultaneous iteration of several vectors a number of smallest or largest eigenvalues with corresponding eigenvectors. The methods are useful if, in addition to the eigenvalues at the end of the spectrum, the pertinent eigenvectors are also needed, for they inevitably turn up as well. The LR transformation and the QD algorithm yield essentially just the eigenvalues, and the subsequent evaluation of eigenvectors often requires additional work before they can be established with certainty.

4-7-1 Classical Vector Iteration. The Power Method

Consider a symmetric matrix A having a *simple* eigenvalue λ_1 that is *largest* in magnitude. Then for the remaining eigenvalues,

$$|\lambda_1| > |\lambda_j| \qquad j = 2, 3, \ldots, n. \tag{4–120}$$

Starting with any arbitrary initial vector $x^{(0)}$, we can form the infinite sequence of iterated vectors

$$x^{(k+1)} = Ax^{(k)} \qquad k = 0, 1, 2, \ldots. \tag{4–121}$$

Clearly $x^{(k)} = A^k x^{(0)}$ is the product of the kth power of A with $x^{(0)}$, and the vector iteration of (4–121) is therefore often designated as the *power method.*

Theorem 4–2 says that matrix A has a complete system of n orthonormal eigenvectors $y_1, y_1, \ldots, y_n$ corresponding to real eigenvalues $\lambda_1, \lambda_2, \ldots, \lambda_n$. The initial vector $x^{(0)}$ can be expanded in eigenvectors

$$x^{(0)} = c_1 y_1 + c_2 y_2 + \cdots + c_n y_n. \tag{4–122}$$

For now assume that $c_1 \neq 0$; that is, the initial vector $x^{(0)}$ is assumed to have a component of the first eigenvector y_1 belonging to the largest eigenvalue λ_1. The remaining expansion coefficients $c_2, c_3, \ldots, c_n$ are arbitrary.

Theorem 4-22. *Given* (4–120), *if in the expansion* (4–122) *we have* $c_1 \neq 0$, *the kth iterated vector* $x^{(k)}$ *of the sequence* (4–121) *asymptotically approaches* $c_1 \lambda_1^k$ *times the first normalized eigenvector* y_1 *of* A *as* $k \longrightarrow \infty$.

Proof: Since y_j is the eigenvector of A pertaining to eigenvalue λ_j, (4–122) yields for the iterated vector $x^{(1)}$ the representation

$$x^{(1)} = c_1 \lambda_1 y_1 + c_2 \lambda_2 y_2 + \cdots + c_n \lambda_n y_n,$$

and, for the general $x^{(k)}$,

$$x^{(k)} = c_1 \lambda_1^k y_1 + c_2 \lambda_2^k y_2 + \cdots + c_n \lambda_n^k y_n. \tag{4–123}$$

If we factor out the term λ_1^k, we have

$$x^{(k)} = \lambda_1^k \left\{ c_1 y_1 + c_2 \left(\frac{\lambda_2}{\lambda_1}\right)^k y_2 + \cdots + c_n \left(\frac{\lambda_n}{\lambda_1}\right)^k y_n \right\}. \tag{4–124}$$

Because of (4–120), the quotients $(\lambda_j/\lambda_1)^k$ converge to zero with increasing k, so that in the bracketed term of (4–124) the first term becomes preponderant for k sufficiently large (recall that $c_1 \neq 0$), and the theorem is thus proven.

For k sufficiently large, two successive iterated vectors $\boldsymbol{x}^{(k)}$ and $\boldsymbol{x}^{(k+1)}$ become proportional to one another by (4–124), and the proportionality factor is the largest-magnitude eigenvalue λ_1. Thus λ_1 is given by the quotients of corresponding components in successive vectors which have been iterated often enough:

$$\lambda_1 \sim \frac{x^{(k+1)}}{x_i^{(k)}} \qquad i = 1, 2, \ldots, n. \tag{4-125}$$

The n component relations of (4–125) generally differ from the exact value λ_1 for a long time. The quotients of *Schwarz constants* (Ref. 54), however, yield better approximations for the eigenvalue λ_1. They are defined as the scalar product of two arbitrary vectors $\boldsymbol{x}^{(\mu)}$ and $\boldsymbol{x}^{(\nu)}$ of the sequence (4–121) as

$$s_{\mu+\nu} = (\boldsymbol{x}^{(\mu)}, \boldsymbol{x}^{(\nu)}) \qquad \mu, \nu = 0, 1, 2, \ldots. \tag{4–126}$$

Through the expansion (4–123) of iterated vectors $\boldsymbol{x}^{(\mu)}$ and the orthonormal character of eigenvectors $\boldsymbol{y}_j$, the Schwarz constants have the representation

$$s_{\mu+\nu} = c_1^2\lambda_1^{\mu+\nu} + c_2^2\lambda_2^{\mu+\nu} + \cdots + c_n^2\lambda_n^{\mu+\nu}. \tag{4–127}$$

The quantities $s_{\mu+\nu}$ of (4–127) are dependent only on the sum $(\mu + \nu)$ of indices of vectors $\boldsymbol{x}^{(\mu)}$ and $\boldsymbol{x}^{(\nu)}$ which are involved, and this fact can be used as a computational check. The quotient of two successive constants s_{m+1} and s_m is then

$$\frac{s_{m+1}}{s_m} = \frac{\sum_{i=1}^{n} c_i^2\lambda_i^{m+1}}{\sum_{i=1}^{n} c_i^2\lambda_i^{m}} = \lambda_1 \cdot \frac{1 + \sum_{i=2}^{n}\left(\frac{c_i}{c_1}\right)^2\left(\frac{\lambda_i}{\lambda_1}\right)^{m+1}}{1 + \sum_{i=2}^{n}\left(\frac{c_i}{c_1}\right)^2\left(\frac{\lambda_i}{\lambda_1}\right)^{m}}. \tag{4–128}$$

Noting the result of (4–120) leads from (4–128) to

$$\lim_{m\to\infty} \frac{s_{m+1}}{s_m} = \lambda_1. \tag{4–129}$$

In particular, the quotients

$$\frac{s_{2k+1}}{s_{2k}} = \frac{(\boldsymbol{x}^{(k+1)}, \boldsymbol{x}^{(k)})}{(\boldsymbol{x}^{(k)}, \boldsymbol{x}^{(k)})} = \frac{(\boldsymbol{A}\boldsymbol{x}^{(k)}, \boldsymbol{x}^{(k)})}{(\boldsymbol{x}^{(k)}, \boldsymbol{x}^{(k)})}$$

are the *Rayleigh quotients* for the kth iterated vector $\boldsymbol{x}^{(k)}$.

If there are m iterated vectors, $(2m + 1)$ Schwarz constants $s_0, s_1, \ldots, s_{2m}$ can be calculated, and thus $2m$ quotients as well. These quotients define the largest eigenvalue significantly better than the component quotients, since the inaccuracies in the individual components are smoothed by the formation of the scalar product.

Convergence of the power method is *linear*, since the components of the initial vector $\boldsymbol{x}^{(0)}$ referred to the eigenvectors $\boldsymbol{y}_2, \boldsymbol{y}_3, \ldots, \boldsymbol{y}_n$ as in (4–124) diminish as a geometric sequence with the quotients $|\lambda_j/\lambda_1| < 1$ ($j = 2, 3, \ldots, n$). The convergence rate is determined by the largest of these quotients. In order to accelerate the rate, methods have been developed which take advantage of the linear convergence and achieve a much improved approximation after a few steps of extrapolation. Worth mentioning are the δ^2-technique of *Aitken* (Ref. 1) and the ϵ-algorithm of *Wynn* (Ref. 78).

The power method demonstrates the property of the iterated vectors $\boldsymbol{x}^{(k)}$ rapidly developing very large components when $|\lambda_1| > 1$, very small components when $|\lambda_1| < 1$, and similarly sized components (having a roughly constant norm) only in the case of $|\lambda_1| = 1$. This behavior must be taken into account, either at every iteration step or at least from time to time, in some manner of normalization of the iterated vector.

If the largest eigenvalue has a multiplicity of p, the iterated vector $\boldsymbol{x}^{(k)}$ becomes progressively more proportional to an eigenvector of the p-dimensional subspace belonging to λ_1. The direction of $\boldsymbol{x}^{(k)}$ is, however, still dependent on the initial vector. If matrix $\boldsymbol{A}$ has two opposing dominant eigenvalues, the square of λ_1 can be found from $\boldsymbol{x}^{(k+2)}$ and $\boldsymbol{x}^{(k)}$, and the corresponding eigenvectors can be found from two successive iterated vectors $\boldsymbol{x}^{(k+1)}$ and $\boldsymbol{x}^{(k)}$ having high enough k (see Refs. 16, 80).

Example 4-13. Find the largest eigenvalue in magnitude, and the corresponding eigenvector, in the symmetric 4×4 matrix below.

$$\boldsymbol{A} = \begin{bmatrix} 2.0 & -0.7 & 0.3 & -0.2 \\ -0.7 & 0.5 & 0.1 & 0.4 \\ 0.3 & 0.1 & 0.3 & 0.1 \\ -0.2 & 0.4 & 0.1 & 0.2 \end{bmatrix},$$

Table 24 gives the initial vector $\boldsymbol{x}^{(0)}$ and the first six iterated vectors. Underneath are shown the corresponding component quotients. The scattering of the component ratios $x_i^{(6)}/x_i^{(5)}$ ($i = 1, 2, 3, 4$) reveals that the vector $\boldsymbol{x}^{(6)}$ is not yet parallel to the eigenvector $\boldsymbol{y}_1$. Finding the eigenvector with greater accuracy requires additional steps. Use of one of the methods mentioned for accelerating convergence would produce a better approximation vector.

The Schwarz constants and quotients which can be assembled from vectors $\boldsymbol{x}^{(0)}, \boldsymbol{x}^{(1)}, \ldots, \boldsymbol{x}^{(6)}$ are shown in Table 25. The improved determination of λ_1 is obvious, and the last quotient s_{12}/s_{11} agrees with the exact eigenvalue within two units of the last decimal place.

Table 24 POWER METHOD: ITERATED VECTORS, COMPONENT RATIOS

$k =$	0	1	2	3	4	5	6
$x^{(k)} =$	1	1.40000	2.73000	6.06300	14.03280	32.86578	77.22787
	1	0.30000	−0.55000	−2.10400	−5.47250	−13.21280	−31.30425
	1	0.80000	0.74000	0.98800	1.83610	4.00412	9.21664
	1	0.50000	0.02000	−0.68800	−2.09300	−5.23055	−12.50397
$\frac{x_1^{(k)}}{x_1^{(k-1)}}$	—	1.40000	1.95000	2.22088	2.31450	2.34207	2.34980
$\frac{x_2^{(k)}}{x_2^{(k-1)}}$	—	0.30000	−1.83333	3.82545	2.60100	2.41440	2.36924
$\frac{x_3^{(k)}}{x_3^{(k-1)}}$	—	0.80000	0.92500	1.33514	1.85840	2.18077	2.30179
$\frac{x_4^{(k)}}{x_4^{(k-1)}}$	—	0.50000	0.04000	−34.40000	3.04215	2.49907	2.39075

Table 25 SCHWARZ CONSTANTS AND QUOTIENTS

k	s_k	$q_k = s_k/s_{k-1}$
0	4.00000	—
1	3.00000	0.75000
2	2.94000	0.98000
3	4.25900	1.44864
4	8.30340	1.94961
5	18.42655	2.21916
6	42.63627	2.31385
7	99.84906	2.34188
8	234.6196	2.34974
9	551.8055	2.35192
10	1298.129	2.35252
11	3054.078	2.35268
12	7185.396	2.35272
		↓
		2.35274

4-7-2 Finding the Second Largest Eigenvalue

The largest eigenvalue λ_1 of A and the corresponding normalized eigenvector $\boldsymbol{y}_1$ have been found with sufficient accuracy by the power method. By taking advantage of the orthogonality of eigenvectors, the next largest eigenvalue λ_2 and its eigenvector $\boldsymbol{y}_2$ can be found by choosing the initial vector $\boldsymbol{x}^{(0)}$ in the $(n - 1)$-dimensional subspace orthogonal to $\boldsymbol{y}_1$. Accordingly, the expansion of $\boldsymbol{x}^{(0)}$ in eigenvectors lacks a $\boldsymbol{y}_1$ component and looks like this:

$$\boldsymbol{x}^{(0)} = c_2\boldsymbol{y}_2 + c_3\boldsymbol{y}_3 + \cdots + c_n\boldsymbol{y}_n. \tag{4–130}$$

Since the subspace is invariant under the linear transformation $\boldsymbol{A}$, theoretically, the vectors iterated by using $\boldsymbol{x}^{(k+1)} = \boldsymbol{A}\boldsymbol{x}^{(k)}$ $(k = 0, 1, 2, \ldots)$ also belong to this subspace, and the expansion of $\boldsymbol{x}^{(k)}$ is

$$\boldsymbol{x}^{(k)} = c_2\lambda_2^k\boldsymbol{y}_2 + c_3\lambda_3^k\boldsymbol{y}_3 + \cdots + c_n\lambda_n^k\boldsymbol{y}_n. \tag{4–131}$$

If in (4–130) $c_2 \neq 0$, and in addition $|\lambda_2| > |\lambda_i|$ $(i = 3, 4, \ldots, n)$, then for $k \longrightarrow \infty$, $\boldsymbol{x}^{(k)}$ approaches asymptotically $c_2\lambda_2^k$ times the second normalized eigenvector $\boldsymbol{y}_2$. For k sufficiently large, vector $\boldsymbol{x}^{(k+1)}$ is proportional to $\boldsymbol{x}^{(k)}$, and the proportionality factor is λ_2. The eigenvalue λ_2 can alternatively be found with the help of Schwarz constants and quotients, just as before.

Because of the inevitable round-off error, the iteration departs from the theory in developing an initially microscopic component of $\boldsymbol{y}_1$ which subsequently grows such that the iteration sequence $\boldsymbol{x}^{(k)}$ will again produce a vector asymptotically which is proportional to $\boldsymbol{y}_1$. In order to prevent this, every iterated vector must immediately shed its undesirable component by orthogonalization with respect to $\boldsymbol{y}_1$. We thus end up with the following modified power method for finding the next to largest eigenvalue λ_2 and its eigenvector $\boldsymbol{y}_2$: Starting with an initial vector $\boldsymbol{x}^{(0)}$ with $(\boldsymbol{x}^{(0)}, \boldsymbol{y}_1) = 0$, produce the vector sequence $z^{(k)}$ and $x^{(k)}$ according to

$$\left.\begin{aligned} \boldsymbol{z}^{(k+1)} &= \boldsymbol{A}\boldsymbol{x}^{(k)} \\ \boldsymbol{x}^{(k+1)} &= \boldsymbol{z}^{(k+1)} - (\boldsymbol{z}^{(k+1)}, \boldsymbol{y}_1)\cdot\boldsymbol{y}_1 \end{aligned}\right\} \quad k = 0, 1, 2, \ldots. \tag{4–132}$$

The generalization of the method for computation of the third largest eigenvalue λ_3 and its eigenvector $\boldsymbol{y}_3$ involves producing a vector sequence $\boldsymbol{x}^{(k)}$ in the $(n-2)$-dimensional subspace, orthogonal to $\boldsymbol{y}_1$ and $\boldsymbol{y}_2$, by orthogonalizing the iterated vector with respect to $\boldsymbol{y}_1$ and $\boldsymbol{y}_2$ analogous to (4–132). For finding several of the largest eigenvalues with their eigenvectors, see also the method of simultaneous vector iteration (Sec. 4–7–4).

4-7-3 Inverse Vector Iteration

Classical vector iteration by 4–7–1 yields in principle the dominant eigenvalue λ_1 with its eigenvector $\boldsymbol{y}_1$. In the case of a nonsingular matrix, we can use the inverse $\boldsymbol{A}^{-1}$ in place of (4–121), start with an arbitrary initial vector $\boldsymbol{x}^{(0)}$, and produce an infinite sequence of iterated vectors

$$\boldsymbol{x}^{(k+1)} = \boldsymbol{A}^{-1}\boldsymbol{x}^{(k)} \qquad k = 0, 1, 2, \ldots. \tag{4–133}$$

Matrix $\boldsymbol{A}^{-1}$ has the same eigenvectors as $\boldsymbol{A}$, but its eigenvalues are reciprocal to those of A. Assume that a symmetric matrix A has a *simple* eigenvalue λ_n which is *smallest* in magnitude, with $|\lambda_n| < |\lambda_j|$ $(j = 1, 2, \ldots, n-1)$, and that the expansion of the initial vector $x^{(0)}$ in eigenvectors $y_1, y_2, \ldots, y_n$ contains a component $c_n y_n$ $(c_n \neq 0)$. Then for $k \longrightarrow \infty$, vector $x^{(k)}$ of the sequence (4–133) asymptotically approaches $c_n y_n^{-k}$ times the normalized eigenvector y_n of eigenvalue λ_n. Approximations of λ_n are given by quotients of corresponding components of iterated vectors $x^{(k)}$ and $x^{(k+1)}$ or, even better, by quotients of successive Schwarz constants as in

$$\lambda_n \sim \frac{x_i^{(k)}}{x_i^{(k+1)}} \qquad (i = 1, 2, \ldots, n), \quad \text{or} \quad \lambda_n \sim \frac{s_m}{s_{m+1}}. \tag{4–134}$$

In practice, inverse iteration does not require the explicit derivation of A^{-1} in order to form the sequence of vectors $x^{(k)}$ by (4–133). On the contrary, $x^{(k+1)}$ can be regarded as the solution of the linear system of equations

$$Ax^{(k+1)} = x^{(k)} \qquad k = 0, 1, 2, \ldots \tag{4–135}$$

with given right-hand side $x^{(k)}$. For every k, the systems (4–135) have the same coefficient matrix A. Since A need not be positive definite, relaxation methods cannot generally be used for the solution of the systems of equations. If, however, matrix A is symmetric definite, the methodical solution of the sequence of linear equation systems is found most effectively through Cholesky's method, where the matrix A is first decomposed in order to find $x^{(k+1)}$ from $x^{(k)}$ by means of forward and backward substitution. In this case, even relaxation methods are admissible, and in some instances even recommended (see Sec. 5–3).

Inverse vector iteration can be extended in a consistent way through a modification given in Sec. 4–7–2 in order to find the second largest eigenvalue and its eigenvector, once the smallest eigenvalue and its eigenvector are known. We should also note in comparison the simultaneous vector iteration of Sec. 4–7–4, which gives the desired eigenvectors and eigenvalues simultaneously.

An extension of the concept of inverse vector iteration is found in the *inverse iteration of Wielandt* (Ref. 72). It serves to improve the approximate value $\bar{\lambda}$ of a simple arbitrary eigenvalue λ_k of a matrix A, and the corresponding eigenvector as well. Matrix $A - \lambda I$ has an eigenvalue which in magnitude is much smaller than all the others, provided that $\bar{\lambda}$ is a good approximation of λ_k, and λ_k is well removed from the other eigenvalues. Inverse vector iteration with the matrix $A - \bar{\lambda} I$ thus produces a rapidly converging vector sequence which often gives a sufficiently accurate eigenvalue and eigenvector after one or two iteration steps. Note, however, that matrix $A - \lambda I$ is almost singular, and special techniques must be employed for the solution of the systems of equations (see Refs. 67, 75, 80).

4-7-4 Simultaneous Vector Iteration

In order to remove a few difficulties, let us confine ourselves to a matrix A of order n that is symmetric and positive definite. The problem of finding the p largest eigenvalues ($p < n$) with their eigenvectors will now be attacked with the *simultaneous iteration* of p linearly independent initial vectors. The eigenvectors of differing eigenvalues form an orthogonal system, but nonetheless the sequences of iterated vectors in the power method exhibit the tendency to approach a multiple of the first eigenvector asymptotically. In order to prevent the p simultaneously iterated vectors from becoming proportional to the same vector, and to do justice to the orthogonality, at every step a system of orthogonal vectors is formed from the iterated vectors. We can visualize this in an n-dimensional space as rotating an orthogonal set of axes by means of iteration into another set which consists of the p eigenvectors of the p largest eigenvalues.

Assume that the p initial vectors $\boldsymbol{x}_1^{(0)}, \boldsymbol{x}_2^{(0)}, \ldots, \boldsymbol{x}_p^{(0)}$ have been orthonormalized and have been assembled into a tall $n \times p$ matrix $\boldsymbol{X}^{(0)}$ whose columns are vectors $\boldsymbol{x}_j^{(0)}$. The method described here can fail if, by accident or unwise selection, the p-dimensional subspace of initial vectors is orthogonal to any one of the eigenvectors of the p largest eigenvalues. The iterated vectors $\boldsymbol{z}_1^{(1)}, \boldsymbol{z}_2^{(1)}, \ldots, \boldsymbol{z}_p^{(1)}$ of $\boldsymbol{z}_j^{(1)} = \boldsymbol{A}\boldsymbol{x}_j^{(0)}$ analogously produce the matrix

$$\boldsymbol{Z}^{(1)} = \boldsymbol{A}\boldsymbol{X}^{(0)}, \tag{4–136}$$

which has maximum rank p. The columns of $\boldsymbol{Z}^{(1)}$ are orthonormalized by the *Schmidt* orthogonalization technique (see Sec. 3–4–1) prior to the next iteration. Matrix $\boldsymbol{Z}^{(1)}$ is decomposed into

$$\boldsymbol{Z}^{(1)} = \boldsymbol{X}^{(1)}\boldsymbol{R}^{(1)}, \tag{4–137}$$

where $\boldsymbol{X}^{(1)}$ indicates an $n \times p$ matrix with p orthonormalized columns, and $\boldsymbol{R}^{(1)}$ a regular upper triangular matrix. The general kth iteration step is then

$$\left.\begin{aligned} \boldsymbol{Z}^{(k+1)} &= \boldsymbol{A}\boldsymbol{X}^{(k)} \\ \boldsymbol{Z}^{(k+1)} &= \boldsymbol{X}^{(k+1)}\boldsymbol{R}^{(k+1)} \end{aligned}\right\} \quad k = 0, 1, 2, \ldots. \tag{4–138}$$

Theorem 4-23. *If the* $(p + 1)$ *largest eigenvalues* λ_j *of a symmetric definite matrix* A *are different, with a suitable choice of* $\boldsymbol{X}^{(0)}$ *the columns of* $\boldsymbol{X}^{(k)}$ *converge to the normalized eigenvectors* $\boldsymbol{y}_1, \boldsymbol{y}_2, \ldots, \boldsymbol{y}_p$ *of the* p *largest eigenvalues* $\lambda_1 > \lambda_2 > \cdots > \lambda_p$. *The sequence of upper triangular matrices* $\boldsymbol{R}^{(k)}$ *converges with increasing* k *to a diagonal matrix whose diagonal elements are the eigenvalues.*[1]

[1]The method converges even if several of the p largest eigenvalues are multiple eigenvalues.

Proof: Let $y_1, y_2, \ldots, y_n$ be the normalized eigenvectors of matrix A with corresponding eigenvalues $\lambda_1 > \lambda_2 > \cdots > \lambda_p > \lambda_{p+1} \geq \lambda_{p+2} \geq \cdots \geq \lambda_n$. The eigenvectors make up an orthogonal matrix Y, which can transform matrix A into diagonal form by

$$Y^T A Y = D, \qquad (4\text{–}139)$$

where D contains the eigenvalues in decreasing order in its diagonals. The individual iterated vectors $x_1^{(k)}, x_2^{(k)}, \ldots, x_p^{(k)}$ can be expanded in eigenvectors:

$$x_l^{(k)} = \sum_{j=1}^{n} c_{jl}^{(k)} y_j \qquad l = 1, 2, \ldots, p;\ k = 0, 1, 2, \ldots.$$

For given k, the quantities $c_{jl}^{(k)}$ form an $n \times p$ matrix $C^{(k)}$, which, in consequence of the orthonormality of the iterated vector systems $x_1^{(k)}, x_2^{(k)}, \ldots, x_p^{(k)}$, itself represents a matrix with orthonormal column vectors. With this observation in mind, matrix $X^{(k)}$ can be written

$$X^{(k)} = YC^{(k)} \qquad k = 0, 1, 2, \ldots. \qquad (4\text{–}140)$$

Next, we will find a relation between the kth expansion matrix $C^{(k)}$ and the matrix $C^{(0)}$ formed by the initial vectors, using the algorithm of (4–138). After the first partial step, by (4–140) and (4–139),

$$Z^{(k)} = AX^{(k-1)} = AYC^{(k-1)} = YDC^{(k-1)}. \qquad (4\text{–}141)$$

Furthermore, in the second partial step of (4–138),

$$Z^{(k)} = X^{(k)}R^{(k)} = YC^{(k)}R^{(k)}. \qquad (4\text{–}142)$$

From (4–141) and (4–142), and the regularity of Y and $R^{(k)}$, we arrive at the recursive relation

$$C^{(k)} = DC^{(k-1)}R^{(k)^{-1}}. \qquad (4\text{–}143)$$

Repeated application of (4–143) produces the desired relation

$$C^{(k)} = D^k C^{(0)}[R^{(k)}R^{(k-1)} \ldots R^{(2)}R^{(1)}]^{-1}. \qquad (4\text{–}144)$$

By Theorem 1–11, the inverse of the product of upper triangular matrices is again an upper triangular matrix:

$$S^{(k)} = [R^{(k)}R^{(k-1)} \ldots R^{(2)}R^{(1)}]^{-1}. \qquad (4\text{–}145)$$

We define $S^{(k)} = (s_{ij}^{(k)})$. The relation $C^{(k)} = D^k C^{(0)} S^{(k)}$ leads to a representation of the general element $c_{\mu\nu}^{(k)}$ of $C^{(k)}$,

$$c_{\mu\nu}^{(k)} = \sum_{l=1}^{\nu} \lambda_\mu^k c_{\mu l}^{(0)} s_{l\nu}^{(k)} \qquad \mu = 1, 2, \ldots, n;\ \nu = 1, 2, \ldots, p. \tag{4–146}$$

The normal property of the first column of $C^{(k)}$ signifies, with $\nu = 1$,

$$\sum_{\mu=1}^{n} (c_{\mu 1}^{(k)})^2 = \sum_{\mu=1}^{n} (\lambda_\mu^k c_{\mu 1}^{(0)} s_{11}^{(k)})^2 = (s_{11}^{(k)})^2 \left\{ \sum_{\mu=1}^{n} \lambda_\mu^{2k} (c_{\mu 1}^{(0)})^2 \right\} = 1,$$

or

$$(s_{11}^{(k)})^2 \lambda_1^{2k} \left[(c_{11}^{(0)})^2 + \sum_{\mu=2}^{n} (c_{\mu 1}^{(0)})^2 \left(\frac{\lambda_\mu}{\lambda_1} \right)^{2k} \right] = 1. \tag{4–147}$$

In case initial vector $x_1^{(0)}$ contains a component of eigenvector y_1 $(c_{11}^{(0)} \neq 0)$, then in (4–147), with k sufficiently large, the sum will become negligible compared to the first term, and we have

$$\lim_{k\to\infty} \lambda_1^k c_{11}^{(0)} s_{11}^{(k)} = 1. \tag{4–148}$$

By (4–146), $c_{11}^{(k)} = \lambda_1^k c_{11}^{(0)} s_{11}^{(k)}$, thus in the limit,

$$\lim_{k\to\infty} c_{11}^{(k)} = 1, \tag{4–149}$$

and since $\sum_{\mu=1}^{n} (c_{\mu 1}^{(k)})^2 = 1$, we have

$$\lim_{k\to\infty} c_{\mu 1}^{(k)} = 0 \qquad \mu = 2, 3, \ldots, n. \tag{4–150}$$

The first iterated vector $x_1^{(k)}$ clearly converges to the first normalized eigenvector y_1. Then, however, through the iteration $z_1^{(k+1)} = Ax_1^{(k)}$, $x_1^{(k)}$ is multiplied by λ_1, thus through the normalization of $z_1^{(k+1)}$ to $x_1^{(k+1)}$ by (4–138), it turns out that $\lim_{k\to\infty} r_{11}^{(k+1)} = \lambda_1$.

The fact that the columns of $C^{(k)}$ are orthonormal leads via the orthogonality condition to

$$\sum_{\mu=1}^{n} c_{\mu 1}^{(k)} c_{\mu\nu}^{(k)} = 0 \qquad \nu = 2, 3, \ldots, p$$

and use of (4–149) and (4–150) to

$$\lim_{k\to\infty} c_{1\nu}^{(k)} = 0 \qquad \nu = 2, 3, \ldots, p. \tag{4–151}$$

In particular, for $\nu = 2$ by (4–146),

$$\lim_{k\to\infty} c_{12}^{(k)} = \lim_{k\to\infty} \lambda_1^k[c_{11}^{(0)}s_{12}^{(k)} + c_{12}^{(0)}s_{22}^{(k)}] = 0. \tag{4–152}$$

From the normal property of the second column of $\boldsymbol{C}^{(k)}$, we have the equation

$$1 = \sum_{\mu=1}^{n} (c_{\mu 2}^{(k)})^2 = \sum_{\mu=1}^{n} \lambda_\mu^{2k}[c_{\mu 1}^{(0)}s_{12}^{(k)} + c_{\mu 2}^{(0)}s_{22}^{(k)}]^2 = \{\lambda_1^k[c_{11}^{(0)}s_{12}^{(k)} + c_{12}^{(0)}s_{22}^{(k)}]\}^2$$
$$+ \lambda_2^{2k}\left\{[c_{21}^{(0)}s_{12}^{(k)} + c_{22}^{(0)}s_{22}^{(k)}]^2 + \sum_{\mu=3}^{n} [c_{\mu 1}^{(0)}s_{12}^{(k)} + c_{\mu 2}^{(0)}s_{22}^{(k)}]^2\left(\frac{\lambda_\mu}{\lambda_2}\right)^{2k}\right\}.$$

In the limit of $k \longrightarrow \infty$, (4–152) tells us that the first term vanishes, and because of $(\lambda_\mu/\lambda_2) < 1$, the sum becomes negligible compared to the first term of the second part. Thus

$$\lim_{k\to\infty} c_{22}^{(k)} = \lim_{k\to\infty} \lambda_2^k[c_{21}^{(0)}s_{12}^{(k)} + c_{22}^{(0)}s_{22}^{(k)}] = 1, \tag{4–153}$$

and, corresponding to (4–150),

$$\lim_{k\to\infty} c_{\mu 2}^{(k)} = 0 \qquad \mu = 3, 4, \ldots, n. \tag{4–154}$$

The second vector $\boldsymbol{x}_2^{(k)}$ clearly converges to the normalized eigenvector $\boldsymbol{y}_2$. Accordingly, $\lim_{k\to\infty} r_{22}^{(k+1)} = \lambda_2$, and the element $r_{12}^{(k)}$ must approach zero, since in the limiting case $\boldsymbol{x}_1^{(k)}$ and $\boldsymbol{x}_2^{(k)}$ are transformed into multiple quantities.

These arguments can be extended to the remaining iteration vectors to complete the proof of the theorem.

Algorithm (4–138) has the drawback that the iterated vectors $z_j^{(k)}$ are orthonormalized by the *Schmidt* technique from left to right, while the first vector is merely normalized. With an unfavorable choice of initial vectors, especially in $\boldsymbol{x}_1^{(0)}$, the first eigenvector may be developed very slowly, thereby slowing the convergence of the p iterated vectors to the system of p eigenvectors which are sought.

A first improvement of this situation is possible in that after the orthonormalization process

$$\boldsymbol{Z}^{(k+1)} = \boldsymbol{X}^{(k+1)}\boldsymbol{R}^{(k+1)},$$

the vectors $\boldsymbol{x}_1^{(k+1)}, \boldsymbol{x}_2^{(k+1)}, \ldots, \boldsymbol{x}_p^{(k+1)}$ are arranged according to the magnitude of the normalization constants $r_{ii}^{(k+1)}$. Thus an iterated vector with a strong component of the first eigenvector $\boldsymbol{y}_1$ leaps into first place.

A second improvement of convergence is achieved through a type of principal axis transformation of the p iterated vectors, from which are assembled more general linear combinations in which the eigenvectors turn up more strongly and in the proper order. With this in mind, for the matrix $\boldsymbol{Z}^{(k+1)} = \boldsymbol{A}\boldsymbol{X}^{(k)}$ we form its *Gauss* transform

$$\boldsymbol{G}^{(k+1)} = \boldsymbol{Z}^{(k+1)\mathrm{T}}\boldsymbol{Z}^{(k+1)}. \tag{4–155}$$

The square matrix $\boldsymbol{G}^{(k+1)}$ is of order p, symmetric and positive definite, since, first, the $n \times p$ matrix $\boldsymbol{X}^{(k)}$, consisting of p orthonormal column vectors, has rank p; second, the matrix $\boldsymbol{A}$ being positive definite is nonsingular; and finally, $Z^{(k+1)}$ is thus also of maximum rank p. The eigenvalues of $\boldsymbol{G}^{(k+1)}$ are greater than zero. By using Jacobi's method, matrix $\boldsymbol{G}^{(k+1)}$ can be transformed to a diagonal form $\boldsymbol{D}_{k+1}$ through an orthogonal matrix $\boldsymbol{U}_{k+1}$

$$\boldsymbol{U}_{k+1}^{\mathrm{T}}\boldsymbol{G}^{(k+1)}\boldsymbol{U}_{k+1} = \boldsymbol{D}_{k+1}, \tag{4–156}$$

where care is taken that the eigenvalues in $\boldsymbol{D}_{k+1}$ appear in diminishing sequence. If necessary, this is done by a permutation. Since the diagonal elements of $\boldsymbol{D}_{k+1}$ are positive, a matrix $\boldsymbol{D}_{k+1}^{1/2}$ can be made up whose diagonal elements are the square roots of the corresponding elements of $\boldsymbol{D}_{k+1}$. Matrix

$$\boldsymbol{X}^{(k+1)} = \boldsymbol{Z}^{(k+1)}\boldsymbol{U}_{k+1}\boldsymbol{D}_{k+1}^{-1/2} \tag{4–157}$$

contains p orthonormal column vectors, since

$$\begin{aligned}\boldsymbol{X}^{(k+1)\mathrm{T}}\boldsymbol{X}^{(k+1)} &= \boldsymbol{D}_{k+1}^{-1/2}\boldsymbol{U}_{k+1}^{\mathrm{T}}\boldsymbol{Z}^{(k+1)\mathrm{T}}\boldsymbol{Z}^{(k+1)}\boldsymbol{U}_{k+1}\boldsymbol{D}_{k+1}^{-1/2}\\ &= \boldsymbol{D}_{k+1}^{-1/2}\boldsymbol{U}_{k+1}^{\mathrm{T}}\boldsymbol{G}^{(k+1)}\boldsymbol{U}_{k+1}\boldsymbol{D}_{k+1}^{-1/2} = \boldsymbol{D}_{k+1}^{-1/2}\boldsymbol{D}_{k+1}\boldsymbol{D}_{k+1}^{-1/2} = \boldsymbol{I}.\end{aligned}$$

By (4–157), each column vector of $\boldsymbol{X}^{(k+1)}$ arises from a linear combination of all the column vectors of $\boldsymbol{Z}^{(k+1)}$, in particular the first one. The orthonormalization of the columns of $\boldsymbol{X}^{(k+1)}$ by (4–157) will not be numerically fulfilled exactly, thus a subsequent orthogonalization of matrix $\boldsymbol{X}^{(k+1)}$ by the Schmidt method is recommended.

The computation time for computing $\boldsymbol{X}^{(k+1)}$ through (4–155), (4–156), and (4–157) with the recommended subsequent orthogonalization is considerable since, in addition to the matrix products, we have the diagonalization of a matrix of (only) order p. The improvement in convergence noted in practice does not justify carrying out the extra transformation in each and every iteration step. On the other hand, a combination of the two variants is a worthwhile method from the point of view of computation time and convergence. This consists of a step using (4–155) through (4–157) introduced periodically after a certain number of iteration steps (4–138).

The simultaneous vector iteration is interrupted as soon as the vector space spanned by the p vectors $\boldsymbol{x}_1^{(k+1)}, \boldsymbol{x}_2^{(k+1)}, \ldots, \boldsymbol{x}_p^{(k+1)}$ agrees with one spanned by $\boldsymbol{x}_1^{(k)}, \boldsymbol{x}_2^{(k)}, \ldots, \boldsymbol{x}_p^{(k)}$ within the limits of tolerance. This can be determined by orthogonalizing the old vectors $\boldsymbol{x}_1^{(k)}$ with respect to the new vectors $x_1^{(k+1)}, x_2^{(k+2)}, \ldots, x_p^{(k+1)}$. The maximum magnitude of the "excess" vectors established by the orthogonalization is a measure of how the p-dimensional space changes under the iteration.

The simultaneous computation of the p *smallest* eigenvalues and their eigenvectors can be done through *simultaneous inverse vector iteration.* In place of the matrix multiplication $Z^{(k+1)} = AX^{(k)}$, we have the solution of a simultaneous system of equations $AZ^{(k+1)} = X^{(k)}$ for matrix $Z^{(k+1)}$. After a Cholesky decomposition of A which must be carried out only once, the p columns of $Z^{(k+1)}$ are found through a p-multiple of forward- and backward-substitution from the corresponding columns of $X^{(k)}$.

PROBLEMS

4-19. For the matrix J of Problem 4-16, find:

(a) the largest eigenvalue and the corresponding eigenvector, employing classical vector iteration;

(b) the smallest eigenvalue and its eigenvector, first through a spectral transformation and subsequent classical vector iteration, and then by inverse vector iteration.

Compare the rates of convergence.

4-8 THE GENERALIZED SYMMETRIC EIGENVALUE PROBLEM

The generalized eigenvalue problem of finding the eigenvalues λ and eigenvectors x of $(A - \lambda B)x = 0$ with matrices A and B given and symmetric can, in principle, be reduced to a special eigenvalue problem $(C - \lambda I)x = 0$ with $C = B^{-1}A$, provided that one of the matrices, say, B, is nonsingular. The eigenvectors of the special eigenvalue problem are exactly the same as those of the generalized eigenvalue problem. Yet there is the serious drawback that matrix C is usually no longer symmetric, despite the symmetry of A, B, and B^{-1}. We have

$$C = B^{-1}A \qquad \text{and} \qquad C^{\mathrm{T}} = (B^{-1}A)^{\mathrm{T}} = A^{\mathrm{T}}(B^{-1})^{\mathrm{T}} = AB^{-1}. \qquad (4\text{–}158)$$

Matrix C is symmetric if and only if matrices A and B^{-1}, or A and B, *commute.* The relation $AB = BA$ means that $A = BAB^{-1}$ and $B^{-1}A = AB^{-1}$; thus C is symmetric by (4–158), and conversely, $C = C^{\mathrm{T}}$ implies that $AB = BA$. In the case of matrices A and B which do not commute, the symmetry is lost.

For all physical problems tackled by *energy methods,* one of the two matrices, representing the positive definite quadratic form of the kinetic energy, is not only nonsingular but actually positive definite. In this case, there are methods taking advantage of symmetry and retaining it.

4-8-1 Transformation to a Special Symmetric Eigenvalue Problem

In the generalized eigenvalue problem

$$(\boldsymbol{A} - \lambda \boldsymbol{B})\boldsymbol{x} = 0, \tag{4–159}$$

the matrices $\boldsymbol{A}$ and $\boldsymbol{B}$ are symmetric, and matrix $\boldsymbol{B}$ is also positive definite. Under these conditions matrix $\boldsymbol{B}$ can be decomposed into the product of a nonsingular upper triangular matrix and its transpose by Cholesky's method:

$$\boldsymbol{B} = \boldsymbol{R}^{\mathrm{T}}\boldsymbol{R}. \tag{4–160}$$

The decomposition of $\boldsymbol{B}$ by (4–160) is substituted into (4–159). After removal of the factor $\boldsymbol{R}$, (4–159) becomes

$$(\boldsymbol{A}\boldsymbol{R}^{-1} - \lambda \boldsymbol{R}^{\mathrm{T}})\boldsymbol{R}\boldsymbol{x} = 0. \tag{4–161}$$

After multiplication of (4–161) from the left by $\boldsymbol{R}^{\mathrm{T}-1}$, we get the special eigenvalue problem

$$(\boldsymbol{R}^{\mathrm{T}-1}\boldsymbol{A}\boldsymbol{R}^{-1} - \lambda \boldsymbol{I})(\boldsymbol{R}\boldsymbol{x}) = 0. \tag{4–162}$$

The matrix

$$\boldsymbol{C} = (\boldsymbol{R}^{-1})^{\mathrm{T}}\boldsymbol{A}\boldsymbol{R}^{-1} \tag{4–163}$$

is symmetric, since the symmetry of $\boldsymbol{A}$ assures that $\boldsymbol{C}^{\mathrm{T}} = [(\boldsymbol{R}^{-1})^{\mathrm{T}}\boldsymbol{A}\boldsymbol{R}^{-1}]^{\mathrm{T}} = (\boldsymbol{R}^{-1})^{\mathrm{T}}\boldsymbol{A}^{\mathrm{T}}\boldsymbol{R}^{-1} = (\boldsymbol{R}^{-1})^{\mathrm{T}}\boldsymbol{A}\boldsymbol{R}^{-1} = \boldsymbol{C}$. With $\boldsymbol{y} = \boldsymbol{R}\boldsymbol{x}$, the generalized eigenvalue problem (4–159) is brought down to the special $(\boldsymbol{C} - \lambda \boldsymbol{I})\boldsymbol{y} = 0$ with symmetric matrix $\boldsymbol{C}$ (4–163) for the eigenvectors $\boldsymbol{y}$. The eigenvalues λ and the eigenvectors $\boldsymbol{y}$ can be computed by one of the methods given above. The eigenvectors $\boldsymbol{x}$ of the original generalized eigenvalue problem may be found by backward substitution from the eigenvectors $\boldsymbol{y}$ out of $\boldsymbol{R}\boldsymbol{x} = \boldsymbol{y}$. Note that the eigenvectors $\boldsymbol{x}$ are not orthogonal in the usual sense of the Euclidean metric, but rather in the more general sense in relation to an inner product $(\boldsymbol{x}, \boldsymbol{B}\boldsymbol{y})$, where $\boldsymbol{B}$ signifies the symmetric definite matrix of the eigenvalue problem.

4-8-2 The Jacobi Method

The positive definite matrix $\boldsymbol{B}$ of the generalized symmetric eigenvalue problem $(\boldsymbol{A} - \lambda \boldsymbol{B})\boldsymbol{x} = 0$ is to be transformed by Jacobi's method to a diagonal form $\boldsymbol{D}$, using the orthogonal matrix $\boldsymbol{U}$:

$$\boldsymbol{D} = \boldsymbol{U}^{\mathrm{T}}\boldsymbol{B}\boldsymbol{U}. \tag{4–164}$$

The generalized eigenvalue problem is transformed with the help of matrix $\boldsymbol{U}$ in that, on one hand, the identity matrix is introduced in the form $\boldsymbol{U}\boldsymbol{U}^{\mathrm{T}}$, and on the other, the eigenvalue equation is multiplied from the left by $\boldsymbol{U}^{\mathrm{T}}$:

$$\boldsymbol{U}^{\mathrm{T}}(\boldsymbol{A} - \lambda\boldsymbol{B})\boldsymbol{U}\boldsymbol{U}^{\mathrm{T}}\boldsymbol{x} = 0.$$

By (4–164), this is the same as

$$(\boldsymbol{U}^{\mathrm{T}}\boldsymbol{A}\boldsymbol{U} - \lambda\boldsymbol{D})(\boldsymbol{U}^{\mathrm{T}}\boldsymbol{x}) = 0. \tag{4–165}$$

The generalized eigenvalue problem is hereby very nearly converted into the special one. In place of the identity matrix $\boldsymbol{I}$, Eq. (4–165) has a diagonal matrix $\boldsymbol{D}$ with only positive elements which are equal to the eigenvalues of $\boldsymbol{B}$. The diagonal matrix $\boldsymbol{D}^{1/2}$, consisting of the square roots of the positive diagonal elements of $\boldsymbol{D}$, exists, as does its inverse $\boldsymbol{D}^{-1/2}$. Thus (4–165) can be further transformed into

$$\boldsymbol{D}^{-1/2}(\boldsymbol{U}^{\mathrm{T}}\boldsymbol{A}\boldsymbol{U} - \lambda\boldsymbol{D})\boldsymbol{D}^{-1/2}\boldsymbol{D}^{1/2}(\boldsymbol{U}^{\mathrm{T}}\boldsymbol{x}) = 0$$

or

$$(\boldsymbol{D}^{-1/2}\boldsymbol{U}^{\mathrm{T}}\boldsymbol{A}\boldsymbol{U}\boldsymbol{D}^{-1/2} - \lambda\boldsymbol{I})(\boldsymbol{D}^{1/2}\boldsymbol{U}^{\mathrm{T}}\boldsymbol{x}) = 0. \tag{4–166}$$

The matrix

$$\boldsymbol{C} = \boldsymbol{D}^{-1/2}\boldsymbol{U}^{\mathrm{T}}\boldsymbol{A}\boldsymbol{U}\boldsymbol{D}^{-1/2} \tag{4–167}$$

must be symmetric since $\boldsymbol{A}$ is symmetric, $\boldsymbol{U}$ is orthogonal, and $\boldsymbol{D}^{-1/2}$ is a diagonal matrix. Through

$$\boldsymbol{y} = \boldsymbol{D}^{1/2}\boldsymbol{U}^{\mathrm{T}}\boldsymbol{x}, \tag{4–168}$$

the generalized eigenvalue problem is reduced to a special one with symmetric matrix $\boldsymbol{C}$ (4–167), or

$$(\boldsymbol{C} - \lambda\boldsymbol{I})\boldsymbol{y} = 0. \tag{4–169}$$

The eigenvalues λ of $\boldsymbol{C}$ agree with those of the generalized eigenvalue problem, and the interrelationship of the eigenvectors of $\boldsymbol{y}$ and $\boldsymbol{x}$ are given by (4–168). The special eigenvalue problem (4–169) is solved by Jacobi's method. Here we encounter the transformation matrix $\boldsymbol{V}$ containing the eigenvectors of $\boldsymbol{C}$ in its columns. Through (4–168), we arrive at matrix $\boldsymbol{X}$ of the eigenvectors of the generalized eigenvalue problem (4–159).

The algorithm for finding the eigenvalues λ_i and eigenvectors $\boldsymbol{x}_i$ of a generalized symmetric eigenvalue problem $(\boldsymbol{A} - \lambda\boldsymbol{B})\boldsymbol{x} = 0$ consists of the following steps:

$$\boxed{\begin{aligned} &\text{(a)}\ \boldsymbol{U}^{\mathrm{T}}\boldsymbol{B}\boldsymbol{U} = \boldsymbol{D} && \text{(Jacobi)} \\ &\text{(b)}\ \boldsymbol{C} = \boldsymbol{D}^{-1/2}\boldsymbol{U}^{\mathrm{T}}\boldsymbol{A}\boldsymbol{U}\boldsymbol{D}^{-1/2} && \\ &\text{(c)}\ \boldsymbol{V}^{\mathrm{T}}\boldsymbol{C}\boldsymbol{V} = \boldsymbol{D}_1 && \text{(Jacobi)} \\ &\text{(d)}\ \boldsymbol{X} = \boldsymbol{U}\boldsymbol{D}^{-1/2}\boldsymbol{V} && \end{aligned}} \tag{4–170}$$

Eigenvalues λ_i are the diagonal elements of $\boldsymbol{D}_1$ and the corresponding eigenvectors are the appropriate columns of matrix $\boldsymbol{X}$. Matrix $\boldsymbol{X}$ is itself not orthogonal. Instead we have $\boldsymbol{X}^{\mathrm{T}}\boldsymbol{B}\boldsymbol{X} = \boldsymbol{I}$, where the generalized metric is expressed.

The calculation procedure (4–170) can be so constructed that the matrices $\boldsymbol{U}$ and $\boldsymbol{V}$ never turn up explicitly. In that case, the successive rotations, which $\boldsymbol{U}$ and $\boldsymbol{V}$ produce, are carried out directly on the appropriate other matrices. In particular, this means that, during the diagonalization of $\boldsymbol{B}$, matrix $\boldsymbol{A}$ is subject to the same rotations. If eigenvectors are required, then $\boldsymbol{U}$ must be found from $\boldsymbol{I}$ recursively as the product of all the rotations. After the formulation of $\boldsymbol{C}$ and of $\boldsymbol{U}\boldsymbol{D}^{-1/2}$, the rotations needed to diagonalize $\boldsymbol{C}$ are applied directly to $\boldsymbol{U}\boldsymbol{D}^{-1/2}$, resulting in $\boldsymbol{X}$. If the eigenvectors are not needed, the problem is correspondingly simplified. Thus executed, the technique is notable for its great uniformity, since by and large only Jacobi rotations need be employed. The computation time for it is greater, however, than in the transformation of Sec. 4–8–1.

4-8-3 Method of Vector Iteration

The techniques described above reduce the generalized eigenvalue problem to a special eigenvalue problem. In contrast, vector iteration solves the generalized eigenvalue problem without transformation to a normal form. Since this method uses the matrix directly, it is especially suitable if we require only a few eigenvalues at one end of the spectrum, with their eigenvectors.

Below, we describe the technique of finding the *smallest* eigenvalue of the general eigenvalue problem $(\boldsymbol{A} - \lambda\boldsymbol{B})\boldsymbol{x} = 0$, where both matrices $\boldsymbol{A}$ and $\boldsymbol{B}$ are positive definite. The generalized eigenvalue problem can, in principle, be reduced to the special problem $(\boldsymbol{B}^{-1}\boldsymbol{A} - \lambda\boldsymbol{I})\boldsymbol{x} = 0$, where $\boldsymbol{B}^{-1}\boldsymbol{A}$ is not necessarily symmetric. Then the inverse vector iteration for $\boldsymbol{B}^{-1}\boldsymbol{A}$ produces the smallest (positive) eigenvalue as a proportionality factor of the vector sequence

$$\boldsymbol{B}^{-1}\boldsymbol{A}\boldsymbol{x}^{(x+1)} = \boldsymbol{x}^{(k)} \qquad k = 0, 1, 2, \ldots. \tag{4–171}$$

In place of (4–171), the iteration law can be written as

$$\boldsymbol{A}\boldsymbol{x}^{(k+1)} = \boldsymbol{B}\boldsymbol{x}^{(k)} \qquad k = 0, 1, 2, \ldots. \tag{4–172}$$

Each iteration step requires, by (4–172), multiplication of vector $\boldsymbol{x}^{(k)}$ by the matrix $\boldsymbol{B}$, and then the solution of the system of equations for $\boldsymbol{x}^{(k+1)}$ containing an unchanging matrix $\boldsymbol{A}$. Vector $\boldsymbol{x}^{(k)}$ of (4–172) asymptotically approaches the direction of $\boldsymbol{y}_n$, the eigenvector corresponding to the smallest eigenvalue.

Vector iteration yields the largest eigenvalue of the generalized eigenvalue problem if in (4–172) matrices $\boldsymbol{A}$ and $\boldsymbol{B}$ are interchanged.

If we want to find additional eigenvalues by either the modified power method (see Sec. 4–7–2) or simultaneous vector iteration, observe that the eigenvectors are orthogonal in the generalized sense, using the metric $(\boldsymbol{x}, \boldsymbol{By})$.

PROBLEMS

4-20. In the generalized eigenvalue problem $(\boldsymbol{A} - \lambda\boldsymbol{B})\boldsymbol{x} = 0$, $\boldsymbol{A}$ and $\boldsymbol{B}$ are symmetric, positive definite band matrices of high order n and of relatively small band width m. In reduction to the special eigenvalue problem $(\boldsymbol{C} - \lambda\boldsymbol{I})\boldsymbol{y} = 0$, the band property is lost! In the special eigenvalue problem $(\boldsymbol{C} - \lambda\boldsymbol{I})\boldsymbol{y} = 0$, use inverse vector iteration for finding the smallest eigenvalues and their eigenvectors in such a way that the band property of matrices $\boldsymbol{A}$ and $\boldsymbol{B}$ can be taken advantage of. (*Hint:* Matrix $\boldsymbol{C}$ should be used formally, while Cholesky decomposition is applied suitably to $\boldsymbol{A}$ and $\boldsymbol{B}$.)

4-9 RÉSUMÉ OF EIGENVALUE METHODS

In the following review of solution methods for the special eigenvalue problem $(\boldsymbol{A} - \lambda\boldsymbol{I})\boldsymbol{x} = 0$, it is assumed that the symmetric matrix $\boldsymbol{A}$ is given as a doubly indexed array of numbers (e.g., as ***array*** a [1 : n, 1 : n]), and that the values of its elements a_{ij} are stored individually in a computer memory.[1] We distinguish between four different cases, depending on the answer sought: whether all the eigenvalues or only a few at one end of the spectrum are to be found, and whether or not the corresponding eigenvectors are required as well. In two problem formulations we also differentiate between whether $\boldsymbol{A}$ is a full matrix or a band matrix. Since the various techniques for the solution of eigenvalue problems are not equally suitable for all types, some indications and recommendations are given for selecting and applying the methods.

1. If *all the eigenvalues and their eigenvectors* are to be found, Jacobi's method (Sec. 4–4) is the most efficient and the safest. Note especially the advantage that the method always yields a system of orthonormal vectors approximating the eigenvectors in columns of the orthogonal transformation

[1]The case where $\boldsymbol{A}$ is an operator, as in operator equations of discretized boundary value problems, is covered in Sec. 5-3.

matrix. Even in the case of a band matrix, the LR transformation (Sec. 4–6–4) would not be recommended, as it serves primarily for eigenvalue determination and proves to be unwieldy for computation of eigenvectors since all the upper triangular matrices would have to be stored (e.g., on tape) and used.

2a. If *all the eigenvalues of a full matrix* are to be found, again Jacobi's method (Sec. 4–4) recommends itself. Nevertheless, the computation time is generally lessened in using Givens' method (Sec. 4–5–1) or Householder's (4–5–2), which first transforms the matrix into a tridiagonal form, and then finds the eigenvalues either by the method of repeated interval bisection (Sec. 4–5–4) or the QD algorithm (Sec. 4–6–5).

2b. If the *matrix* is *banded*, from the standpoint of computation time it is best to reduce it to a tridiagonal matrix without destroying its form (see Refs. 47, 56) and then to apply the bisection method (4–5–4) or the QD algorithm (4–6–5).

3a. If only *a few eigenvalues at an end of the spectrum* are required, as in vibration problems, then a *full matrix* suggests the use of Givens' method (4–5–1) or Householder's (4–5–2), each of which immediately transforms the matrix to tridiagonal form. In this step no particular advantage is gained from the specific aspects of the problem. Thereafter, either the method of repeated interval bisection (4–5–4) or the QD algorithm (4–6–5) produces exactly the eigenvalues desired.

3b. For a *band matrix* where *several eigenvalues* are required at *one end of the spectrum*, the LR transformation is best (Sec. 4–6–4), yielding the desired eigenvalues with certainty and a minimum of effort.

4. If *several eigenvalues at the end of the spectrum and their eigenvectors* are required, it is best to solve the problem with the help of simultaneous vector iteration (4–7–4), since this method leads immediately to the eigenvectors and eigenvalues and constantly produces, as an approximation of the eigenvectors, a system of orthonormalized vectors. It differs significantly from the unnatural procedure whereby eigenvalues are found first by one method and eigenvectors are subsequently found by a completely different technique.

5 BOUNDARY VALUE PROBLEMS, RELAXATION

Many problems of mathematical physics lead to boundary value problems for partial differential equations of elliptic type. The *Dirichlet problem* with its foundation built on the *Laplace differential equation* $\Delta u = 0$ belongs to this class of problems. These problems have the property of being *self-adjoint*. It is thus expedient to use special numerical methods which take advantage of self-adjointness. The key lies in the fact that self-adjoint problems can be formulated as *variational problems* where a certain integral expression of the function sought is to be made an extremum. For the Dirichlet problem, that means a *Dirichlet integral.* In elasticity applications, the variational integral can be interpreted as energy, and thus we speak of an *energy method.* In practice, it is preferable to attack the variational problem directly and to forget about the boundary value problem. For the purpose of *discretization* of a continuous problem, the variational integral is approximated and the resultant quadratic function of discretized function values is made extremal. The quadratic terms produce a *quadratic positive definite form*, and through it we have access to *relaxation* methods of solution.

Below, we will be concerned with problems of *Dirichlet* and *Poisson* type. The methods can logically be extended to more general applications.

5-1 BOUNDARY VALUE PROBLEMS

5-1-1 The Energy Method

Let G be a bounded connected region of the (x, y) plane, enclosed by the boundary C, which may consist of several pieces. In this two-dimensional region G, we seek a continuous and twice piecewise continuously differentiable function $u(x, y)$ of the two independent variables x and y, such that the *variational integral*

$$J = \iint_G [\tfrac{1}{2}(\text{grad } u)^2 - \tfrac{1}{2}\rho(x, y)\cdot u^2 + f(x, y)\cdot u]\, dx\, dy + \oint_C [\tfrac{1}{2}\alpha(s)u^2 - \gamma(s)u]\, ds \tag{5–1}$$

takes on a stationary value subject to the side condition that the function $u(x, y)$ take on certain prescribed values $\phi(s)$ on a portion C_1 of the boundary, while on the remainder of the boundary C_2 it is free. Thus the *boundary condition* is

$$u = \phi(s) \text{ on } C_1. \tag{5–2}$$

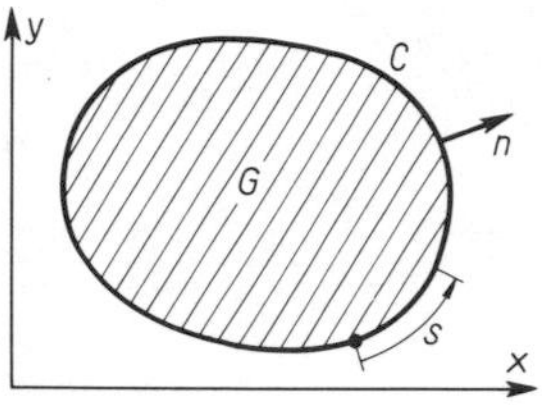

Figure 20. Domain for variational problem.

The arc length along the boundary C is denoted by s, measured from a fixed point. Functions $\rho(x, y)$ and $f(x, y)$ are given in G, and $\alpha(s)$ and $\gamma(s)$ are given functions of arc length s on the boundary C (Fig. 20).

The problem formulated in (5–1) and (5–2) is equivalent to a *boundary value problem* for a partial differential equation of *elliptic type*. A necessary condition for the integral (5–1) to take on a stationary value, according to classical methods of variational calculus, is that its first variation vanish (see Ref. 11):

$$\delta J = \iint_G [\text{grad } u\cdot\text{grad }(\delta u) - \rho(x, y)\cdot u\cdot\delta u + f(x, y)\,\delta u]\, dx\, dy + \oint_C [\alpha\cdot u\cdot\delta u - \gamma\cdot\delta u]\, ds = 0. \tag{5–3}$$

The first term of the first integral is rewritten, using *Green's formula*,

$$\iint_G \text{grad } u\cdot\text{grad } v\, dx\, dy = -\iint_G \Delta u\cdot v\, dx\, dy + \oint_C \frac{\partial u}{\partial n}\cdot v\, ds. \tag{5–4}$$

Here Δ indicates the Laplace operator and $\partial u/\partial n$ the derivative of u in the direction of the outward normal on C. Thus (5–3) becomes

$$\delta J = -\iint_G [\Delta u + \rho(x, y)\cdot u - f(x, y)]\delta u\, dx\, dy + \oint \left[\frac{\partial u}{\partial n} + \alpha\cdot u - \gamma\right] \delta u\, ds = 0. \tag{5–5}$$

Through an initial restriction to admissible functions u with variation δu vanishing over the entire boundary, there follows, according to the general variational principles, the *Euler differential equation*

$$\Delta u + \rho(x, y) \cdot u = f(x, y) \quad \text{in} \quad G. \tag{5–6}$$

The function $u(x, y)$ must necessarily satisfy (5–6) in G in order for the integral to take on a stationary value. If the values of $u = \phi(s)$ are specified by Eq. (5–2) over the entire boundary C (Dirichlet boundary conditions), then $\delta u = 0$ on C, and the second integral in (5–5) vanishes. In this case, the boundary integral in (5–1) has a fixed value and plays no role in the determination of the stationary value of J. In the other case, the variation δu can be arbitrarily selected on the portion of the boundary C_2 having no conditions on u. By a second restriction in which the class of functions are required to satisfy the Euler differential equation in G and take on the prescribed values on the boundary C_1, the vanishing of the boundary integral of (5–5) leads to the *natural boundary condition*

$$\frac{\partial u}{\partial n} + \alpha \cdot u = \gamma \quad \text{on} \quad C_2. \tag{5–7}$$

The variational problem (5–1) and (5–2) is thereby linked to the boundary value problem with the elliptic differential equation (5–6) having the boundary conditions (5–2) and (5–7). Every function $u(x, y)$, which among all twice piecewise continuously differentiable functions fulfilling boundary condition (5–2) makes integral (5–1) stationary, will assuredly fulfill differential equation (5–6) in G and satisfy condition (5–7) on the portion of the boundary C_2. Clearly the inverse is also true, in that every solution of the boundary value problem (5–6), (5–2), and (5–7) makes the integral (5–1) stationary. There is a complete equivalence between the two formulations, since the solution of one problem satisfies the other as well.

The formulation of the variational problem points out a significant distinction in the boundary conditions of the boundary value problem. The requirement that function $u(x, y)$ take on prescribed values on the boundary segment C_1 belongs inseparably to the variational problem, and it is referred to as a *constraint condition*. However, a function $u(x, y)$ making the integral (5–1) stationary under the prescribed conditions automatically fulfills the *natural boundary conditions* (5–7). They are a natural consequence of the variational problem. Therein lies an *important advantage in the formulation of a boundary value problem as a variational one: we need worry only about the constraint conditions.*

Note that the coefficient of the normal derivative in the natural boundary condition (5–7) is unity; thus this term always appears. Furthermore, no term in $\partial u/\partial s$ (derivative in the tangential direction of C) turns up in the boundary conditions. This is a consequence of the variational problem leading to a boundary value problem with the Laplacian operator.

The partial differential equation (5–6) is the normal form of an *elliptic differential equation*. Particular choices of the functions $\rho(x, y)$ and $f(x, y)$ in the variational problem (5–1) yield particular cases:

$\Delta u = 0$ *Laplace Equation* $(\rho(x, y) = f(x, y) = 0)$ (5–8)

$\Delta u = f(x, y)$ *Poisson Equation* $(\rho(x, y) = 0)$ (5–9)

$\Delta u + \rho(x, y) \cdot u = 0$ *Vibration Equation* $(f(x, y) = 0)$ (5–10)

In like manner, the natural boundary conditions (5–7) are dependent on the functions $\alpha(s)$ and $\gamma(s)$. If $\alpha(s)$ vanishes, we have the natural boundary condition

$$\frac{\partial u}{\partial n} = \gamma(s) \quad \text{on} \quad C_2. \tag{5–11}$$

If the boundary integral does not turn up in (5–1) at all, then (5–7) reduces to

$$\frac{\partial u}{\partial n} = 0 \quad \text{on} \quad C_2. \tag{5–12}$$

For every boundary value problem with a partial differential equation (5–6) and corresponding constraints and natural boundary conditions of type (5–2) and (5–7), the pertinent variational integral (5–1) can be constructed. The functions therein are defined through the differential equation and/or the boundary conditions.

5-1-2 Self-Adjointness

The concept of self-adjointness is inseparably linked with boundary value problems stemming from a variational problem. The differential equation (5–6) in its special form is clearly not the most general form of a partial differential equation of second order.

Consider now the more general boundary value problem

$$L(u) = f(x, y) \quad \text{in} \quad G, \tag{5–13}$$

where $L(u)$ indicates an arbitrary linear *differential operator*. Boundary conditions are given by

$$au + b\frac{\partial u}{\partial n} + c\frac{\partial u}{\partial s} = d \quad \text{on} \quad C, \tag{5–14}$$

which can be either constraints of the problem or natural boundary conditions.

Definition 5–1

A boundary value problem $L(u) = f(x, y)$ *in a domain* G *with boundary conditions* (5–14) *on the boundary* C *is called self-adjoint if for two arbitrary, sufficiently continuously differentiable functions* $u(x, y)$ *and* $v(x, y)$ *fulfilling the homogeneous boundary conditions*

$$au + b\frac{\partial u}{\partial n} + c\frac{\partial u}{\partial s} = 0 \quad \text{and} \quad av + b\frac{\partial v}{\partial n} + c\frac{\partial v}{\partial s} = 0 \qquad (5\text{–}15)$$

appropriate to (5–14) *on the boundary* C, *it always follows that*

$$\iint_G [v \cdot L(u) - u \cdot L(v)]\, \mathrm{d}x\, \mathrm{d}y = 0. \qquad (5\text{–}16)$$

The self-adjointness of the boundary value problem (5–6) with boundary conditions (5–2) and (5–7) ensues as follows: For $L(u) = \Delta u + \rho \cdot u$, after simplification and rearranging by using *Green's* formula (5–4), the integral expression (5–16) becomes

$$\begin{aligned} &\iint_G [v(\Delta u + \rho \cdot u) - u(\Delta v + \rho \cdot v)]\, \mathrm{d}x\, \mathrm{d}y \\ &= \iint_G [v \cdot \Delta u - u \cdot \Delta v]\, \mathrm{d}x\, \mathrm{d}y = \oint_C \left[\frac{\partial u}{\partial n} v - \frac{\partial v}{\partial n} u\right] \mathrm{d}s. \end{aligned} \qquad (5\text{–}17)$$

It remains to be shown that the boundary integral (5–17) vanishes for two arbitrary functions u and v which satisfy the homogeneous boundary conditions

$$u = 0 \quad \text{on} \quad C_1, \quad \frac{\partial u}{\partial n} + \alpha \cdot u = 0 \quad \text{on} \quad C_2. \qquad (5\text{–}18)$$

For the portion of the boundary C_1 this is trivial, since both u and v vanish there. For the portion C_2, the vanishing of the appropriate line integral follows from the fact that after substitution of the homogeneous boundary conditions, the integrand vanishes identically:

$$\int_{C_2} \left[\frac{\partial u}{\partial n} v - \frac{\partial v}{\partial n} u\right] \mathrm{d}s = \int_{C_2} [-\alpha \cdot uv + \alpha vu]\, \mathrm{d}s = 0.$$

Many problems of mathematical physics can be formulated as variational problems, with possible constraint conditions (utilizing *Hamilton's*, *Rayleigh's*, or *Fermat's* principles). In order to achieve a numerical solution, it is then unnecessary to transfer them into a boundary value problem with a partial differential equation having, in addition, natural boundary conditions. Besides the hardly insignificant advantage that only constraint conditions need be examined, we also have the certainty that after discretization, the ensuing equations will be symmetric. The self-adjointness of a boundary value problem coming from a variational problem corresponds to the symmetry of a real matrix operator.

If, however, the problem is given in the form of a partial differential equation with boundary conditions, in the case of a self-adjoint boundary value problem we can always try to determine the matching variational problem. For this, we would have to conduct the calculations in reverse in the example of Sec. 5–1–1, and use Green's formula or, in more general cases, integration by parts on a plane, applied to the sought function u and the variation δu (Ref. 17).

5-1-3 Discretization

Discretization of the problem is usually achieved by laying a regular square *grid* or *network* over the domain G and regarding as unknowns merely the values u_i of the function $u(x, y)$ at the nodes of the grid. The variational integral (5–1) will be approximated in the following example by sums taken over the unknowns u_i, constraint conditions (if any) will be approximated by expressions in the unknowns which may yield side conditions for admissible u_i's.

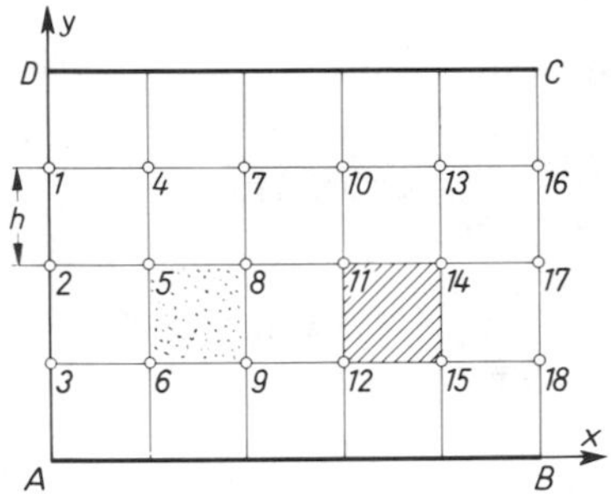

Figure 21. Grid region in variational problem.

Example 5-1. Consider the domain G shown in Fig. 21, forming a rectangle in the (x, y) plane, with a ratio of side lengths 5: 4. A function $u(x, y)$ is to be found in G such that for a given function $f(x, y)$, the integral

$$J = \iint_G [\tfrac{1}{2}(\text{grad } u)^2 + f(x, y)\cdot u]\, dx\, dy + \tfrac{1}{2}\int_B^C u^2\, dy \tag{5–19}$$

is stationary, under the constraint conditions

$$u = 0 \quad \text{on} \quad AB \quad \text{and} \quad CD. \tag{5–20}$$

This variational problem corresponds to a boundary value problem with a

Poisson differential equation

$$\left.\begin{aligned} &\Delta u = f(x, y) \quad \text{in} \quad G \\ &\text{having boundary conditions} \\ &u = 0 \quad \text{on} \quad AB \quad \text{and} \quad CD \\ &\frac{\partial u}{\partial n} + u = 0 \quad \text{on} \quad BC, \qquad \frac{\partial u}{\partial n} = 0 \quad \text{on} \quad DA. \end{aligned}\right\} \tag{5–21}$$

The given problem has the physical interpretation of determining the equilibrium position of an elastic *membrane* or diaphragm covering the rectangular region G when unloaded, having both horizontal boundaries fixed, the left vertical one free and the right vertical one elastically constrained. A specified continuous loading $f(x, y)$ normal to the plane is applied to the membrane. The sought function $u(x, y)$ represents the deflection of the membrane under this load. The physical interpretation of the variational integral (5–19) is the sum of the linearized deformation energy in the membrane, in the work done during loading, and in the elastic loading at the right-hand boundary. By the *extremum principle of mechanics* (Ref. 11), the actual deflection $u(x, y)$ of the membrane minimizes the expression J.

A square grid with mesh width h is laid over domain G (see Fig. 21). There are 18 nodes within the region and on the vertical boundaries which are numbered by columns, as in Fig. 21. At these points, the function values $u_1, u_2, \ldots, u_{18}$ of the desired function $u(x, y)$ are to be found. The function values in the grid points of the horizontal boundaries are specified by the constraints and thus do not turn up as unknowns.

By the *direct method*, the integral (5–19) is approximated by the unknowns $u_1, u_2, \ldots, u_{18}$. After this discretization, a minimum can be sought. We begin with the approximation of the first term of the integral sum

$$J_1 = \tfrac{1}{2} \iint_G (\text{grad}\, u)^2 \, \mathrm{d}x \, \mathrm{d}y = \tfrac{1}{2} \iint_G (u_x^2 + u_y^2) \, \mathrm{d}x \, \mathrm{d}y,$$

and compute the approximate contribution of the cross-hatched cell in Fig. 21. At the mid-point of the upper horizontal edge, rules of numerical differentiation dictate the approximation

$$u_x \sim \frac{u_{14} - u_{11}}{h}, \tag{5–22}$$

and, similarly, at the midpoint of the lower horizontal edge

$$u_x \sim \frac{u_{15} - u_{12}}{h}. \tag{5–23}$$

As an approximation for the value of u_x^2 at the center of the cell, we take the arithmetic mean of the squares of the two quantities (5–22) and (5–23):

$$u_x^2 \sim \frac{1}{2h^2}[(u_{14} - u_{11})^2 + (u_{15} - u_{12})^2]. \tag{5–24}$$

For differentiation in the y direction, we have the analogous result for the value of u_y^2 at the center,

$$u_y^2 \sim \frac{1}{2h^2}[(u_{14} - u_{15})^2 + (u_{11} - u_{12})^2]. \tag{5–25}$$

The cross-hatched cell with area h^2 contributes to the integral J_1 the quantity

$$J_1 \approx \tfrac{1}{4}[(u_{14} - u_{11})^2 + (u_{15} - u_{12})^2 + (u_{14} - u_{15})^2 + (u_{11} - u_{12})^2]. \tag{5–26}$$

As the contribution of the same cell to the integral component

$$J_2 = \iint_G f(x, y) \cdot u \, dx \, dy, \tag{5–27}$$

we use the approximate value

$$J_2 \approx \frac{h^2}{4}[f_{11} \cdot u_{11} + f_{12} \cdot u_{12} + f_{14} \cdot u_{14} + f_{15} \cdot u_{15}], \tag{5–28}$$

where f_{11} signifies the value of the given function $f(x, y)$ at point 11. This expression, which is essentially the arithmetic mean of the values of the integrand at the four corners of the cell, can be regarded as a sort of two-dimensional trapezoid rule. Consistent with the coarse approximation by rectangles, the boundary integral

$$J_3 = \tfrac{1}{2} \oint_B^C u^2 \, dy \tag{5–29}$$

is represented by

$$J_3 \approx \frac{h}{2}[u_{16}^2 + u_{17}^2 + u_{18}^2]. \tag{5–30}$$

The integral J is approximately equal to the sum of all components (5–26) and (5–28) of the individual cells, plus the expression (5–30). The discretized expression for $J = J_1 + J_2 + J_3$ is a quadratic function in the variables $u_1, u_2, \ldots, u_{18}$. It is assembled from the *quadratic form* $J_1 + J_3$ and the linear function J_2. Since it is a sum of squares, the quadratic form $J_1 + J_2$ is positive, vanishing only when all variables $u_1, u_2, \ldots, u_{18}$ vanish identically. The quadratic part, then, is a *positive definite quadratic form*, and the linear system of equations resulting from the minimization is *symmetric definite*. This is an important point in the energy method. If, for instance, we were to discretize a boundary value problem, it is not at all certain beforehand that the resultant system of equations would be symmetric; that is, such as to retain the self-adjointness. Only an especially careful procedure will retain symmetry in this case.

The linear system of equations results from the minimum requirement in the integral J by differentiation of its discretized expression with respect to individual variables $u_1, u_2, \ldots, u_{18}$ and by setting equal to zero the individual partial derivatives. In the above example, distinction must be made among three cases, depending on whether a point is in the interior of the rectangle or on either of the vertical boundaries. Let us start arbitrarily with the interior point 11. Variable u_{11} influences only the cells having a corner at point 11. In differentiation with respect to u_{11}, therefore, only the four surrounding cells need be considered. The cross-hatched cell shown in Fig. 21 has J_1 and J_2 contributions, by (5–26) and (5–28), of

$$\frac{1}{2}(-u_{14} + 2u_{11} - u_{12}) + \frac{h^2}{4}f_{11}. \tag{5–31}$$

Analogously, the three other squares have contributions

$$\frac{1}{2}(-u_{10} + 2u_{11} - u_{14}) + \frac{h^2}{4}f_{11}, \quad \frac{1}{2}(-u_8 + 2u_{11} - u_{10}) + \frac{h^2}{4}f_{11},$$
$$\frac{1}{2}(-u_8 + 2u_{11} - u_{12}) + \frac{h^2}{4}f_{11}. \tag{5–32}$$

After addition of these four partial results, we have for interior point 11 the *discretized equation*

$$4u_{11} - u_{14} - u_{10} - u_8 - u_{12} + h^2 f_{11} = 0. \tag{5–33}$$

An equation similar to (5–33) holds for each of the interior points of the domain, as shown symbolically in Fig. 22, and it is called an *operator equation*. The cross shown is called the *operator* of the problem. At its points, the coefficients of the corresponding linear equation are entered. The operator equation is to be interpreted such that the function values u_i at the individual nodes are to be multiplied by the corresponding weighting factors and then added, and the function value f is to be taken at the *center point* of the operator.

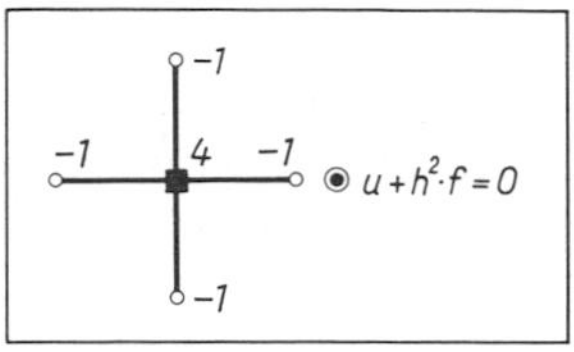

Figure 22. Operator equation for an interior point.

Obviously, in an interior point of the upper row, as at point 7, the value of u to be used in the upper arm of the operator is determined by the boundary condition. Since the boundary value in this case must be zero, the upper arm drops out of the operator equation; the case of the lower row is analogous. The operator equation (5–33) could also have been derived by replacing the two partial derivatives with the corresponding second differences. Up to this point, the energy method provides nothing new. But the boundary points are a different story. Consider point 2 of the left boundary. Here only two squares of the grid meet, producing in differentiation of J with respect to u_2 the contributions analogous to (5–32):

$$\frac{1}{2}(-u_5 + 2u_2 - u_1) + \frac{h^2}{4}f_2, \quad \frac{1}{2}(-u_3 + 2u_2 - u_5) + \frac{h_2}{4}f_2. \qquad (5\text{–}34)$$

The unknown u_2 does not turn up elsewhere; thus the sum of these quantities must vanish. Point 2 thus has the equation

$$2u_2 - u_5 - \frac{1}{2}u_1 - \frac{1}{2}u_3 + \frac{h^2}{2}f_2 = 0, \qquad (5\text{–}35)$$

which is depicted in operator form in Fig. 23. This operator equation is applicable for the three boundary points of the left vertical. For points 1 and 3 we invoke the modifications discussed above for boundary conditions. Note that the value of the partial derivative of J with respect to u_2 was still set equal to zero. The tempting multiplication of (5–35), or of the operator form of Fig. 23, by the factor 2 must be suppressed in order to retain the symmetry of the resultant system of equations.

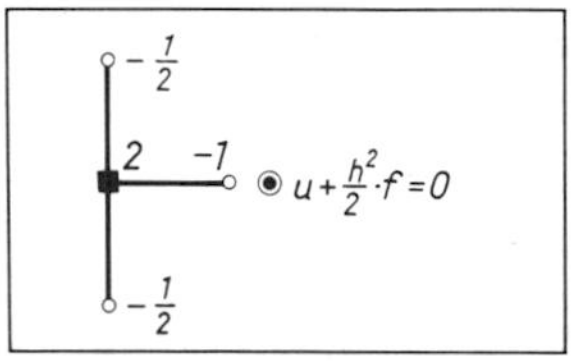

Figure 23. Operator equation for a point at left boundary.

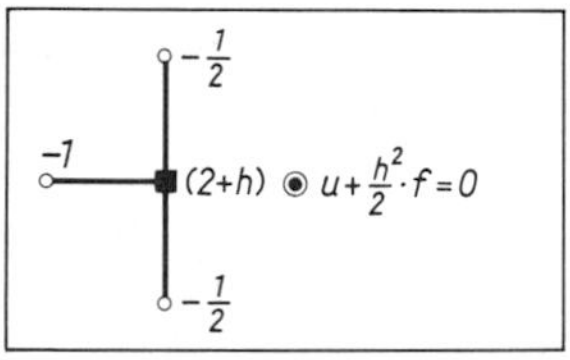

Figure 24. Operator equation for a point at right boundary.

For a point on the right-hand boundary, the calculation is entirely analogous. Let us examine point 17. The two grid cells meeting there produce, after differentiation with respect to u_{17}, the contributions

$$\frac{1}{2}(-u_{14} + 2u_{17} - u_{16}) + \frac{h^2}{4}f_{17}, \quad \frac{1}{2}(-u_{14} + 2u_{17} - u_{18}) + \frac{h^2}{4}f_{17}, \qquad (5\text{–}36)$$

while the approximation of the integral J_3 on the right boundary adds on the term hu_{17} by (5–30). Thus for point 17 we have the equation

$$2u_{17} - u_{14} - \frac{1}{2}u_{16} - \frac{1}{2}u_{18} + \frac{h^2}{2}f_{17} + hu_{17} = 0, \qquad (5\text{–}37)$$

which is depicted in Fig. 24 in operator form. It applies to the three points on the right-hand boundary, with appropriate modifications for points 16 and 18. Here, too, a multiplication by the factor 2 must be suppressed on the grounds of maintaining symmetry of the resultant system of equations. In contrast to the operator equation (5–35), the central point 17 exhibits a different weighting factor stemming from the boundary integral. It manifests the influence of the *natural boundary conditions* of the boundary value problem. The natural boundary conditions which were nowhere considered in the derivation are implicitly contained in the operator equations of Figs. 23 and 24.

Up to now only the simplest approximations of the partial derivatives using differences have been employed to approximate the variational integral. The discretization is thus very coarse, and the local discretization error is large. Better techniques exist for approximating partial differential equations by difference equations. The *Hermitian finite-difference operators** (Ref. 10) are especially applicable. In the same fashion the energy method can give better approximations by operator equations in that more exact numerical integration formulas are used. This is carried out concretely in Ref. 13, and the energy method is also applied to a simple elasticity problem.

If domain G is not rectangular but is of a more general form, following straight lines of the network, the energy method experiences no significant modification; but with an arbitrary domain G having a curved boundary, the case is quite different. As before, a regular square grid with mesh width h is imposed on the domain (see Fig. 25). One coarse method involves replacing the given domain G by another, G', consisting entirely of grid squares lying completely within G. The constraints on C are piecewise approximated by linear interpolation along grid lines. Say that the boundary value $u(R_1)$ at point R_1 on boundary C is to be satisfied. Linear interpolation along grid lines of R_1 to points 1 and 2 yields

$$u_1 = \frac{a \cdot u_2 + h \cdot u(R_1)}{a + h}.$$

Here a is the distance from point 1 to point R_1, and $u(R_1)$ is a given quantity. On the strength of this equation, the variable u_1 is eliminated from the

*In German, "Mehrstellenoperatoren." The translation given in this text is consistent with the English edition of Ref. 10. (Translator's note.)

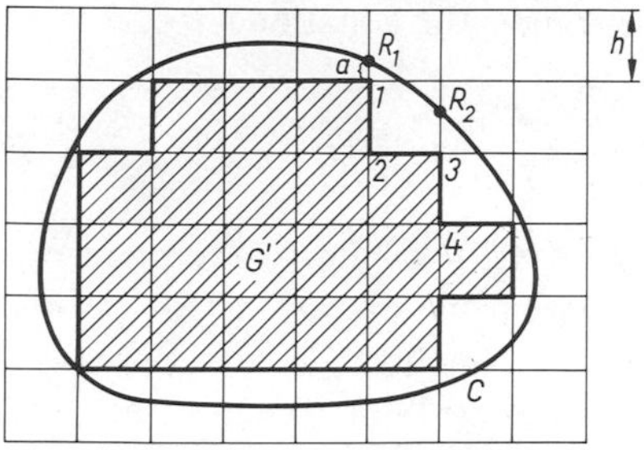

Figure 25. Curved boundary.

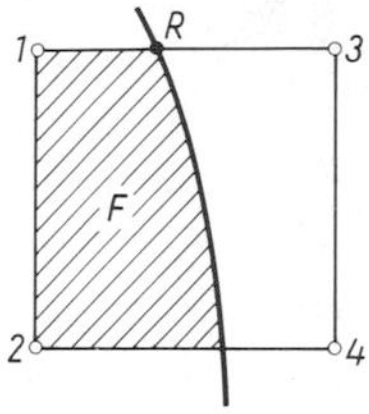

Figure 26. Cell on a curved boundary.

energy integrals and thus from the whole problem. Therefore, point 1 is called an *elimination point*, and point 2 the corresponding *auxiliary point*. If other boundary points R_i are chosen at which the given boundary conditions are to be fulfilled, an auxiliary point may never become an elimination point.

A better approximation of the variational integral results if, in addition to the cells of G', all cells cut by G are also included. Figure 26 shows a typical case of an intersected cell with corner points 1, 2, 3, and 4. Even though points 3 and 4 do not belong to G, the corresponding function values u_3 and u_4 nevertheless turn up as variables. Let F be the area of the part of the cell within the domain G. The diminished contribution of the mesh to the energy integral is most easily represented by the expression

$$J_1 \approx \frac{F}{4h^2}[(u_1 - u_2)^2 + (u_2 - u_4)^2 + (u_3 - u_4)^2 + (u_1 - u_3)^2].$$

Constraint equations are adapted by linear interpolation and subsequent elimination. For instance, in Fig. 26 the elimination point 3 with the auxiliary point 1 can be linked to the boundary point R such that u_3 vanishes in the energy expression. If there are no constraint equations, the function values u_3 and u_4 are retained as variables. Boundary integrals are approximated with interpolated values used on the boundary curve. Recent efforts have been made to automate this time-consuming preparatory step (see Ref. 14).

5-1-4 Structure of the Linear Equations

The totality of the 18 operator equations of Example 5–1 forms a symmetric definite linear system of equations in the unknowns $u_1, u_2, \ldots, u_{18}$. The system of equations exhibits a special structure as a result of its origin. The operator equations link, at most, five of the unknowns u_i, so that the coefficient matrix of the linear system of equations is only sparsely occupied by nonzero elements.

Suppose that it is required that function $f(x, y)$ in the speckled square of Fig. 21 have the constant value 4, and a value zero in the rest of the domain G. Then the square in question would make a contribution to the integral J_2 by

$$J_2 \approx \frac{h^2}{4}[f_5 u_5 + f_6 u_6 + f_8 u_8 + f_9 u_9] = h^2(u_5 + u_6 + u_8 + u_9).$$

This must be considered in differentiation of J with respect to u. If the dimensionless mesh width is set $h = 1$, the system acquires the form (5–38) because of the numbering of Fig. 21. Only coefficients are shown, corresponding to the unknown at the top of each column:

u_1	u_2	u_3	u_4	u_5	u_6	u_7	u_8	u_9	u_{10}	u_{11}	u_{12}	u_{13}	u_{14}	u_{15}	u_{16}	u_{17}	u_{18}	1
2	$-\frac{1}{2}$	0	−1															0
$-\frac{1}{2}$	2	$-\frac{1}{2}$	0	−1														0
0	$-\frac{1}{2}$	2	0	0	−1													0
−1	0	0	4	−1	0	−1												0
	−1	0	−1	4	−1	0	−1											1
		−1	0	−1	4	0	0	−1										1
			−1	0	0	4	−1	0	−1									0
				−1	0	−1	4	−1	0	−1								1
					−1	0	−1	4	0	0	−1							1
						−1	0	0	4	−1	0	−1						0
							−1	0	−1	4	−1	0	−1					0
								−1	0	−1	4	0	0	−1				0
									−1	0	0	4	−1	0	−1			0
										−1	0	−1	4	−1	0	−1		0
											−1	0	−1	4	0	0	−1	0
												−1	0	0	3	$-\frac{1}{2}$	0	0
													−1	0	$-\frac{1}{2}$	3	$-\frac{1}{2}$	0
														−1	0	$-\frac{1}{2}$	3	0

(5–38)

The obvious characteristic is the *band form* of the coefficient matrix. The band width $m = 3$ is a consequence of the numbering scheme by columns of the grid points. Numbering by rows of grid points would produce an analogous system of equations with band width $m = 6$, since then the largest difference in the numberings of points linked by one operator equation would be six. The band width is dependent on the numbering of grid points.

The band property, together with the symmetry and the definiteness of the system, permit its solution by Cholesky's method (Sec. 1–4–5). With a relatively coarse grid and a correspondingly small number of unknowns, this way is appropriate. With a significantly finer grid than in Fig. 21, the operator equations of Figs. 22 to 24 remain valid, but the number of grid points grows rapidly, and also the band width increases. With half as large a mesh width (see Fig. 27), there are 77 grid points with a similar number of

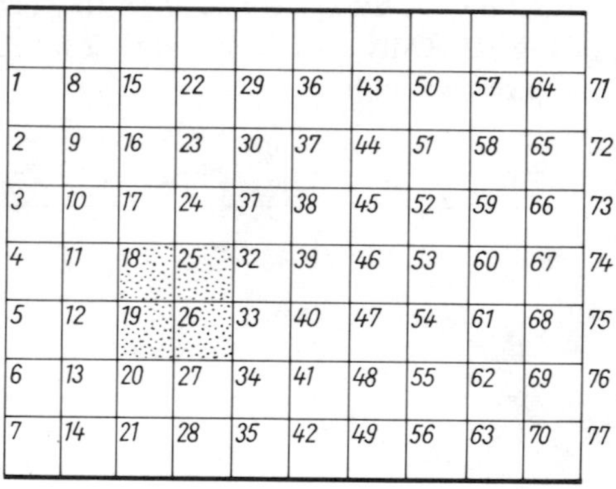

Figure 27. Finer subdivision of the network.

function quantities u_i. With numbering by columns, the band width of the system is $m = 7$, which is the smallest possible band width for this case. Nonetheless, the ratio of nonzero elements to the total number of matrix elements has diminished. In addition, the many zero elements within the band cannot be exploited in Cholesky's method, as they are disturbed in the process of reduction. For the reasons given, iterative relaxation methods are usually used in place of direct methods for the solution of operator equations with a larger number of grid points. The system of equations was given explicitly just to study its structure.

In addition, matrix (5–38) decomposes into three-rowed submatrices, as shown in (5–39),

$$A = \begin{bmatrix} \boldsymbol{B}_1 & \boldsymbol{D}_1 & & & & \\ \boldsymbol{D}_1 & \boldsymbol{B}_2 & \boldsymbol{D}_1 & & & \\ & \boldsymbol{D}_1 & \boldsymbol{B}_2 & \boldsymbol{D}_1 & & \\ & & \boldsymbol{D}_1 & \boldsymbol{B}_2 & \boldsymbol{D}_1 & \\ & & & \boldsymbol{D}_1 & \boldsymbol{B}_2 & \boldsymbol{D}_1 \\ & & & & \boldsymbol{D}_1 & \boldsymbol{B}_3 \end{bmatrix}, \qquad (5\text{–}39)$$

where $\boldsymbol{B}_1$, $\boldsymbol{B}_2$, $\boldsymbol{B}_3$, and $\boldsymbol{D}_1$ are defined as the three-rowed matrices

$$\boldsymbol{B}_1 = \begin{bmatrix} 2 & -\frac{1}{2} & 0 \\ -\frac{1}{2} & 2 & -\frac{1}{2} \\ 0 & -\frac{1}{2} & 2 \end{bmatrix}, \quad \boldsymbol{B}_2 = \begin{bmatrix} 4 & -1 & 0 \\ -1 & 4 & -1 \\ 0 & -1 & 4 \end{bmatrix},$$
$$\boldsymbol{B}_3 = \begin{bmatrix} 3 & -\frac{1}{2} & 0 \\ -\frac{1}{2} & 3 & -\frac{1}{2} \\ 0 & -\frac{1}{2} & 3 \end{bmatrix}, \quad \boldsymbol{D}_1 = \begin{bmatrix} -1 & 0 & 0 \\ 0 & -1 & 0 \\ 0 & 0 & -1 \end{bmatrix}, \tag{5–40}$$

and where $\boldsymbol{B}_1$, $\boldsymbol{B}_2$, and $\boldsymbol{B}_3$ are tridiagonal, and $\boldsymbol{D}_1$ is actually a diagonal matrix.

The inner regularity of the system of equations now is manifested even more clearly. Several of the equations exhibit a repetitive structure because of the similar structure of operator equations and the regular numbering of the grid points. In the example given, nine different types of operator equations appear. With 18 operator equations, this is hardly exciting. With a finer subdivision of the grid as in Fig. 27, the structure of the system of equations remains as in (5–39), only the submatrices are of order seven, and nine matrices having a structure like $\boldsymbol{B}_2$ appear. Since, as before, only nine differently constructed operator equations turn up, it is inefficient to store the system by coefficients. Because of the extremely simple constuction of the operator equations, it is more advantageous to formulate them as rules of arithmetic computation applicable to the different groups of nodes.

PROBLEMS

5-1. Formulate the boundary value problem corresponding to the variational integral

$$J = \tfrac{1}{2} \iint_G (\text{grad } u)^2 \, \mathrm{d}x \, \mathrm{d}y - \int_C^D u \cdot \mathrm{d}s$$

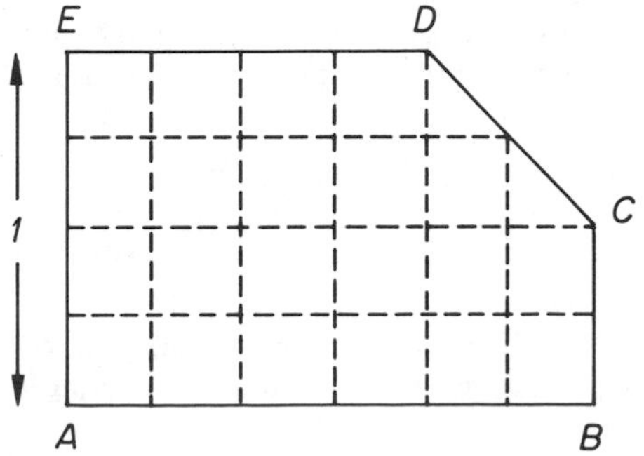

with the boundary conditions $u = 0$ on BC, $u = 1$ on DE and EA, and where G indicates the domain shown in the figure. Side AE has the length unity.

5-2. Construct the variational integral for the boundary value problem of the domain shown in the figure, with $\Delta u + 2 = 0$ in G, $u = 0$ on AB, CD, and DA, and $\partial u/\partial n = 0$ on BC.

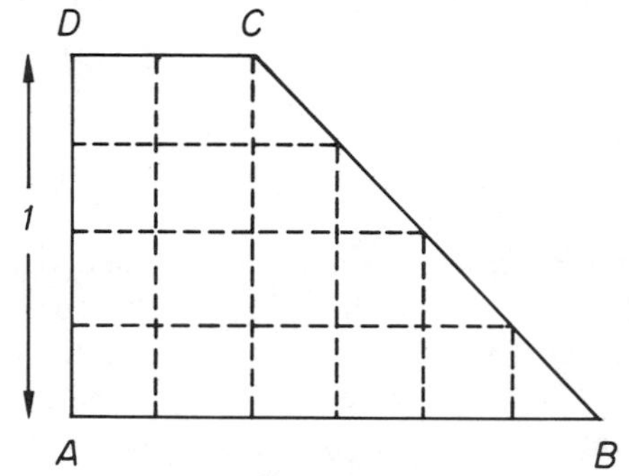

5-3. Find the variational integral for the boundary value problem involving the domain shown in the figure, with $\Delta u = 1$ in G, $u = 1$ on AB, $\partial u/\partial n = 0$ on BC, $u = 0$ on CD, and $2u + (\partial u/\partial n) = 1$ on DA.

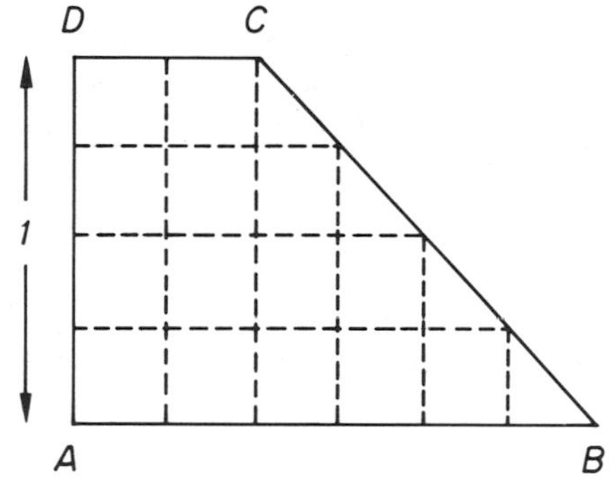

5-4. For boundary value Problems 5-1 to 5-3, derive the operator equations for the various nodes of the network shown, based on the variational integral (energy method). The integrals are to be approximated in the simplest possible fashion (trapezoid rule). For the approximation of an integral

$$\iint (u_x^2 + u_y^2)\,dx\,dy$$

for a triangular cell, both u_x and u_y are to be approximated by a single difference quotient.

5-5. For Problems 5-1 to 5-3, using the operator equations of Problem 5-4, set up the systems of equations by using various numberings of nodes (rowwise;

columnwise; diagonally), and examine the structure of the systems. Are the systems symmetric definite?

5-2 OPERATOR EQUATIONS AND RELAXATION

In this section appropriate *iterative treatment* of operator equations is taken up. Besides the general relaxation methods of Chapter 2, the unique structure of operator equations permits additional avenues of approach.

5-2-1 Elementary Relaxation Methods

In the early stages of relaxation computations, the structure of operator equations was accommodated through the use of a worksheet as in Fig. 28, where the domain and the imposed grid with its nodes are depicted. As the first step, underneath and to the left of each node, a plausible approximation for u_i is entered. The corresponding residuals are computed, using the operator equations, and are written underneath and to the right of each

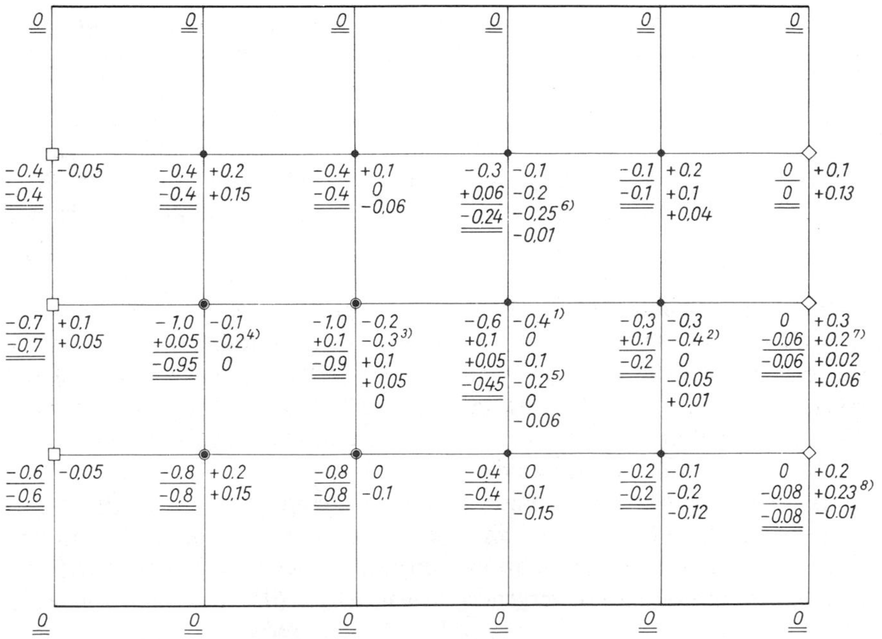

Figure 28. Worksheet for hand relaxation.
● Operator equation as in Fig. 22.
□ Operator equation as in Fig. 23.
◇ Operator equation as in Fig. 24.
⊙ Points where f is set equal to one, consistent with the contribution to J_2 as given in Sec. 5–1–4.

point. By the *hand relaxation* method of Sec. 2–2–1, the largest residual in absolute value is made to vanish through a correction to the function value. In addition, the residual must be divided by the negative of the weight assigned to the central point in the pertinent operator equation. The *correction*, which is rounded off to one significant figure for convenience, is written underneath the function value. The residual is thus made to vanish, if only in the predominant decimal places. The diminishing of the residual in the course of computations causes a reduction in the corrections and thus an increasing accuracy in the approximation values. A single correction influences only the neighboring points whose operator equations encompass the corrected function value. In the case of the *five-point operator* (Fig. 22), this means, in general, the four adjacent points. In order to help depict the computation, *residual operators* are defined for the operator equations with weights assigned, showing how the residuals of the points in the "star" are changed with a unit correction of the function value at the central point.

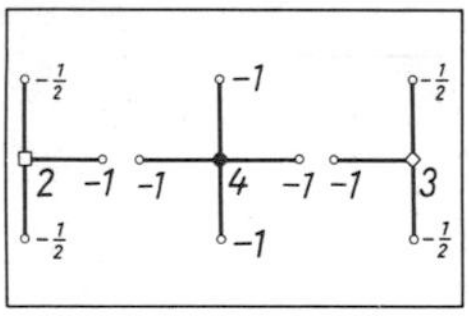

Figure 29. Residual operators for hand relaxation ($h=1$).

For Example 5–1, the residual operators are shown in Fig. 29. They are constructed analogous to the operator equations; only the terms with f have dropped out. In the worksheet of Fig. 28, the first eight steps have been carried out, where not only the corrections to the function values but also the actual residual have been entered. To assist in tracing the sequence, the residuals are numbered in the order in which they were attacked. The total corrections added to the initial values yield the doubly underlined approximate values.

This iterative method for the solution of operator equations is attributable to *Southwell* (Ref. 59). He interpreted the residuals r_i at the nodes as the negatives of additional external forces needed to make the membrane deflect to the approximation values u_i. Each step of hand relaxation reduces the absolute value of the largest applied force to zero. This interpretation gave the name *relaxation* to the iterative technique of gradual annihilation of residuals.

Southwell's relaxation method is notable for its great simplicity in calculational procedure, such that an experienced mathematician could develop great skill in repeated computational operations. And yet in larger systems, phenomena inhibiting convergence turn up (Ref. 61).

5-2-2 Overrelaxation and Property A

In computer usage, methods with a cyclic handling of nodes (*Gauss-Seidel*, Sec. 2–2–2; overrelaxation, Sec. 2–2–3) are superior. Because of the improved convergence, only overrelaxation comes into question, practically speaking.

The theory governing the optimal choice of relaxation factor ω_{opt} of Sec. 2–2–4 is applicable only for a *symmetric definite, diagonally block tridiagonal* coefficient matrix of the equation system. By Eq. (5–38), matrix A is *block tridiagonal*, but the submatrices on the diagonal are not diagonal in form. The theory cannot then be applied to Eq. (5–38). In the concrete Example 5–1, it is shown that the requirement of Theorem 2–7 can be satisfied by a skillful numbering of nodal points.

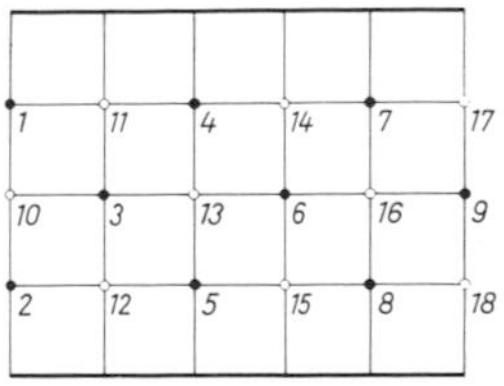

Figure 30. Separation of grid points into two groups.

Figure 30 shows a pattern of white and black circles at the nodes such that each line segment connects points of unlike color. Each operator equation of Figs. 22, 23, and 24 binds to the functional value at the central point only functional values of points of opposite marking. Unknowns u_i may thus be divided into two groups of white and black nodes.

The operator equations link the unknown at the central point with those of the opposite group. If one entire group is numbered completely, and then the other, and if the operator equations are appropriately rearranged, the system of equations for the numbering of nodes shown in Fig. 30 takes on the form depicted in (5–41):

u_1	u_2	u_3	u_4	u_5	u_6	u_7	u_8	u_9	u_{10}	u_{11}	u_{12}	u_{13}	u_{14}	u_{15}	u_{16}	u_{17}	u_{18}	1
2									$-\frac{1}{2}$	-1								0
	2								$-\frac{1}{2}$	0	-1							0
		4							-1	-1	-1	-1						1
			4							-1	0	-1	-1					0
				4							-1	-1	0	-1				1
					4							-1	-1	-1	-1			0
						4							-1	0	-1	-1		0
							4							-1	-1	0	-1	0
								3							-1	$-\frac{1}{2}$	$-\frac{1}{2}$	0
$-\frac{1}{2}$	$-\frac{1}{2}$	-1							2									0
-1	0	-1	-1							4								0
	-1	-1	0	-1							4							1
		-1	-1	-1	-1							4						1
			-1	0	-1	-1							4					0
				-1	-1	0	-1							4				0
					-1	-1	-1	-1							4			0
						-1	0	$-\frac{1}{2}$								3		0
							-1	$-\frac{1}{2}$									3	0

(5–41)

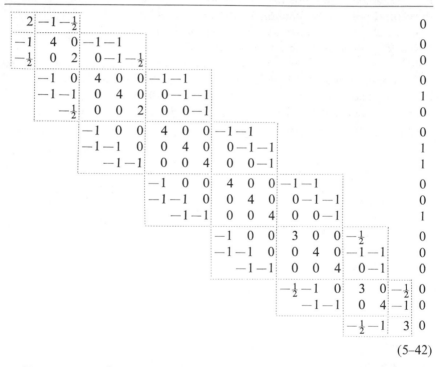

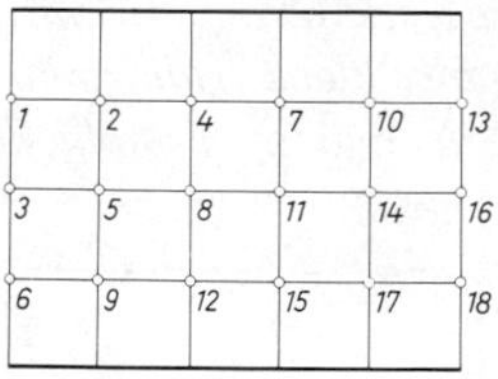

Figure 31. Numbering by diagonals.

Matrix A is now actually *diagonally block tridiagonal.* It decomposes into four square submatrices of order 9, and the two submatrices on the diagonal are themselves diagonal.

In practice, numbering of nodes by diagonals (see Fig. 31) is easier. This, too, leads to a *diagonally block tridiagonal* matrix A of the linear system of equations (5–42). But the submatrices on the diagonal are of unequal order, and some of the others are rectangular. The subdivision is shown in (5–42):

u_1	u_2	u_3	u_4	u_5	u_6	u_7	u_8	u_9	u_{10}	u_{11}	u_{12}	u_{13}	u_{14}	u_{15}	u_{16}	u_{17}	u_{18}	1
2	−1	$-\frac{1}{2}$																0
−1	4	0	−1	−1														0
$-\frac{1}{2}$	0	2	0	−1	$-\frac{1}{2}$													0
	−1	0	4	0	0	−1	−1											0
	−1	−1	0	4	0	0	−1	−1										1
		$-\frac{1}{2}$	0	0	2	0	0	−1										0
			−1	0	0	4	0	0	−1	−1								0
			−1	−1	0	0	4	0	0	−1	−1							1
				−1	−1	0	0	4	0	0	−1							1
						−1	0	0	4	0	0	−1	−1					0
						−1	−1	0	0	4	0	0	−1	−1				0
							−1	−1	0	0	4	0	0	−1				1
									−1	0	0	3	0	0	$-\frac{1}{2}$			0
									−1	−1	0	0	4	0	−1	−1		0
										−1	−1	0	0	4	0	−1		0
												$-\frac{1}{2}$	−1	0	3	0	$-\frac{1}{2}$	0
													−1	−1	0	4	−1	0
															$-\frac{1}{2}$	−1	3	0

(5–42)

For systems of equations which can be arranged to the forms (5–41) or (5–42), a particular designation was introduced in Ref. 79.

Definition 5–2

A square matrix A possesses property A *if, through simultaneous permutation of columns and rows,*[1] *it can be brought to a diagonally block tridiagonal form.*

[1]A simultaneous permutation of rows and columns means that any interchanging of two columns must be accompanied by a similar interchange of two rows such that the matrix remains symmetric.

The coefficient matrix of the system of equations (5–38) possesses property A, as the transition to the system (5–41) or (5–42) consists of a simultaneous permutation of rows and columns, which is done by an appropriate numbering of points. A matrix with property A generally has several different diagonally block tridiagonal representations. The representation (5–41) with just two submatrices in the diagonal is typical, and it is even used at times to characterize a matrix with property A. An arbitrary permutation of the first nine rows and columns of (5–41) brings about an analogous representation in which, however, the well-ordered structure of the nondiagonal elements is lost.

Having a meaning identical to Definition 5–2 is:

Definition 5–3

A square matrix $A = (a_{ij})$ *of order n possesses property* A *if two nonempty, disjunct subsets S and T of the first n positive integers* $W = (1, 2, \ldots, n)$ *exist such that* $S \cup T = W$ *and* $S \cap T = \phi$, *and if* $a_{ij} \neq 0$, *then either* $i = j$, *or* $i \in S$ *and* $j \in T$, *or* $i \in T$ *and* $j \in S$.

By this definition, matrix A of (5–38) exhibits property A, since the set $S = (1, 3, 5, 7, 9, 11, 13, 15, 17)$ and $T = (2, 4, 6, 8, 10, 12, 14, 16, 18)$. The indices i and j of every nonzero nondiagonal element a_{ij} are such that either $i \in S$ and $j \in T$, or $i \in T$ and $j \in S$. For the diagonally block tridiagonal matrices A of (5–41) and (5–42), it is easy to select the two subsets S and T.

The representation of the coefficient matrix A of a system of operator equations having property A is strongly dependent on the numbering scheme of nodes; in other words, on the *sequence* in which the nodal points are taken up during overrelaxation.

Definition 5–4

The sequences of nodes whose corresponding representations of the matrix A having property A *are diagonal block tridiagonal, are called consistent.*

For the five-point operator of the Laplace differential operator, the sequences as shown in Figs. 30 and 31 are consistent. The relatively similar numbering sequence by rows or columns, however, is not consistent. And in problems of *elasticity theory* involving deformation of plates and shells, property A cannot usually be established.

Theorem 2–7 governing the optimum choice of relaxation factor ω_{opt} is applicable for consistent sequences. Then the dominant eigenvalue λ_1 of matrix $-\boldsymbol{D}^{-1}(\boldsymbol{E} + \boldsymbol{F})$ or the symmetric matrix $-\boldsymbol{D}^{-1/2}(\boldsymbol{E} + \boldsymbol{F})\boldsymbol{D}^{-1/2}$, both pertaining to matrix $\boldsymbol{A}$, must be found in order to establish

$$\omega_{\text{opt}} = \frac{2}{1 + \sqrt{1 - \lambda_1^2}}.$$

Two different consistent sequences exhibit different representations of matrix A; nonetheless, we should note:

Theorem 5-1. *The optimal relaxation factor* ω_{opt} *is independent of the particular consistent sequence. The optimal asymptotic convergence of all consistent sequences is the same.*

Proof: The two different representations of matrix A for two different sequences are similar to one another through a permutation matrix. The corresponding matrices $-D^{-1/2}(E + F)D^{-1/2}$ are similar, thanks to the same permutation matrix, and thus their dominant eigenvalues are identical. This leads to the same optimal relaxation factor, and thus the convergence factors and the asymptotic convergence rate actually coincide.

Example 5-2. With a consistent sequence of 18 nodes as in Example 5-1, the dominant eigenvalue of matrix $-D^{-1/2}(E + F)D^{-1/2}$ is found by calculation to be $\lambda_1 = 0.837319$. For the method of successive displacements ($\omega = 1$), the convergence radius and convergence rate are

$$\rho(M(1) = \lambda_1^2 = 0.70110, \quad R(M(1)) = -\log_{10}\rho(M(1)) = 0.1542.$$

Asymptotically, $k \approx 6.5$ iteration cycles are needed in order to gain an additional decimal place of accuracy in the approximate solution. At the start, several iterations are necessary before the asymptotic convergence becomes applicable. The determination of the solution to six decimal places requires about 50 iteration cycles.

The optimal relaxation factor, the corresponding convergence radius, and the convergence rate are

$$\omega_{\text{opt}} = 1.29306, \quad \rho(M(\omega_{\text{opt}})) = 0.29306, \quad R(M(\omega_{\text{opt}})) = 0.5330,$$

so that with $k \approx 1.88$, roughly two iterations are needed asymptotically to gain an additional decimal place. The solution accurate to six decimals is found after 16 iterations. It is given in Table 26 by nodal point locations as depicted in Figs. 21, 30, and 31.

Table 26 Solution of the Boundary Value Problem with a Coarse Grid

−0.405937	−0.442200	−0.388185	−0.227890	−0.116926	−0.052551
−0.739347	−0.974678	−0.882650	−0.406449	−0.187264	−0.081455
−0.602094	−0.834516	−0.761289	−0.327991	−0.144227	−0.061651

For a finer subdivision with 77 grid points, the corresponding quantities become

$$\lambda_1 = 0.957686, \quad \rho(M(1)) = \lambda_1^2 = 0.917162$$
$$R(M(1)) = 0.03756, \quad k \approx 26.6.$$

The successive displacements method requires about 200 iterations to produce a solution accurate to six places. Exploiting property A produces the quantities $\omega_{\text{opt}} = 1.5530$, $\rho(M(\omega_{\text{opt}})) = 0.5530$, $R(M(\omega_{\text{opt}})) = 0.2573$, $k \approx 3.9$. With an optimum choice of relaxation factor, the six-place solution is done after 35 iterations, involving a negligible increment of computational work compared to the successive displacements method. Further subdivision of the grid forces the dominant eigenvalue λ_1 ever closer to unity, and the successive displacements method converges in a hopelessly leisurely fashion, and only overrelaxation holds out a promise of success (see Table 4).

ALGOL Procedure of an Iteration Cycle for the Standard Example. With the worksheet of Fig. 28 as a guidepost, the functional values (and their approximations) at the nodes are treated as doubly indexed variables, with rows numbered from top to bottom and columns from left to right. For convenience, the given boundary values are included as additional rows which, however, may not be altered. By this approach there are actually only three operator equations to be formulated. The values $f(x, y)$ at the nodes are given as doubly indexed quantities just like the function values. The nodal points of the consistent sequence are run along diagonals (see Fig. 31). By using the appropriate operator equation, the new approximations are computed by Eq. (2–32).

The procedure is kept sufficiently general to handle problems of our standard type with corresponding boundary conditions, but with completely different integer ratios of side lengths.

The procedure parameters are defined as:

- *n* — Number of points per row.
- *m* — Number of interior rows.
- *h* — Mesh width of a cell square.
- *omega* — Relaxation factor.
- *f* — Functional values f_{ik} at the nodal points $(i = 1, 2, \ldots, m;\ k = 1, 2, \ldots, n)$.
- *u* — Functional values u_{ik} at nodal points before and after over-relaxation $(i = 0, 1, \ldots, m + 1;\ k = 1, 2, \ldots, n)$.

ALGOL Procedure No. 15

```
procedure overrelax (n, m, h, omega, f, u);
          value n, m, h, omega;
          integer n, m;  real h, omega;  array f, u;
begin integer i, k, z, max, min;  real delta;
   for z: = 1 step 1 until n + m - 1 do
   begin max: = if z > n then z - n + 1 else 1;
      min: = if z < m then z else m;
     for i: = max step 1 until min do
```

```
        begin k: = z - i + 1;
          if k = 1 then
            delta: = (f[i, k] - 0.5 × (u[i - 1, k] + u[i + 1, k])
                     - u[i, k + 1])/2
          else
          if k = n then
            delta: = (f[i, k] - 0.5 × (u[i - 1, k] + u[i + 1, k])
                     -u[i, k - 1])/(2 + h)
          else
            delta: = (f[i, k] - u[i, k + 1] - u[i, k - 1]
                     - u[i - 1, k] - u[i + 1, k])/4;
          u[i, k]: = u[i, k] - omega × (u[i, k] + delta)
        end i
      end z
    end overrelax
```

In order to find the optimal relaxation factor ω_{opt}, it was assumed that the dominant eigenvalue λ_1 of the matrix $-\boldsymbol{D}^{-1/2}(\boldsymbol{E}+\boldsymbol{F})\boldsymbol{D}^{-1/2}$ was known. For the two grid subdivisions, λ_1 was actually evaluated. In general, however, the evaluation of the dominant eigenvalue is too time consuming, and then the choice of optimal relaxation factor cannot be based on knowing λ_1. The value of λ_1 can be found approximately in the course of calculation. Not knowing ω_{opt}, we can begin iterating with $\omega = 1$ (*successive displacements*). The *error vector* $\boldsymbol{f}^{(k)} = \boldsymbol{x} - \boldsymbol{v}^{(k)}$ exhibits asymptotic convergence to the zero vector like a geometric sequence with the quotient $\rho(\boldsymbol{M}(1))$. The residual vector $\boldsymbol{r}^{(k)}$, however, exhibits the same asymptotic behavior as the error vector $\boldsymbol{f}^{(k)}$, since

$$\boldsymbol{r}^{(k)} = \boldsymbol{A}\boldsymbol{v}^{(k)} + \boldsymbol{b}, \quad 0 = \boldsymbol{A}\boldsymbol{x} + \boldsymbol{b} \tag{5–43}$$

leads, after subtraction, to

$$\boldsymbol{r}^{(k)} = \boldsymbol{A}(\boldsymbol{v}^{(k)} - \boldsymbol{x}) = -\boldsymbol{A}\boldsymbol{f}^{(k)}. \tag{5–44}$$

For a consistent sequence, $\rho(\boldsymbol{M}(1)) = \lambda_1^2$ with real λ_1. Thus for the asymptotic behavior of the error vector, it follows that $\boldsymbol{f}^{(k+1)} \approx \lambda_1^2 \boldsymbol{f}^{(k)}$ and thus, by (5–44),

$$\boldsymbol{r}^{(k+1)} = -\boldsymbol{A}\boldsymbol{f}^{(k+1)} \approx \lambda_1^2 \boldsymbol{A}\boldsymbol{f}^{(k)} = \lambda_1^2 \boldsymbol{r}^{(k)}. \tag{5–45}$$

The quotient of norms of successive residual vectors in the successive displacements method provides asymptotic approximations to λ_1^2, and, therefore, the optimal relaxation factor ω_{opt} can be located in the right ballpark.

Example 5-3. The quotient $q_k = \|\boldsymbol{r}^{(k)}\|/\|\boldsymbol{r}^{(k-1)}\|$ of Euclidean norms of successive residual vectors is shown as a function of k in Fig. 32 for the case of a fine grid (77 points). After a few steps q_k is in the neighborhood of $\rho(M(1)) = 0.91716$. The asymptotic value is approached from below, and this results in too *small* an optimal relaxation factor ω_{opt}. After 20 iterations with $\omega = 1$, we have $q_{20} = 0.9162$, and thus $\omega'_{opt} = 1.551$. With this value of ω'_{opt}, overrelaxation reaches the solution quickly. The linear convergence of q_k makes possible use of the *Wynn* ϵ-algorithm (Ref. 78) for finding a better approximate value.

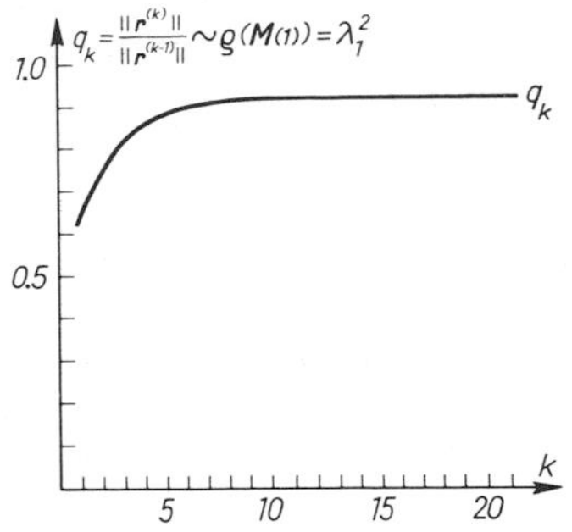

Figure 32. Approximations for the spectral radius λ_1^2 ($\omega = 1$).

In order to locate the optimal relaxation factor ω_{opt}, the iteration could also be started with a value $1 < \omega < \omega_{opt}$ so that the initial relaxation steps could contribute better to the solution. In that case, the quotients of norms of successive residual vectors converge to the spectral radius of the matrix $M(\omega)$; that is, to the value $\rho(M(\omega)) = \mu_1^{(1)}$ (see Fig. 4). The relation (2–63) linking λ_1^2, ω and $\mu_1^{(1)}$ yields the result

$$\lambda_1^2 = \frac{(\mu_1^{(1)} + \omega - 1)^2}{\omega^2 \mu_1^{(1)}}. \tag{5–46}$$

With $\mu_1^{(1)}$ approximately known, an approximate λ_1^2 can be found from (5–46), and thus ω_{opt} can again be found approximately.

Example 5-4. If the relaxation of Example 5-3 is started with $\omega = 1.5$, we find approximations for $\mu_1^{(1)}$ as quotients of successive residual vectors and the ensuing quantities λ_1^2 by Eq. (5–46) (see Fig. 33). The decaying oscillation of $\mu_1^{(1)}$ is explained by the partially complex eigenvalues of $M(\omega)$, whose magnitude is only slightly under unity. Convergence is unmistakable, and the optimal relaxation factor can be found approximately.

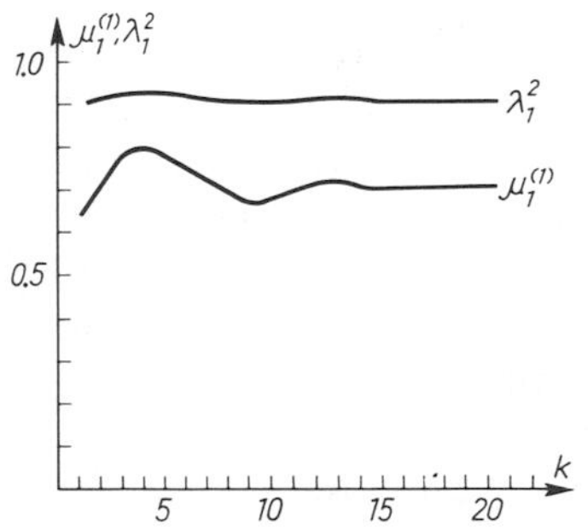

Figure 33. Approximations for $\mu_1^{(1)}$ and λ_1^2 ($\omega = 1.5$).

For $\omega > \omega_{opt}$, $M(\omega)$ has nothing but complex eigenvalues. The quotients of norms of successive residuals exhibit an irregular pattern and fail to converge to an eigenvalue. In such a case, ω should be reduced.

These statements indicate that ω_{opt} can be found from the residuals after a complete iteration cycle. In conventional computations, however, the residuals are not immediately available and they must be explicitly calculated. The computation time is comparable with that for one iteration cycle, and is thus too costly. This can be reduced by not computing residuals until after every p (say, 5 to 10) iterations, and then by finding mean convergence radii by (5–47):

$$\left[\frac{\|\boldsymbol{r}^{[(m+1)\cdot p]}\|}{\|\boldsymbol{r}^{(m\cdot p)}\|}\right]^{1/p} \approx \rho(\boldsymbol{M}(\omega)) = \mu_1^{(1)}. \tag{5–47}$$

Ultimately, the residual calculation could be omitted altogether. The sequence of difference vectors $\boldsymbol{y}^{(k)} = \boldsymbol{v}^{(k)} - \boldsymbol{v}^{(k-1)}$ of two approximations after complete cycles exhibits the same asymptotic behavior as the sequence of error vectors themselves. A norm of the vector $\boldsymbol{y}^{(k)}$ can be trivially evaluated during an iteration cycle; however, the values are not very accurate, generally.

Not every discretization of a boundary value problem (as in the case of the energy integral) leads to a system of operator equations having property A. The more accurate approximation of the Poisson equation $\Delta u = f(x, y)$ by the *nine-point* operator or the *Hermitian finite-difference* operator (see Refs. 10, 13), as shown in Fig. 34 for a regular interior point of the domain, leads to a system of equations lacking property A.

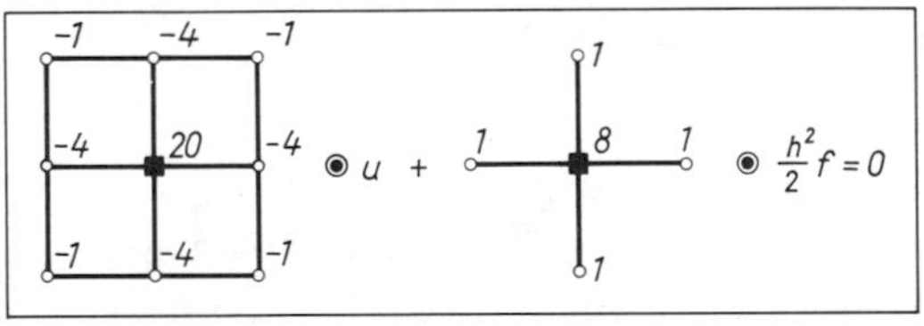

Figure 34. Nine-point operator equation for $u = f(x, y)$. Regular interior point.

The system of equations is, to be sure, symmetric definite, but the result of Theorem 2–7 cannot be applied. The method of overrelaxation does converge for every value of ω in the interval $0 < \omega < 2$ by Theorem 2–6, but nothing is known about the best choice of ω. Reference 35 examines the class of operator equations with a property such that in the operator "star," the weights of all function values are negative, except for the central point. The operator equations of Figs. 22, 23, 24, and 34 are in this category. First, it is shown that ω values in the interval $0 < \omega < 1$ exhibit a poorer convergence than that of $\omega = 1$, and thus underrelaxation is out of the question. Furthermore, let λ_1 be the dominant eigenvalue of $-\boldsymbol{D}^{-1/2}(\boldsymbol{E} + \boldsymbol{F})\boldsymbol{D}^{-1/2}$, and

$$\omega_b = \frac{2}{1 + \sqrt{1 - \lambda_1^2}}.$$

For systems of operator equations of the specified category lacking property

A, then, the spectral radius $\rho(M(\omega))$ for all ω in the interval $0 < \omega < 2$ is larger than the corresponding spectral radius for a system with property A in a consistent sequence and with the same λ_1. In particular,

$$\rho(M(\omega)) \geq \omega_b - 1. \tag{5–48}$$

The convergence in overrelaxation without property A is definitely not better than with it in a consistent sequence. For the choice of ω, either experiment or experience should serve as the guide. The best possible value, incidentally, need not coincide with ω_b.

5-2-3 Implicit Block Relaxation

In the relaxation schemes examined so far (successive displacements, overrelaxation), the value of each individual component of the new approximation is found from an explicit linear formula, using the best currently accessible approximate values of unknowns. We thus speak of *explicit relaxation methods. Implicit relaxation methods*, on the other hand, evaluate simultaneously whole groups or blocks of new approximations, requiring the solution of linear systems of equations.

Using a numbering of points by columns or rows, the system of operator equations exhibits a tridiagonal block structure (5–39), where each square block matrix on the diagonal corresponds to a group of nodal points belonging to a column or a row. It thus seems logical to group the nodal points of a column or row together and to relax the pertinent functional values as a block. With the subdividing of the approximation vector $\boldsymbol{v}^{(m)}$ and the constant vector $\boldsymbol{b}$ corresponding to that of the partitioned coefficient matrix $\boldsymbol{A}$, with p blocks as shown,

$$\boldsymbol{A} = \begin{bmatrix} \boldsymbol{D}_1 & \boldsymbol{F}_1 & & & & \\ \boldsymbol{E}_1 & \boldsymbol{D}_2 & \boldsymbol{F}_2 & & & \\ & \boldsymbol{E}_2 & \boldsymbol{D}_3 & \boldsymbol{F}_3 & & \\ & & \cdot & \cdot & \cdot & \\ & & & \boldsymbol{E}_{p-2} & \boldsymbol{D}_{p-1} & \boldsymbol{F}_{p-1} \\ & & & & \boldsymbol{E}_{p-1} & \boldsymbol{D}_p \end{bmatrix}, \quad \boldsymbol{v}^{(m)} = \begin{bmatrix} \boldsymbol{v}_1^{(m)} \\ \boldsymbol{v}_2^{(m)} \\ \boldsymbol{v}_3^{(m)} \\ \cdot \\ \boldsymbol{v}_{p-1}^{(m)} \\ \boldsymbol{v}_p^{(m)} \end{bmatrix}, \quad \boldsymbol{b} = \begin{bmatrix} \boldsymbol{b}_1 \\ \boldsymbol{b}_2 \\ \boldsymbol{b}_3 \\ \cdot \\ \boldsymbol{b}_{p-1} \\ \boldsymbol{b}_p \end{bmatrix}, \tag{5–49}$$

the iteration procedure for block relaxation runs as follows, analogous to the successive displacements method:

$$\left.\begin{aligned} \boldsymbol{D}_1\boldsymbol{v}_1^{(m+1)} + \boldsymbol{F}_1\boldsymbol{v}_2^{(m)} + \boldsymbol{b}_1 &= 0 \\ \boldsymbol{E}_{i-1}\boldsymbol{v}_{i-1}^{(m+1)} + \boldsymbol{D}_i\boldsymbol{v}_i^{(m+1)} + \boldsymbol{F}_i\boldsymbol{v}_{i+1}^{(m)} + \boldsymbol{b}_i &= 0 \quad (i = 2, 3, \ldots, p-1). \\ \boldsymbol{E}_{p-1}\boldsymbol{v}_{p-1}^{(m+1)} + \boldsymbol{D}_p\boldsymbol{v}_p^{(m+1)} + \boldsymbol{b}_p &= 0 \end{aligned}\right\} \tag{5–50}$$

Matrices $\boldsymbol{D}_i$ usually are not diagonal matrices; thus finding the components of the ith group $\boldsymbol{v}_i^{(m+1)}$ requires solution of a linear system of equations. The coefficient matrices, nevertheless, are symmetric definite and banded, and they do not change during the iteration. In the concrete example having numbering by columns, only three distinct types of *tridiagonal* matrices turn up. These band matrices are decomposed by Cholesky's method before the start of relaxation, so that for the evaluation of $\boldsymbol{v}_i^{(m+1)}$ only forward and backward substitution need be carried out on the vector

$$\boldsymbol{c}_i = \boldsymbol{b}_i + \boldsymbol{E}_{i-1}\boldsymbol{v}_{i-1}^{(m+1)} + \boldsymbol{F}_i\boldsymbol{v}_{i+1}^{(m)}$$

along with the corresponding decomposition of $\boldsymbol{D}_i$. Components of vector $\boldsymbol{c}_i$ are, of course, computed with the help of operator equations.

Under the *block representation* of $\boldsymbol{A} = \boldsymbol{D} + \boldsymbol{E} + \boldsymbol{F}$, the iteration law (5–50) reads

$$\boldsymbol{E}\boldsymbol{v}^{(m+1)} + \boldsymbol{D}\boldsymbol{v}^{(m+1)} + \boldsymbol{F}\boldsymbol{v}^{(m)} + \boldsymbol{b} = 0, \tag{5–51}$$

or, when solved,

$$\boldsymbol{v}^{(m+1)} = -(\boldsymbol{E} + \boldsymbol{D})^{-1}\boldsymbol{F}\boldsymbol{v}^{(m)} - (\boldsymbol{E} + \boldsymbol{D})^{-1}\boldsymbol{b}. \tag{5–52}$$

The law (5–52) represents a general iteration method [see Eq. (2–22)], with iteration matrix $\boldsymbol{M}_B$ of block relaxation given by

$$\boldsymbol{M}_B = -(\boldsymbol{E} + \boldsymbol{D})^{-1}\boldsymbol{F}. \tag{5–53}$$

The spectral radius of matrix $\boldsymbol{M}_B$ determines asymptotic convergence. For a symmetric definite matrix $\boldsymbol{A}$, it is less than unity.

The *method of block overrelaxation* furnishes better convergence. The correction factor, which turns up after the block relaxation for the transition from $\boldsymbol{v}_i^{(m)}$ to $\boldsymbol{v}_i^{(m+1)}$, is multiplied by the relaxation factor ω, which remains constant. The iteration laws thus read:

$$\left.\begin{aligned} \boldsymbol{E}\boldsymbol{v}^{(m+1)} + \boldsymbol{D}\boldsymbol{v}^{(m+1/2)} + \boldsymbol{F}\boldsymbol{v}^{(m)} + \boldsymbol{b} &= 0 \\ \boldsymbol{v}^{(m+1)} = \boldsymbol{v}^{(m)} + \omega(\boldsymbol{v}^{(m+1/2)} - \boldsymbol{v}^{(m)}) & \end{aligned}\right\}, \tag{5–54}$$

where $\boldsymbol{v}^{(m+1/2)}$ is the interim vector.

Upon elimination of the auxiliary vector $\boldsymbol{v}^{(m+1/2)}$, we arrive formally at the same law as (2–50):

$$(\boldsymbol{E} + \omega^{-1}\boldsymbol{D})\boldsymbol{v}^{(m+1)} + [\boldsymbol{F} + (1 - \omega^{-1})\boldsymbol{D}]\boldsymbol{v}^{(m)} + \boldsymbol{b} = 0. \tag{5–55}$$

In Eq. (5–55) we understand the matrices to be in partitioned block form.

The convergence rate is dependent on the spectral radius of the matrix

$$M_B(\omega) = -(E + \omega^{-1}D)^{-1}[F + (1 - \omega^{-1})D]. \tag{5–56}$$

Index B refers to the block overrelaxation. A proof procedure which is formally identical to *point overrelaxation* shows that $\rho(M_B(\omega)) < 1$ for $0 < \omega < 2$ if matrix A is symmetric definite. The fact that matrix A of the system (5–49) is block tridiagonal establishes the link to Theorem 2–7, which makes possible finding the optimal relaxation factor for just such tridiagonal matrices. To a large extent, it is merely a formality whether the elements of a tridiagonal matrix are numbers or matrices. Consequently, the same relationship stands between the eigenvalues of M_B of Eq. (5–53) as between those of $M_B(\omega)$ of Eq. (5–56), and we are led to:

Theorem 5-2. *The optimal relaxation factor ω_{opt} in block overrelaxation of a symmetric definite system of equations whose coefficient matrix is block tridiagonal is given by*

$$\omega_{\text{opt}} = \frac{2}{1 + \sqrt{1 - \lambda_1^2}}. \tag{5–57}$$

Here λ_1^2 indicates the dominant eigenvalue of the iteration matrix $M_B = (-D + E)^{-1} F$ of the corresponding block relaxation, and λ_1 represents the largest eigenvalue of the matrix $-D^{-1}(E + F)$.

The importance of Theorem 5–2 is that the block tridiagonal form of a system of operator equations manifests itself also in difference operators more general than the 5-point operator. For instance, a system of operator equations of the 9-point operator in Fig. 34 does not exhibit property A. Nonetheless, in block relaxation by columns or rows, the coefficient matrix is block tridiagonal. Theorem 5–2 also makes possible a statement on the best possible choice of relaxation factor. The operator equations of Fig. 34 combine in one row (column) three neighboring values of functions of the same row (column); thus the block matrices along the diagonal are tridiagonal. The blocks in the adjacent diagonals contain up to three nonzero elements in each row. In the event of a rectangular domain, they are likewise tridiagonal. The particular form of block matrices is in no way a requirement for Theorem 5–2, but it does simplify calculations appreciably.

The optimal relaxation factor ω_{opt} for block overrelaxation can be found from the residual vectors as in the explicit method, whether the computation is begun with $\omega = 1$ or with some value $1 < \omega < \omega_{\text{opt}}$.

Block overrelaxation by rows (columns) exhibits a greater convergence rate $R(M_B(\omega_{\text{opt}}))$ than in point overrelaxation; therefore, fewer iterations are needed for achieving a given accuracy (Ref. 17). The total computational time, however, remains essentially unchanged, since the gain in iterations is largely compensated by additional solution of implicit systems of equations.

Example 5-5. In block relaxation by rows in the concrete boundary value problem of the rectangle, the square submatrices $\boldsymbol{D}_i$ along the diagonal are identical. For the coarse grid (three rows of six points), the coefficient matrix $\boldsymbol{A}$ has the block tridiagonal form

$$\boldsymbol{A} = \begin{bmatrix} \boldsymbol{D}_1 & \boldsymbol{F}_1 & 0 \\ \boldsymbol{E}_1 & \boldsymbol{D}_2 & \boldsymbol{F}_2 \\ 0 & \boldsymbol{E}_2 & \boldsymbol{D}_3 \end{bmatrix}$$

with the three six-rowed submatrices

$$\boldsymbol{D}_i = \begin{bmatrix} 2 & -1 & & & & \\ -1 & 4 & -1 & & & \\ & -1 & 4 & 1 & & \\ & & -1 & 4 & -1 & \\ & & & -1 & 4 & -1 \\ & & & & -1 & 3 \end{bmatrix}, \quad \boldsymbol{E}_i = \boldsymbol{F}_i = \begin{bmatrix} -\frac{1}{2} & & & & & \\ & -1 & & & & \\ & & -1 & & & \\ & & & -1 & & \\ & & & & -1 & \\ & & & & & -\frac{1}{2} \end{bmatrix}.$$

For each row, the same matrix $\boldsymbol{D}_i$ applies, and thus initially only one Cholesky decomposition $\boldsymbol{D}_1 = \boldsymbol{R}^{\mathrm{T}}\boldsymbol{R}$ is needed. For the block relaxation we have the quantities

$$\rho(\boldsymbol{M}_B) = \lambda_1^2 = 0.46726, \quad R(\boldsymbol{M}_B) = 0.3304, \quad k \approx 3.$$

Compared to the explicit method (Example 5–2), block relaxation converges twice as fast. The optimal relaxation factor, by Theorem 5–2, and the corresponding related quantities are

$$\omega_{\text{opt}} = 1.1562, \quad \rho(\boldsymbol{M}_B(\omega_{\text{opt}})) = 0.1562, \quad R(\boldsymbol{M}_B(\omega_{\text{opt}})) = 0.8063,$$
$$k \approx 1.24.$$

Block overrelaxation, using optimal factor ω_{opt}, gives a 33% better convergence than point overrelaxation.

The corresponding results for the finer 77-grid-point subdivision are

$$\rho(\boldsymbol{M}_B) = \lambda_1^2 = 0.8390, \quad R(\boldsymbol{M}_B) = 0.0762, \quad k \approx 13.1.$$
$$\omega_{\text{opt}} = 1.4273, \quad \rho(\boldsymbol{M}_B(\omega_{\text{opt}}) = 0.4273, \quad R(\boldsymbol{M}_B(\omega_{\text{opt}})) = 0.3693, \quad k \approx 2.7.$$

Compared to point relaxation or overrelaxation, block relaxation exhibits twice as good a convergence, while for block overrelaxation it is 30% better.

One variant of block overrelaxation by rows (or columns) lies in following one pass through the rows (columns) by a second one through rows in reverse sequence. One such "double" traverse will be considered a single iteration step. The iteration rule for this overrelaxation procedure with constant relaxation factor ω becomes, for the two half-steps, in analogy to (5–54),

1. Half-Step (forward):

$$\left.\begin{cases} \boldsymbol{E}\boldsymbol{v}^{(m+1/2)} + \boldsymbol{D}\boldsymbol{v}^{(m+1/4)} + \boldsymbol{F}\boldsymbol{v}^{(m)} + \boldsymbol{b} = 0, \\ \boldsymbol{v}^{(m+1/2)} - \boldsymbol{v}^{(m)} - \omega(\boldsymbol{v}^{(m+1/4)} - \boldsymbol{v}^{(m)}) = 0; \end{cases}\right\} \tag{5–58}$$

2. Half-Step (backward):

$$\left.\begin{cases} \boldsymbol{E}\boldsymbol{v}^{(m+1/2)} + \boldsymbol{D}\boldsymbol{v}^{(m+3/4)} + \boldsymbol{F}\boldsymbol{v}^{(m+1)} + \boldsymbol{b} = 0, \\ \boldsymbol{v}^{(m+1)} - \boldsymbol{v}^{(m+1/2)} - \omega(\boldsymbol{v}^{(m+3/4)} - \boldsymbol{v}^{(m+1/2)}) = 0. \end{cases}\right\} \tag{5–59}$$

Elimination of $\boldsymbol{v}^{(m+1/4)}$ from (5–58) and of $\boldsymbol{v}^{(m+3/4)}$ from (5–59) leads directly to the relations

$$(\boldsymbol{E} + \omega^{-1}\boldsymbol{D})\boldsymbol{v}^{(m+1/2)} + [\boldsymbol{F} + (1 - \omega^{-1})\boldsymbol{D}]\boldsymbol{v}^{(m)} + \boldsymbol{b} = 0, \tag{5–60}$$

$$[\boldsymbol{E} + (1 - \omega^{-1})\boldsymbol{D}]\boldsymbol{v}^{(m+1/2)} + (\boldsymbol{F} + \omega^{-1}\boldsymbol{D})\boldsymbol{v}^{(m+1)} + \boldsymbol{b} = 0. \tag{5–61}$$

Elimination of $\boldsymbol{v}^{(m+1/2)}$ from (5–60) and (5–61) finally leaves the relation (5–62) between the approximations $\boldsymbol{v}^{(m+1)}$ and $\boldsymbol{v}^{(m)}$ after one double traverse:

$$\left.\begin{aligned} &[\boldsymbol{E} + (1 - \omega^{-1})\boldsymbol{D}]^{-1}(\boldsymbol{F} + \omega^{-1}\boldsymbol{D})\boldsymbol{v}^{(m+1)} \\ &\quad - (\boldsymbol{E} + \omega^{-1}\boldsymbol{D})^{-1}[\boldsymbol{F} + (1 - \omega^{-1})\boldsymbol{D}]\boldsymbol{v}^{(m)} \\ &\quad + \{[\boldsymbol{E} + (1 - \omega^{-1})\boldsymbol{D}]^{-1} - (\boldsymbol{E} + \omega^{-1}\boldsymbol{D})^{-1}\}\boldsymbol{b} = 0. \end{aligned}\right\} \tag{5–62}$$

Convergence rate is determined by the matrix

$$\boldsymbol{M}_{\mathrm{BS}}(\omega) = (\boldsymbol{F} + \omega^{-1}\boldsymbol{D})^{-1}[\boldsymbol{E} + (1 - \omega^{-1})\boldsymbol{D}](\boldsymbol{E} + \omega^{-1}\boldsymbol{D})^{-1}[\boldsymbol{F} + (1 - \omega^{-1})\boldsymbol{D}]. \tag{5–63}$$

Theorem 5-3. *Given that matrix* $\mathbf{A} = \mathbf{D} + \mathbf{E} + \mathbf{F}$ *is symmetric definite, the eigenvalues of matrix* $\boldsymbol{M}_{BS}(\omega)$ *in Eq.* (5–63) *are real, nonnegative, and smaller than unity for all values of* ω *in the interval* $0 < \omega < 2$.

Proof: The positive definiteness of matrix $\boldsymbol{A}$ assures that the block diagonal matrix $\boldsymbol{D}$ is also positive definite. Then there exists a symmetric definite matrix $\boldsymbol{D}^{1/2}$ such that $\boldsymbol{D}^{1/2}\boldsymbol{D}^{1/2} = \boldsymbol{D}$. Furthermore, we introduce

$$\boldsymbol{U} = \boldsymbol{D}^{-1/2}\boldsymbol{F}\boldsymbol{D}^{-1/2}, \quad \boldsymbol{L} - \boldsymbol{D}^{-1/2}\boldsymbol{E}\boldsymbol{D}^{-1/2}. \tag{5–64}$$

Through the symmetry of $D^{1/2}$ and thus of $D^{-1/2}$, and since $F = E^T$, it follows that $U^T = L$. With this substitution, matrix $M_{BS}(\omega)$ becomes

$$
\begin{aligned}
M_{BS}(\omega) &= (D^{1/2}UD^{1/2} + \omega^{-1}D)^{-1}\cdot[D^{1/2}LD^{1/2} + (1-\omega^{-1})D] \\
&\quad \times (D^{1/2}LD^{1/2} + \omega^{-1}D)^{-1}\cdot[D^{1/2}UD^{1/2} + (1-\omega^{-1})D] \\
&= [D^{1/2}(U+\omega^{-1}I)D^{1/2}]^{-1}\cdot D^{1/2}\cdot[L+(1-\omega^{-1})I]D^{1/2} \\
&\quad \times [D^{1/2}(L+\omega^{-1}I)D^{1/2}]^{-1}\cdot D^{1/2}\cdot[U+(1-\omega^{-1})I]D^{1/2} \\
&= D^{-1/2}(U+\omega^{-1}I)^{-1}[L+(1-\omega^{-1})I](L+\omega^{-1}I)^{-1} \\
&\quad \times [U+(1-\omega^{-1})I]D^{1/2}.
\end{aligned}
$$

Let us set

$$
P = L + \omega^{-1}I, \quad P^T = U + \omega^{-1}I, \tag{5–65}
$$

such that after some additional rearrangement of matrices we get for $M_{BS}(\omega)$

$$
\begin{aligned}
M_{BS}(\omega) &= D^{-1/2}P^{T^{-1}}\cdot[P+(1-2\omega^{-1})I]\cdot P^{-1}\cdot[P^T+(1-2\omega^{-1})I]D^{1/2} \\
&= (P^T\cdot D^{1/2})^{-1}\cdot[I+(1-2\omega^{-1})P^{-1}]\cdot[I+(1-2\omega^{-1})P^{T^{-1}}]\cdot(P^T\cdot D^{1/2}) \\
&= (P^T\cdot D^{1/2})^{-1}\cdot\{[I+(1-2\omega^{-1})P^{-1}]\cdot[I+(1-2\omega^{-1})P^{-1}]^T\}\cdot(P^T\cdot D^{1/2}).
\end{aligned}
$$

Within the pair of curly brackets (braces) we have a symmetric, semidefinite matrix as a product of a matrix with its own transpose. Matrix $M_{BS}(\omega)$ is formed through the similarity transformation of this matrix, using matrix $(P^T\cdot D^{1/2})$. Consequently, the eigenvalues of $M_{BS}(\omega)$ are real and nonnegative. The spectral radius $\rho(M_{BS}(\omega))$ is smaller than unity, since the result of the two half-steps can be looked on as a combination of two convergent block relaxations.

Because of the similarity of the iteration matrix $M_{BS}(\omega)$ to a symmetric semidefinite matrix, the relaxation procedure is called *symmetric block relaxation.* Unfortunately, there is no simple relation between the eigenvalues of $M_{BS}(\omega)$ and other more simply constructed matrices; thus no direct statement about convergence rate can be deduced. Practice shows that symmetric block overrelaxation exhibits poorer convergence rate than double steps of conventional block overrelaxation. On the other hand, the spectral radius $\rho(M_{BS}(\omega))$ is less sensitive to the choice of ω in the vicinity of the optimal value, since at that point $\rho(M_{BS}(\omega))$ as a function of ω has a horizontal tangent instead of an inverted peak. Since the eigenvalues of $M_{BS}(\omega)$ are real, *Chebyshev's* convergence improvement methods may be used to speed up convergence so that with this modification the symmetric block overrelaxation is actually better than block overrelaxation (see Refs. 58, 13).

Example 5-6. The spectral radius $\rho(\mathbf{M}_{BS}(\omega))$ of symmetric block overrelaxation for the rectangle having a fine grid (77 points) is shown as a function of ω in Fig. 35, for a row-by-row procedure. For comparison, the spectral radius $\rho(M_B(\omega))$ of block overrelaxation by rows is also depicted. Note that the computational effort for one step of symmetric block overrelaxation corresponds to that of two iterations in ordinary block overrelaxation. By this observation, the apparently better convergence for $\omega < 1.37$ is actually poorer, and in the vicinity of the optimal relaxation factor, symmetric block overrelaxation in its basic form clearly drops out of consideration.

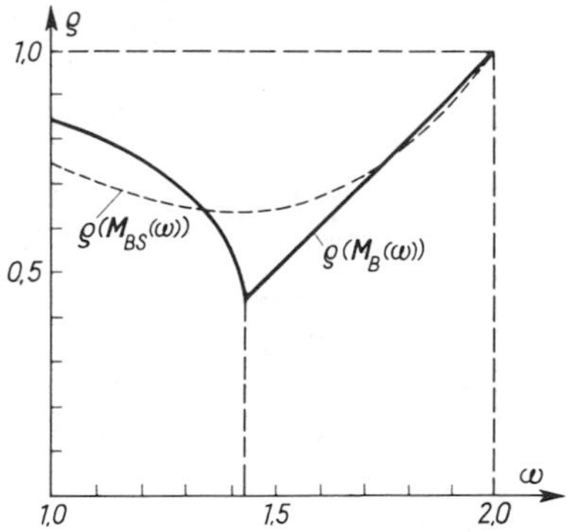

Figure 35. Spectral radii of ordinary and symmetric block overrelaxation by rows (fine division, 77 points).

The fact that the optimal relaxation factors for both block relaxations are essentially identical is a coincidence. In more general boundary value problems, the values can diverge appreciably (see Ref. 13).

5-2-4 Method of Alternating Directions

Recently *Peaceman* and *Rachford* (Ref. 41) have developed a method originally intended for the solution of *parabolic differential equations*, which is related to block relaxation but fundamentally different. It is applicable to boundary value problems having a partial differential equation of elliptical type of the sort

$$\Delta u - \rho(x, y)u = f(x, y) \quad \text{in} \quad G. \tag{5-66}$$

Note that this time the sign has changed, and in addition we require that $\rho(x, y) \geq 0$ in G. The operator equations are decomposed by *central differences* in the x- and y-directions. At a regular interior point with the coordinates (x_i, y_i), the operator equation is

$$\left.\begin{aligned} &\{-u(x_i - h, y_i) + 2u(x_i, y_i) - u(x_i + h, y_i)\} \\ &+ \{-u(x_i, y_i - h) + 2u(x_i, y_i) - u(x_i, y_i + h)\} \\ &+ h^2 \cdot \rho(x_i, y_i)u(x_i, y_i) + h^2 f(x_i, y_i) = 0. \end{aligned}\right\} \tag{5-67}$$

The *matrix operators* $\boldsymbol{H}$, $\boldsymbol{V}$, and $\boldsymbol{\Sigma}$ are defined:

$$\boldsymbol{H}u(x_i, y_i) = -u(x_i - h, y_i) + 2u(x_i, y_i) - u(x_i + h, y_i) \tag{5-68}$$

$$Vu(x_i, y_i) = -u(x_i, y_i - h) + 2u(x_i, y_i) - u(x_i, y_i + h) \tag{5–69}$$

$$\Sigma u(x_i, y_i) = h^2 \cdot \rho(x_i, y_i)u(x_i, y_i). \tag{5–70}$$

Operators $\boldsymbol{H}$ and $\boldsymbol{V}$ correspond to the horizontal and vertical second central difference operators, respectively. At boundary points and at grid points close to the boundary, the operators are defined analogously. If we lump the terms coming from $h^2 f(x_i, y_i)$ and from the boundary values, which are all independent of the unknowns $u(x_i, y_i)$, into a constant vector $\boldsymbol{b}$, then (5–67) is replaced by a system of linear equations in $\boldsymbol{u}$ with the operators $\boldsymbol{H}$, $\boldsymbol{V}$, and, $\boldsymbol{\Sigma}$ as shown:

$$(\boldsymbol{H} + \boldsymbol{V} + \boldsymbol{\Sigma})\boldsymbol{u} + \boldsymbol{b} = 0. \tag{5–71}$$

Matrices $\boldsymbol{H}$, $\boldsymbol{V}$, and $\boldsymbol{\Sigma}$ are real, symmetric matrices. $\boldsymbol{\Sigma}$ is a diagonal matrix whose diagonal elements are nonnegative in light of the requirement $\rho(x, y) \geq 0$. Matrices $\boldsymbol{H}$ and $\boldsymbol{V}$ have, at most, three nonzero elements in each row, the diagonal one being positive and the two others usually being dominated in the weak sense. $\boldsymbol{H}$ and $\boldsymbol{V}$ are thus generally positive definite, or at least positive semidefinite. By means of appropriate simultaneous row and column permutations, they can be brought into a tridiagonal form.

For any real r, Eq. (5–71) is clearly equivalent with the two vector equations

$$(\boldsymbol{H} + \tfrac{1}{2}\boldsymbol{\Sigma} + r\boldsymbol{I})\boldsymbol{u} + (\boldsymbol{V} + \tfrac{1}{2}\boldsymbol{\Sigma} - r\boldsymbol{I})\boldsymbol{u} + \boldsymbol{b} = 0, \tag{5–72}$$

$$(\boldsymbol{V} + \tfrac{1}{2}\boldsymbol{\Sigma} + r\boldsymbol{I})\boldsymbol{u} + (\boldsymbol{H} + \tfrac{1}{2}\boldsymbol{\Sigma} - r\boldsymbol{I})\boldsymbol{u} + \boldsymbol{b} = 0. \tag{5–73}$$

The *method of alternating directions* is defined by the iteration procedure[1]

$$(\boldsymbol{H} + \tfrac{1}{2}\boldsymbol{\Sigma} + r_{m+1}\boldsymbol{I})\boldsymbol{u}^{(m+1/2)} + (\boldsymbol{V} + \tfrac{1}{2}\boldsymbol{\Sigma} - r_{m+1}\boldsymbol{I})\boldsymbol{u}^{(m)} + \boldsymbol{b} = 0, \tag{5–74}$$

$$(\boldsymbol{V} + \tfrac{1}{2}\boldsymbol{\Sigma} + r_{m+1}\boldsymbol{I})\boldsymbol{u}^{(m+1)} + (\boldsymbol{H} + \tfrac{1}{2}\boldsymbol{\Sigma} - r_{m+1}\boldsymbol{I})\boldsymbol{u}^{(m+1/2)} + \boldsymbol{b} = 0. \tag{5–75}$$

Here the parameter r_m indicates a sequence of positive numbers serving to accelerate the convergence. In Eq. (5–74) the components of $\boldsymbol{u}^{(m+1/2)}$ belonging to one row are coupled to one another by the operator $(\boldsymbol{H} + \tfrac{1}{2}\boldsymbol{\Sigma} + r_{m+1}\boldsymbol{I})$, and thus their evaluation requires solution of a linear system of equations. Therefore, the method is an *implicit* one.

Furthermore, in $(\boldsymbol{V} + \tfrac{1}{2}\boldsymbol{\Sigma} - r_{m+1}\boldsymbol{I})\boldsymbol{u}^{(m)}$, the approximations after the mth step are always used, and not, for instance, the currently best available values. The sequence in which we solve for values $\boldsymbol{u}^{(m+1/2)}$ of the individual rows is immaterial. The same is true in Eq. (5–75) for the columns. The two half-

[1]We use a variant of the original *Peaceman-Rachford* method as used by *Sheldon* and *Wachspress* (Refs. 69, 6).

steps, (5–74) and (5–75), represent a sort of simultaneous displacement iteration by rows or columns, but, of course, not in the sense of the block relaxation of Sec. 5–2–3, since different matrix operators appear in place of operator equations. The designation of the method is attributable to the alternating grouping of nodal points by rows and columns.

It is useful to define the matrices $\boldsymbol{H}_1 = \boldsymbol{H} + \frac{1}{2}\boldsymbol{\Sigma}$ and $\boldsymbol{V}_1 = \boldsymbol{V} + \frac{1}{2}\boldsymbol{\Sigma}$, having the same properties as $\boldsymbol{H}$ and $\boldsymbol{V}$, respectively. Upon elimination of the vector $\boldsymbol{u}^{(m+1/2)}$, Eqs. (5–74) and (5–75) lead to the relation

$$(\boldsymbol{V}_1 + r_{m+1}\boldsymbol{I})\boldsymbol{u}^{(m+1)} - (\boldsymbol{H}_1 - r_{m+1}\boldsymbol{I})(\boldsymbol{H}_1 + r_{m+1}\boldsymbol{I})^{-1}(\boldsymbol{V}_1 - r_{m+1}\boldsymbol{I})\boldsymbol{u}^{(m)} + \{\boldsymbol{I} - (\boldsymbol{H}_1 - r_{m+1}\boldsymbol{I})(\boldsymbol{H}_1 + r_{m+1}\boldsymbol{I})^{-1}\}\boldsymbol{b} = 0. \qquad (5\text{–}76)$$

This is a general iteration method, as in Eq. (2–22), having in the mth step the iteration matrix dependent on r_m,

$$\boldsymbol{M}_A(r_m) = (\boldsymbol{V}_1 + r_m\boldsymbol{I})^{-1}(\boldsymbol{H}_1 - r_m\boldsymbol{I})(\boldsymbol{H}_1 + r_m\boldsymbol{I})^{-1}(\boldsymbol{V}_1 - r_m\boldsymbol{I}). \qquad (5\text{–}77)$$

The convergence of the *Peaceman-Rachford* method is established by the selected sequence of the *iteration parameter* r_m. In general, it follows for the error vectors $\boldsymbol{f}^{(m)} = \boldsymbol{u} - \boldsymbol{u}^{(m)}$ that

$$\boldsymbol{f}^{(m)} = \left(\prod_{i=1}^{m} \boldsymbol{M}_A(r_i)\right)\boldsymbol{f}^{(o)}. \qquad (5\text{–}78)$$

For arbitrarily shaped domains G, no generally applicable, theoretically based statements on the optimal choice of iteration parameter have so far been established. In order to explore convergence in principle, let us consider the special case where all the iteration parameters r_i have a fixed value $r > 0$. In this case, $\boldsymbol{M}_A(r)$ is a matrix dependent on r, but not on the step m:

$$\boldsymbol{M}_A(r) = (\boldsymbol{V}_1 + r\boldsymbol{I})^{-1}(\boldsymbol{H}_1 - r\boldsymbol{I})(\boldsymbol{H}_1 + r\boldsymbol{I})^{-1}(\boldsymbol{V}_1 - r\boldsymbol{I}). \qquad (5\text{–}79)$$

Theorem 5-4. *If the matrices* $\mathbf{H}_1 = \mathbf{H} + \frac{1}{2}\boldsymbol{\Sigma}$ *and* $\mathbf{V}_1 = \mathbf{V} + \frac{1}{2}\boldsymbol{\Sigma}$ *are symmetric and positive semidefinite, and one of them is positive definite, the method of alternating directions of Peaceman-Rachford, given by* (5–74) *and* (5–75), *converges for any constant positive iteration parameter* r.

Proof: Matrix $\boldsymbol{M}_A(r)$ in Eq. (5–79) is similar to

$$\begin{aligned}\boldsymbol{M}_A^*(r) &= (\boldsymbol{V}_1 + r\boldsymbol{I})\boldsymbol{M}_A(r)(\boldsymbol{V}_1 + r\boldsymbol{I})^{-1} \\ &= (\boldsymbol{H}_1 - r\boldsymbol{I})(\boldsymbol{H}_1 + r\boldsymbol{I})^{-1}(\boldsymbol{V}_1 - r\boldsymbol{I})(\boldsymbol{V}_1 + r\boldsymbol{I})^{-1}, \end{aligned} \qquad (5\text{–}80)$$

thus their eigenvalues are the same. Since the spectral radius of a matrix is also a matrix norm, it follows that

$$\rho(\boldsymbol{M}_A(r)) \leq \rho\{(\boldsymbol{H}_1 - r\boldsymbol{I})(\boldsymbol{H}_1 + r\boldsymbol{I})^{-1}\} \cdot \rho\{(\boldsymbol{V}_1 - r\boldsymbol{I})(\boldsymbol{V}_1 + r\boldsymbol{I})^{-1}\}. \qquad (5\text{–}81)$$

The given information assures that the eigenvalues λ_i and ν_i of the matrices $\boldsymbol{H}_1$ and $\boldsymbol{V}_1$ are real and nonnegative. Matrices $(\boldsymbol{H}_1 - r\boldsymbol{I})(\boldsymbol{H}_1 + r\boldsymbol{I})^{-1}$ and $(\boldsymbol{V}_1 - r\boldsymbol{I})(\boldsymbol{V}_1 + r\boldsymbol{I})^{-1}$ are also symmetric for real positive r, and their eigenvalues are given by

$$\frac{\lambda_i - r}{\lambda_i + r} \quad \text{and} \quad \frac{\nu_i - r}{\nu_i + r},$$

as is evident from inspection of the eigenvalues and corresponding eigenvectors of $\boldsymbol{H}_1$ and $\boldsymbol{V}_1$, respectively. Since $\lambda_i \geq 0$ and $\nu_i \geq 0$, there results for the spectral radii on the right-hand side of (5–81)

$$\rho\{(\boldsymbol{H}_1 - r\boldsymbol{I})(\boldsymbol{H}_1 + r\boldsymbol{I})^{-1}\} = \max_i \left|\frac{\lambda_i - r}{\lambda_i + r}\right| \leq 1 \quad \text{for} \quad r > 0 \qquad (5\text{–}82)$$

and

$$\rho\{(\boldsymbol{V}_1 - r\boldsymbol{I})(\boldsymbol{V}_1 + r\boldsymbol{I})^{-1}\} = \max_i \left|\frac{\nu_i - r}{\nu_i + r}\right| \leq 1 \quad \text{for} \quad r > 0. \qquad (5\text{–}83)$$

By the given information, at least one of the matrices $\boldsymbol{H}_1$ and $\boldsymbol{V}_1$ is positive definite. Suppose it is $\boldsymbol{H}_1$. Then its eigenvalues are greater than zero, and thus $\rho\{(\boldsymbol{H}_1 - r\boldsymbol{I})(\boldsymbol{H}_1 + r\boldsymbol{I})^{-1}\} < 1$. From (5–81) and the fact that $\rho(\boldsymbol{M}_A(r)) < 1$, the convergence of the method of alternating directions for any arbitrary positive value of r follows directly.

The method of alternating directions converges under rather general requirements, according to Theorem 5–4. But the question of the optimum choice of parameter r for minimizing the spectral radius $\rho(\boldsymbol{M}_A(r))$ remains unanswered. The relation (5–81) provides merely an *inequality* for the spectral radius $\rho(\boldsymbol{M}_A(r))$. Even if the upper bound is minimal, the corresponding $\rho(\boldsymbol{M}_A(r))$ need not be the smallest one. In order to be able to solve the problem quite strictly, the matrices $\boldsymbol{H}_1$ and $\boldsymbol{V}_1$ must be *commutative*:

$$\boldsymbol{H}_1\boldsymbol{V}_1 = \boldsymbol{V}_1\boldsymbol{H}_1. \qquad (5\text{–}84)$$

With the additional requirement (5–84), the symmetric matrices $\boldsymbol{H}_1$ and $\boldsymbol{V}_1$ possess a common complete system of orthonormal eigenvectors $\boldsymbol{x}_i$ ($i = 1, 2, \ldots, n$), so that we have, simultaneously,

$$\boldsymbol{H}_1\boldsymbol{x}_i = \lambda_i\boldsymbol{x}_i, \quad \boldsymbol{V}_1\boldsymbol{x}_i = \nu_i\boldsymbol{x}_i \qquad (i = 1, 2, \ldots, n). \qquad (5\text{–}85)$$

From this fact, it follows that the matrices $(\boldsymbol{H}_1 - r\boldsymbol{I})$, $(\boldsymbol{H}_1 + r\boldsymbol{I})^{-1}$, $(\boldsymbol{V}_1 - r\boldsymbol{I})$ and $(\boldsymbol{V}_1 + r\boldsymbol{I})^{-1}$ possess these same eigenvectors $\boldsymbol{x}_i$, and this applies, too, to $\boldsymbol{M}_A^*(r)$. The eigenvalues μ_i of $\boldsymbol{M}_A(r)$, and thus of $\boldsymbol{M}_A^*(r)$, are thus

$$\mu_i = \frac{(\lambda_i - r)(\nu_i - r)}{(\lambda_i + r)(\nu_i + r)}.$$

Under the assumption that $\boldsymbol{H}_1$ or $\boldsymbol{V}_1$ is not just semidefinite but positive definite, there follows for the spectral radius that

$$\rho(\boldsymbol{M}_A(r)) = \max_i \left| \frac{(\lambda_i - r)(\nu_i - r)}{(\lambda_i + r)(\nu_i + r)} \right| < 1 \quad \text{for} \quad r > 0. \tag{5–86}$$

The optimal value of r is given by the solution of the minimum-maximum problem,

$$\min_{r>0} \left\{ \max_i \left| \frac{(\lambda_i - r)(\nu_i - r)}{(\lambda_i + r)(\nu_i + r)} \right| \right\}.$$

An explicit solution exists for a square or rectangular domain, for in these cases, the eigenvalues λ_i and ν_i are known. For such domains, Ref. 68 shows that Peaceman-Rachford's method of alternating directions with a constant optimal parameter r exhibits the same convergence rate as *point overrelaxation* with optimal ω_{opt}. The computation effort per iteration step, however, is appreciably greater, and, therefore, the Peaceman-Rachford method in this form is less advantageous than the variants of overrelaxation.

The method of alternating directions becomes superior to other methods if a succession of differing positive iteration parameters r_m is used. The spectral radius of a sequence of p iteration steps with varying parameters $r_j > 0$ is, for the commutative case (5–84),

$$\rho\Big(\prod_{j=1}^{p} \boldsymbol{M}_A(r_j)\Big) = \max_i \prod_{j=1}^{p} \left| \frac{(\lambda_i - r_j)(\nu_i - r_j)}{(\lambda_i + r_j)(\nu_i + r_j)} \right|. \tag{5–87}$$

In the event that all the eigenvalues λ_i of $\boldsymbol{H}_1$ are known (and/or all the eigenvalues ν_i of $\boldsymbol{V}_1$), then with $p = n$, the sequence of r_j's could be chosen equal to the set of positive eigenvalues, so that

$$\rho\Big(\prod_{j=1}^{n} \boldsymbol{M}_A(r_j)\Big) = 0.$$

The procedure would theoretically be terminated after n steps and would thus be finite. In general, however, the eigenvalues are not known, and we confine ourselves to a few parameters which turn up cyclically in the application.

If both matrices $\boldsymbol{H}_1$ and $\boldsymbol{V}_1$ are symmetric definite, and at least some of the bounds are known for the eigenvalues such that $0 < a \leq \lambda_i \leq b$ and $a \leq \nu_i \leq b$, the r_j's are so determined that the expression

$$\max_{a \leq x, y \leq b} \prod_{j=1}^{p} \left| \frac{(x - r_j)(y - r_j)}{(x + r_j)(y + r_j)} \right|$$

is minimal. This problem is an approximation problem for rational functions, in the sense of Chebyshev. For $p = 1$, simplification in an elementary way yields $r_1 = r = \sqrt{ab}$. For $p = 2^q$ (q being a nonnegative integer), *Wachspress* (Refs. 70, 71) gives the optimal values of r_j. On the other hand, we are often content with reasonably good parameters which can guarantee a very good convergence for the Peaceman-Rachford method. According to Wachspress, when $p > 1$, the values

$$r_j^{(W)} = a\left(\frac{b}{a}\right)^{(j-1)/(p-1)} \qquad j = 1, 2, \ldots, p \tag{5–88}$$

produce a set of good iteration parameters, for which he provides an upper bound for the spectral radius:

$$\rho\left(\prod_{j=1}^{p} \boldsymbol{M}_A(r_j^{(W)})\right) \leq \left(\frac{d-1}{d+1}\right)^4, \quad \text{with} \quad d = \left(\frac{b}{a}\right)^{1/(2p-2)}. \tag{5–89}$$

A better choice of parameters is suggested in Ref. 8. With such good sets of parameters, we can produce such an extraordinary convergence that the total computational effort is notably less than in overrelaxation methods.

Despite these remarkable properties of the Peaceman-Rachford method, it should be used with a few reservations. The statements on the excellent convergence hold only in the case of *commutative operators* $\boldsymbol{H}_1$ and $\boldsymbol{V}_1$. This case occurs, however, only for rectangular domains (Ref. 68). The strict theory cannot as yet be extended to more general domains, and thus we are driven to numerical experiments. Numerical explorations have been promising; nevertheless, the overrelaxation methods, whose theory encompasses a larger class of boundary value problems over general domains, appear more favorable. In addition, the method of alternating directions in the commutative case shows computational economies in relation to overrelaxation only with a very fine mesh width ($h \longrightarrow 0$). With a given general domain and given mesh width h, theoretical results favoring one or the other method are still wanting.

More thorough descriptions and discussion of the Peaceman-Rachford method and its variants are given in Refs. 7, 68, and 71 and in other works cited therein. Reference 7 in particular lists results of extensive numerous experiments for various domains and compares the methods with each other.

Example 5-7. In Example 5–1, $\rho(x, y) \equiv 0$, thus operator $\boldsymbol{\Sigma}$ vanishes. Furthermore, the schematic representations of operators $\boldsymbol{H}$ and $\boldsymbol{V}$ are given for the three types of points in Figs. 36 and 37.

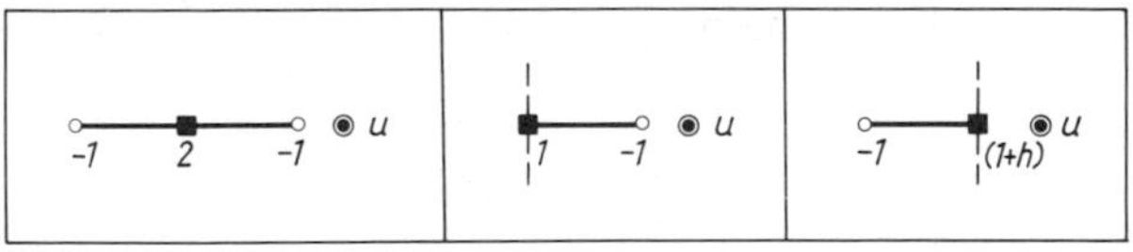

Figure 36. Operator $\boldsymbol{H}$. (a) Interior point. (b) Left boundary point. (c) Right boundary point.

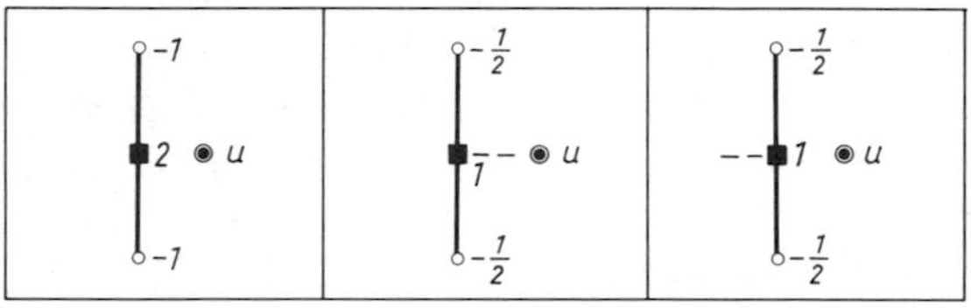

Figure 37. Operator $\boldsymbol{V}$. (a) Interior point. (b) Left boundary point. (c) Right boundary point.

For the practical execution of computations, we should use these operators without setting up the corresponding matrices. The positive definiteness of $\boldsymbol{H}$ follows from the weak diagonal dominance by means of Theorem 1–5 since in the right-hand boundary points (Fig. 36c), at least, there is a true inequality. For operator $\boldsymbol{V}$, positive definiteness follows from the fact that for interior nodes on the upper and lower boundaries, one arm drops out, and thus weak diagonal dominance is again established. The operators $\boldsymbol{H}$ and $\boldsymbol{V}$ are positive definite and commutative.

Let us examine the coarse network of the domain G. The operator matrices for the first half-step are the same for every line, so that with numbering of nodal points by rows, matrix $\boldsymbol{H}$ is made up of three identical six-rowed tridiagonal matrices $\boldsymbol{H}^{(i)}$:

$$\boldsymbol{H}^{(i)} = \begin{bmatrix} 1 & -1 & & & & \\ -1 & 2 & -1 & & & \\ & -1 & 2 & -1 & & \\ & & -1 & 2 & -1 & \\ & & & -1 & 2 & -1 \\ & & & & -1 & 1+h \end{bmatrix} \qquad i = 1, 2, 3,$$

arranged along the diagonal. Matrix $\boldsymbol{H}$ itself is tridiagonal and reduceable.

With a numbering of nodal points by rows, matrix V would not be tridiagonal. A numbering of nodes by columns is thus in order. Matrix V decomposes into six three-rowed tridiagonal matrices $V^{(i)}$, lying on the diagonal. Two types occur, corresponding to a column either on the right or left boundary, or in the interior of the domain:

$$V^{(i)} = \begin{bmatrix} 1 & -\frac{1}{2} & \\ -\frac{1}{2} & 1 & -\frac{1}{2} \\ & -\frac{1}{2} & 1 \end{bmatrix}, \qquad V^{(i)} = \begin{bmatrix} 2 & -1 & \\ -1 & 2 & -1 \\ & -1 & 2 \end{bmatrix}$$

$$(i = 1, 6); \qquad\qquad (i = 2, 3, 4, 5).$$

With variable iteration parameters r_j, only three decompositions are thus to be taken per step, and in addition only forward and backward substitution need be carried out. For a good choice of iteration parameters tending to improve convergence, bounds on the eigenvalues of $\boldsymbol{H}$ and V are needed. Since the matrices decompose, it suffices to find bounds on the eigenvalues of the three submatrices. By *Gerschgorin's* theorem, $b = 4$ is an upper bound on all eigenvalues. Lower bounds are easily found by the method of *bisection* or the *QD algorithm*.

For the coarse mesh width ($h = 1$), the smallest eigenvalue of matrices $\boldsymbol{H}$ and V is 0.058116. Table 27 shows the iteration parameters $r_j^{(W)}$ of Wachspress from (5–88), the resultant asymptotic convergence radii by means of p steps, and the estimate (5–89) of Wachspress, for various values of the lower bound a and of the quantity p.

Table 27 METHOD OF ALTERNATING DIRECTIONS; ITERATION PARAMETERS FOR COARSE GRID; CONVERGENCE RADII

a	b	p	$r_j^{(W)}$	ρ	d	$\left(\frac{d-1}{d+1}\right)^4$
0.01	4.00	2	0.01000, 4.00000	0.55	20	0.670
0.01	4.00	4	0.01000, 0.07368 0.54288, 4.00000	0.033	2.714	0.045
0.05	4.00	2	0.05000, 4.00000	0.38	8.944	0.407
0.05	4.00	4	0.05000, 0.21544 0.92832, 4.00000	0.014	2.076	0.015
0.05	4.00	8	0.05000, 0.09351 0.17487, 0.32702 0.61158, 1.14372 2.13890, 4.00000	0.00011	1.368	0.00058

The quality of convergence is better than indicated by the estimate. In the last example ($p = 8$), the error vector is multiplied roughly by the factor 10^{-4} for every eight iterations, so that after 16 steps the solution is good to six-figure accuracy. The analogous numbers for the finer grid are shown in Table 28. The smallest eigenvalue is 0.01581.

Table 28 METHOD OF ALTERNATING DIRECTIONS; ITERATION PARAMETERS FOR FINE GRID; CONVERGENCE RADII

a	b	p	$r_j^{(W)}$	ρ	d	$\left(\frac{d-1}{d+1}\right)^4$
0.015	4.00	2	0.01500, 4.00000	0.55	16.33	0.612
0.015	4.00	4	0.01500, 0.09655 0.62145, 4.00000	0.016	2.537	0.036
0.015	4.00	8	0.01500, 0.03332 0.07400, 0.16436 0.36506, 0.81082 1.80091, 4.00000	0.00044	1.490	0.0015

5-2-5 Method of Conjugate Gradients

The method of conjugate gradients is especially applicable if the calculation of the matrix vector product $z = Ap$ can be based on arithmetic expressions such that matrix A never enters explicitly. The operator equations of a discretized boundary value problem are of this type. Besides, this method is usable without qualifications for biharmonic problems of the theory of elasticity. In contrast to the other relaxation methods, it requires no estimates of eigenvalues or of optimal relaxation parameters, and thus its use offers the fewest complications. In addition, the method of conjugate gradients yields some information about the eigenvalues of operator A through the q and e values that turn up (see Sec. 4–6–6 and Ref. 13).

Example 5-8. For the standard Example 5–1 with a fine grid, the procedure *cg* of Sec. 2–4–3 can be developed in detail by defining the procedure *op*(n, p, z) necessary to compute the vector $z = Ap$ from a given vector p. The vectors are designated as simply subscripted variables. Thus the nodal points and their functional values must constantly be numbered. In the numbering scheme of Fig. 27, nine types of definition equations must be utilized, as noted in the following procedure.

Table 29 SOLUTION OF THE BOUNDARY VALUE PROBLEM WITH A FINE MESH

−0.210656	−0.213297	−0.216977	−0.210935	−0.188135	−0.153765
−0.416029	−0.425557	−0.443677	−0.438627	−0.387841	−0.309147
−0.602349	−0.629223	−0.693547	−0.712055	−0.615454	−0.463792
−0.734919	−0.795439	−0.989232	−1.100592	−0.898127	−0.598781
−0.746450	−0.828382	−1.117352	−1.302952	−1.027683	−0.632196
−0.594116	−0.654288	−0.848840	−0.966184	−0.777456	−0.497729
−0.321436	−0.345815	−0.407535	−0.435487	−0.368228	−0.259969

ALGOL Procedure No. 16

```
procedure op (n, p, z);  value n;
          integer n;  array p, z;
begin integer i, k;
op1:  z[1]: = 2 × p[1] − 0.5 × p[2] − p[8];
    for i: = 2 step 1 until 6 do
op2:   z[i] = 2 × p[i] − 0.5 × (p[i − 1] + p[i + 1]) − p[i + 7];
op3:  z[7]: = 2 × p[7] − 0.5 × p[6] − p[14];
    for i: = 8 step 7 until 64 do
    begin
op4:   z[i]: = 4 × p[i] − p[i − 7] − p [i + 7] − p[i + 1];
        for k: = i + 1 step 1 until i + 5 do
op5:  z[k]: = 4 × p[k] − p[k − 1] − p[k − 7] − p[k + 1] − p[k + 7];
op6:  z[i + 6]: = 4 × p[i + 6] − p[i + 5] − p[i − 1] − p[i + 13]
    end i;
op7:  z[71]: = 2.5 × p[71] − 0.5 × p[72] − p[64];
    for i: = 72 step 1 until 76 do
op8:  z[i]: = 2.5 × p[i] − 0.5 × (p[i − 1] + p[i + 1]) − p[i − 7];
op9:  z[77]: = 2.5 × p[77] − 0.5 × p[76] − p[70]
end op
```

The length of the residual vector $\boldsymbol{r}^{(k)}$ as a function of the number of steps k, as manifested in procedure *cg* in conjunction with procedure *op*, is shown in Fig. 38 on a logarithmic scale. It does not decrease monotonically with increasing iteration index k. The irregular decrease with occasional increases of the residual vector length is a widely observed phenomenon (Ref. 13). After $n = 77$ steps the length of the residual vector has decreased to approximately 4.7×10^{-22}. The approximate solution found for the boundary value problem is given in Table 29. The decimal places shown are all significant figures. Via a graphic display of functional values along horizontal lines of the network, we can observe that the natural boundary conditions of the boundary value problem are well adhered to at the two vertical boundaries.

Table 29 (Cont'd)

−0.117779	−0.086159	−0.060804	−0.041351	−0.026530
−0.231191	−0.166052	−0.115707	−0.078070	−0.049948
−0.331786	−0.231153	−0.157900	−0.105273	−0.067071
−0.401008	−0.268873	−0.179467	−0.118053	−0.074862
−0.404591	−0.263864	−0.173043	−0.112607	−0.071135
−0.321297	−0.208950	−0.136235	−0.088198	−0.055598
−0.173917	−0.114404	−0.074747	−0.048351	−0.030460

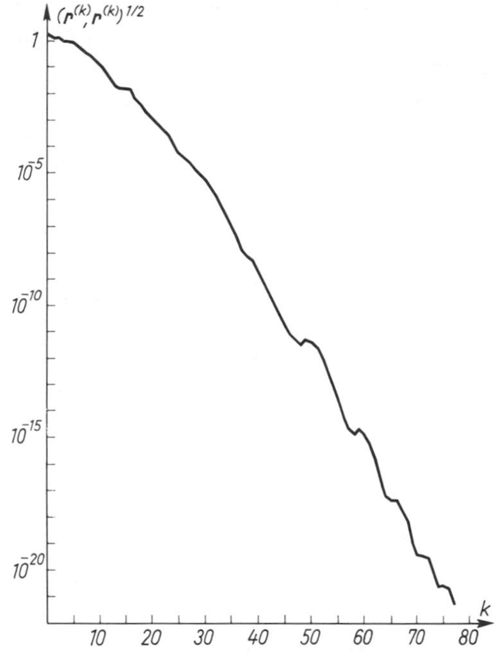

Figure 38. Length of the residual vector.

PROBLEMS

5-6. Solve the boundary value of Problem 5-2, using a consistent sequence of nodal points, by the method of overrelaxation. How large are the convergence radius and convergence rate for the successive displacement method and for optimal overrelaxation? (A computer is required.)

5-7. The boundary value Problem 5-2 is to be solved with a mesh width that is twice and four times as fine ($h = \frac{1}{8}$ and $h = \frac{1}{16}$), where the optimal relaxation factor is to be found experimentally, using the residual vector in the successive displacement method, or using a selected factor in the overrelaxation method. (A computer is necessary.)

5-8. For boundary value Problem 5-2, find the convergence radius of implicit block relaxation and of optimal blockwise overrelaxation, if nodal points are grouped by rows. (A computer is necessary.)

5-9. For boundary value Problem 5-2, find matrix operators **H**, **V**, and **Σ** of the method of alternating directions. Are the operators **H** and **V** commutative?

5-3 THE EIGENVALUE PROBLEM

Let us examine the eigenvalue problem of partial differential equations and consider the solution techniques. Physically, this eigenvalue problem corresponds to the task of finding the lowest frequencies of vibration and, if desired, the corresponding vibration mode shapes of, say, a vibrating linear elastic membrane or plate. Mathematically, the task consists of computing several of the *smallest eigenvalues and the corresponding eigenvectors* of $(A - \lambda I)x = 0$, where A is defined by operator equations as derived and discussed in Sec. 5–1. The coefficient matrix A exhibits numerous zero elements. Furthermore, with a fine mesh width, the order of the matrix becomes so very large that listing and storage of its coefficients in computer memories is often impossible or undesirable. Even if the matrix should turn out to be banded as a result of an opportune numbering of nodal points, the LR transformation (Sec. 4–6–4) will not destroy the zeros outside the banded region, but it will certainly do so to the zeros within. As a result, we will now confine ourselves to problem formulations where A is given as an *operator*, in the form of operator expressions.[1]

(1) If only *several of the smallest eigenvalues of operator A* are sought, in principle, we can solve the appropriate boundary value problem under an arbitrary load function by the method of conjugate gradients (Sec. 2–4). Then q and e values dependent on the loading will turn up and, after application of the QD algorithm (Secs. 4–6–5 and 4–6–6), these will furnish initial information about the eigenvalues. This procedure, moreover, has the property of utilizing the many zeros in the problem matrix without destroying them. This solution method, however, is subject to numerical instability, and it is even possible that one of the smallest eigenvalues sought is totally omitted. Note Ref. 13, which describes how to sidestep this pitfall by a refinement of the technique. The approach described is, in any event, recommendable if the boundary value problem must be solved for a given loading, in order to acquire initial information about the smallest eigenvalues. An improvement of these approximate values by invoking the *Wielandt* technique (Sec. 4–7–3) is, however, impossible since in the spectral transformation the definiteness of operator A generally disappears. Wielandt's inverse vector iteration requires at each step the solution of such an indefinite boundary value problem, but relaxation methods are no longer applicable.

[1]The case of a given explicit matrix is treated in Sec. 4–9.

(2) For finding the *p smallest eigenvalues with the corresponding eigenvectors*, the method of vector iteration (Sec. 4–7) is the only one to consider. It can be used in two ways:

(a) Through a norm, an upper bound m of the eigenvalues of operator $\boldsymbol{A}$ is always known. For the operator $\boldsymbol{B} = m\boldsymbol{I} - \boldsymbol{A}$, where $\boldsymbol{I}$ is the identity operator, or identity matrix, the smallest eigenvalues of $\boldsymbol{A}$ are carried over as the largest eigenvalues of $\boldsymbol{B}$ by the spectral transformation. Ordinary simultaneous vector iteration (Sec. 4–7–4) with operator $\boldsymbol{B}$ yields the dominant eigenvalues and thus, indirectly, the smallest eigenvalues of $\boldsymbol{A}$, as well as the corresponding eigenvectors, which are identical to $\boldsymbol{A}$ and $\boldsymbol{B}$. This procedure has the notable advantage of simplicity. Unfortunately, however, it has the drawback of a very slow convergence in general, since the key convergence quotient of the eigenvalues of $\boldsymbol{B}$ lies very close to unity. For the case of a plate in Ref. 13, the quotients for various values of p lie in the vicinity of 0.9998! We should also bear in mind that at the end of the calculations, the smallest eigenvalues of $\boldsymbol{A}$ turn up as the difference of almost equal quantities because of the spectral transformation, and thus significant decimal places are lost. Accordingly, the eigenvalues of $\boldsymbol{B}$ should be computed more accurately than is required for those of $\boldsymbol{A}$.

(b) Simultaneous inverse vector iteration (Sec. 4–7–4), applied directly to operator $\boldsymbol{A}$, immediately produces the p eigenvalues and eigenvectors that are sought. Each iteration step requires the simultaneous solution of a boundary value problem for p different "loadings," given by the instantaneous eigenvector approximations. Because of the positive definiteness of $\boldsymbol{A}$, this problem can be solved by any relaxation method. It should be emphasized that *for all iteration steps, the same unaltered operator is involved.* For instance, in the use of the overrelaxation method, the same optimal relaxation factor can always be used, once it is found experimentally in the first iterations. The convergence of simultaneous inverse vector iteration is significantly better, compared to the alternate method in (a) above, since the quotient of important eigenvalues determining convergence is much smaller. For the example of the plate in Ref. 13, the quotients of the first eight successive smallest eigenvalues lie between 0.303 and 0.965. Another advantage is that the eigenvalues are found directly, without added loss of significant places. For this reason, despite the complicated structure and the significantly greater computational effort per iteration step, simultaneous inverse vector iteration is recommended, since the faster convergence prevents the total computational effort from growing unduly large.

APPENDIX A

FORTRAN Subroutine Counterparts of ALGOL Procedures

The 16 ALGOL procedures appearing in the body of this book have been converted into their FORTRAN IV subroutine counterparts in this Appendix. The closest possible structural correlation has been maintained and, where-ever possible, the same subprogram names, variable names, index numberings, and operations have been retained. Only in the Subprograms Nos. 4 and 14 did subtle indexing differences arise which could confuse the unwary. Table 30 lists all name counterparts which involve changes, as well as some others which could be ambiguous.

Most of the alterations appearing in the FORTRAN counterpart programs are attributable to:

(a) FORTRAN integer variables having to start with letters I through N;

(b) names being allowed no more than six letters;

(c) limits on the types and sequences of integer expressions allowed in a DO statement; and

(d) rules regarding arguments appearing in a subroutine statement. In most cases, only very minor changes had to be made. In some cases, however, variable dimensioning had to be abandoned, and argument lists in subroutine statements had to be altered.

The following alterations should be particularly noted:

Subprogram No. 2. The final argument in the subroutine statement, INDEF, serves a role not as a statement number but as an indicator. If the array A is not positive definite, the subprogram sets INDEF = 1 and returns immediately to the main program. Otherwise, INDEF = 0 and computations are carried out normally.

Table 30 ALGOL TO FORTRAN EQUIVALENTS

Program Number	ALGOL *Variable*		FORTRAN *Variable*
1	*linverse*	=	LINVRS
2	*cholesky*	=	CHLSK
2, 4, 6	*p*	=	IP
2, 4, 14	*indef*		(redefined)
3	*choleskysol*	=	CHLSKS
4	*k*, *m*	=	K, M
4, 14	*n*	=	N and-or NDIM
4, 14	*choleskyband*	=	CHLSKB
7	*fehlerorth*	=	FELORT
8, 9, 10, 11	*rotation*	=	ROTATN
8, 9, 10	*p*	=	IP
8, 9, 10	*q*	=	IQ
9	*jacobi* 1	=	JACOB1
9	*max*	=	XMAX
10	*jacobi* 2	=	JACOB2
12	*householder*	=	HSHLDR
12, 13	*t*	=	IT
13	*bisection*	=	BISECT
13	*min*	=	EIGMIN
13	*max*	=	EIGMAX
13	*lambda*	=	ALAMDA
13	*v*	=	IV
14	*lrcholband*	=	LRCHLB
14, 15	*m*, *n*	=	M, N
15	*overrelax*	=	OVRLAX
15	*z*	=	IZ

Subprogram No. 3. Variable dimension array size cannot be used for indexed variables not appearing in the argument list of the subroutine statement. For those indexed variables, a maximum array size of 200 was used, which should be more than ample for most applications.

Subprogram No. 4. The third argument in the subroutine statement, MPLS1 = M + 1, is one greater than the corresponding argument in the ALGOL program. Furthermore, the column numbers of both R and A run from 1 to M + 1, whereas in the ALGOL procedures, *r* and *a* had columns numbered 0 to *m*. See also the note to Subprogram No. 2. Also, observe that in some versions of FORTRAN, complicated subscript expressions such as R(IP, I − IP + 1) are inadmissible, and some dummy variable like IDUMMY = I − IP + 1 must then be introduced in a previous statement. Finally, note that ALGOL variable *n* splits into two different FORTRAN variables N and NDIM. NDIM is used to indicate the storage space set aside for an indexed variable by the dimension statement. N, which may not exceed NDIM, signifiies the maximum number of elements of a variable on which operations will be carried out. The last (NDIM — N) elements are totally ignored.

Subprograms Nos. 5, 7, 12, 13. See note to Subprogram No. 3.

Subprogram No. 14. The second argument in the subroutine statement, MPLS1 = M + 1, is one greater than the corresponding argument in the ALGOL program. The doubly indexed variable R has also been added to the argument list of the subroutine statement. Just as for Subprogram No. 4, the columns of R and A both run from 1 to $m + 1$, whereas in the ALGOL procedures, r and a had columns numbered 0 to m. See also other notes to Subprograms Nos. 2 and 4 above.

Subprogram No. 15. The second argument in the subroutine statement, MPLS2 = M + 2, is two greater than the corresponding argument in the ALGOL procedure statement. And while doubly indexed variables U and F correspond to u and f, their row numberings differ. The rows of U run from 1 to M + 2, while for u they ran 0 to $m + 1$. The rows of F are numbered 2 to M + 1, while for f they ran 1 to m.

All of these FORTRAN IV subprograms have been successfully compiled on a digital computer, and several test problems have been run as a check.

```
C        FORTRAN SUBROUTINE NO.1, LINVERSE...
C        THIS SUBROUTINE FINDS THE INVERSE MATRIX B OF A
C         GIVEN LOWER TRIANGULAR MATRIX A...
         SUBROUTINE  LINVRS(N,A,B)
         DIMENSION  A(N,N),  B(N,N)
         DO  50   I=1,N
          IF (I-1)  45,45,10
   10     ILSS1= I-1
C         COMPUTATION OF I-TH ROW OF B.
          DO 40  K=1, ILSS1
           S=0.
           DO 30  J=K,ILSS1
            S= S + A(I,J)*B(J,K)
   30      CONTINUE
           B(I,K) = -S/A(I,I)
   40     CONTINUE
   45     B(I,I) = 1./A(I,I)
   50    CONTINUE
         RETURN
         END
```

```
C     FORTRAN SUBROUTINE  NO. 2. CHOLESKY.
C     THIS SUBROUTINE COMPUTES THE UPPER TRIANGULAR MATRIX R WHICH,
C      WHEN MULTIPLIED BY ITS OWN TRANSPOSE, YIELDS THE GIVEN
C      SYMMETRIC MATRIX A, BY THE CHOLESKY METHOD...
      SUBROUTINE  CHLSK(N,A,R, INDEF)
      DIMENSION  A(N,N) , R(N,N)
      INDEF = 0
      DO  30  IP= 1,N
       IF  (A(IP,IP))   05,05,10
C     MATRIX A NOT POSITIVE DEFINITE, ABORT COMPUTATION...
   05 PRINT 06
   06 FORMAT (53H ARRAY NOT POSITIVE DEFINITE.COMP≠S ABORTED.          )
      INDEF = 1
      GO TO 40
C     MATRIX A LOOKS O.K., COMPUTE R(IP,K).
   10  R(IP,IP) = SQRT(A(IP,IP))
       IF  (IP-N) 12, 40, 40
   12  IPPLS1= IP + 1
       DO 20  K=IPPLS1, N
        R(IP,K) = A(IP,K)/R(IP,IP)
   20  CONTINUE
C      REDUCTION OF ELEMENTS A(I,K).
       DO 30  I=IPPLS1, N
        DO 30  K= I, N
         A(I,K) = A(I,K) - R(IP,I)*R(IP,K)
   30 CONTINUE
   40 RETURN
      END
```

```
C      FORTRAN SUBROUTINE NO, 3. CHOLESKYSOL.
C     THIS SUBROUTINE SOLVES VECTOR EQUATIONS AX+B=0 FOR VECTOR X,
C      WHERE VECTOR B IS GIVEN AND WHERE R, THE CHOLESKY DECOMPO-
C      SITION OF THE POSITIVE DEFINITE MATRIX A, IS KNOWN....
      SUBROUTINE  CHLSKS(N,R,B,X)
      DIMENSION R(N,N), B(N), X(N), Y(200)
C     NOTE.IF N GREATER THAN 200, ALTER DIMENSION STATEMENT,RECOMPILE.
C     FORWARD SUBSTITUTION.
      DO 30  K= 1, N
       S= B(K)
       IF (K-1) 25,25, 10
   10  KLSS1 = K-1
       DO 20  I = 1, KLSS1
        S = S + R(I,K)*Y(I)
   20  CONTINUE
   25  Y(K) = -S/R(K,K)
   30 CONTINUE
C         BACKWARD SUBSTITUTION.
      DO 50  IDUM = 1, N
       I = N + 1 - IDUM
       S = Y(I)
       IF  (N-I)  45, 45, 35
   35  IPLS1 = I + 1
       DO 40  K = IPLS1, N
        S = S - R(I,K)*X(K)
   40  CONTINUE
   45  X(I) = S/R(I,I)
   50 CONTINUE
      RETURN
      END
```

```
C              FORTRAN SUBROUTINE NO. 4. CHOLESKYBAND
C        THIS SUBROUTINE FINDS MATRIX R WHICH WHEN MULTIPLIED BY
C        TRANSPOSE YIELDS GIVEN BAND MATRIX A.  NOTE SHORTHAND
C        STORAGE OF MATRICES.....
      SUBROUTINE  CHLSKB(A,NDIM, N, MPLS1, R, INDEF)
      DIMENSION  A(NDIM, MPLS1), R(NDIM, MPLS1)
      M = MPLS1 - 1
      INDEF = 0
C            PRELIMINARY TRANSFER.
      DO 20  I = 1, N
       DO 20  J = 1, MPLS1
        R(I,J) = A(I,J)
   20 CONTINUE
C          ACTUAL CHOLESKY DECOMPOSITION WITH ELEMENTS R(I,J).
      DO  70  IP = 1,N
       IF  (R(IP,1))  25, 25, 30
C        FAILURE, ARRAY R IS NOT POSITIVE DEFINITE.
   25 PRINT 26
   26 FORMAT ( 53H ARRAY IS NOT POSITIVE DEFINITE.RETURN TO MAIN.        )
      INDEF = 1
      GO TO 80
C        ARRAY R LOOKS O.K., PROCEED WITH SOLUTION...
   30  R(IP,1) = SQRT(R(IP,1))
       DO  40  J = 2, MPLS1
        R(IP,J) = R(IP,J)/R(IP,1)
   40  CONTINUE
       MIN   = MIN0(N, IP + M)
        IF  (MIN-IP-1)  70, 50, 50
   50  IPPLS1 = IP + 1
       DO  60  I = IPPLS1, MIN
        MIPI1 = M + IP - I + 1
        DO  60  J = 1, MIPI1
         R(I,J) = R(I,J)-R(IP,I-IP+1)*R(IP,I-IP+J)
   60  CONTINUE
   70 CONTINUE
   80 RETURN
      END
```

```
C        FORTRAN SUBROUTINE NO. 5.  CG.
C       THIS SUBROUTINE USES METHOD OF CONJUGATE GRADIENTS, A RELAXATION
C       METHOD FOR FINDING SOLUTION VECTOR X OF EQUATION  AX+B=0.
C         ALWAYS USE SUBROUTINE OP(N,P,Z)  IN CONJUNCTION WITH THIS ONE.
        SUBROUTINE  CG(N,N1, B, OP, X, Q, E, NCG)
C       NOTE.IF N GREATER THAN 200,ALTER DIMENSION STATEMENT,RECOMPILE.
        DIMENSION  B(N), X(N), Q(N1), E(N1), Z(200),P(200),  R(200)
        DO  10  I = 1,N
         X(I) = 0.
         R(I) = B(I)
         P(I) = -R(I)
     10 CONTINUE
        DO 70  K = 1, N1
C       CONJUGATE GRADIENT STEP...
         RR = 0.
         NCG = K - 1
         DO 20  I = 1, N
          RR = RR + R(I)**2
     20  CONTINUE
C       IF RR=0., BREAK OFF COMP≠S.OTHERWISE,CONTINUE RELAXATION.
         IF (RR)  30, 100, 30
     30  IF  (K-1)  40, 40, 31
     31  E(K-1) = RR/RR1
         DO  35  I = 1, N
          P(I) = E(K-1)*P(I) - R(I)
     35  CONTINUE
C       CALL SUBROUTINE OP TO FIND VECTOR Z=AP.
     40  CALL  OP(N,P,Z)
         H = 0.
         DO 50  I = 1, N
          H = H + P(I)*Z(I)
     50  CONTINUE
         Q(K) = RR/H
         DO  60  I = 1,N
          X(I) = X(I) + Q(K)*P(I)
          R(I) = R(I) + Q(K)*Z(I)
     60  CONTINUE
         RR1 = RR
     70 CONTINUE
C       N1 STEPS COMPLETED, BREAK OFF COMPUTATIONS...
        NCG = N1
    100 RETURN
        END
```

```
C          FORTRAN SUBROUTINE NO. 6.  ORTH.
C       THIS SUBROUTINE APPLIES SCHMIDT ORTHOGONALIZATION  TO A TALL
C       N X IP MATRIX A, STORING THE ORTHONORMAL VECTORS IN THE COLUMNS
C       OF MATRIX A...
        SUBROUTINE ORTH(N,IP,A,R)
        DIMENSION  A(N,IP),  R(IP,IP)
        DO 60  K = 1, IP
```

```
C          COMPUTATION OF R(J,K) VALUES,ORTHOGONALIZATION OF K-TH COLUMN.
         IF  (K-1)  30, 30, 5
   05    KLSS1 = K-1
         DO  20  J = 1, KLSS1
           R(J,K) = 0.
           DO  10  I = 1, N
             R(J,K) = R(J,K) + A(I,J)*A(I,K)
   10      CONTINUE
           DO  20  I = 1,N
             A(I,K) = A(I,K) - R(J,K)*A(I,J)
   20    CONTINUE
C          NORMALIZATION OF K-TH COLUMN.
   30    S = 0.
         DO  40  I = 1, N
           S = S + A(I,K)**2
   40    CONTINUE
         R(K,K) = SQRT(S)
         DO  60  I = 1, N
           A(I,K) = A(I,K)/R(K,K)
   60 CONTINUE
      RETURN
      END
```

```
C          FORTRAN SUBROUTINE NO. 7. FEHLERORTH.
C       THIS SUBROUTINE USES ORTHOGONALIZATION TO SOLVE ERROR
C       EQUATIONS AND EVALUATE VECTOR X...
C          ALWAYS USE SUBROUTINE ORTH WHEN USING THIS SUBROUTINE.
       SUBROUTINE  FELORT(N,M,D,C,X,R,RR)
C      IF M GREATER THAN 200, ALTER DIMENSION STATEMENT,RECOMPILE.
       DIMENSION  D(N), C(N,M),  X(M), R(N), RR(M,M), F(200)
C          COMPUTATION OF QUANTITIES F(K), AND SUCCESSIVE
C           ORTHOGONALIZATION OF THE CONSTANT VECTOR.
       CALL  ORTH(N,M,C,RR)
       DO  10  I = 1, N
         R(I) = D(I)
    10 CONTINUE
       DO  30  K = 1,M
         F(K) = 0.
         DO  20  I = 1,N
           F(K) = F(K) + C(I,K)*R(I)
    20   CONTINUE
         DO  30  I = 1,N
         R(I) = R(I) - F(K)*C(I,K)
    30 CONTINUE
C          FIND SOLUTION BY BACKWARD SUBSTITUTION.
       DO  50  IDUM= 1, M
         I = M + 1 - IDUM
         S = F(I)
         IF  (M-I)  45, 45, 35
    35   IPLS1 = I + 1
         DO  40  K = IPLS1,M
           S = S + X(K)*RR(I,K)
    40   CONTINUE
    45   X(I) = -S/RR(I,I)
    50 CONTINUE
       RETURN
       END
```

```
C         FORTRAN SUBROUTINE NO. 8. ROTATION.
C      THIS SUBROUTINE APPLIES JACOBI ROTATIONS TO A GIVEN MATRIX A,
C      WITH FINAL RESULTS STORED ALSO IN MATRIX A...
       SUBROUTINE  ROTATN(N,C,S,IP,IQ,A)
       DIMENSION A(N,N)
       H = C*C*A(IP,IP) - 2.*C*S*A(IP,IQ) + S*S*A(IQ,IQ)
       G = S*S*A(IP,IP) + 2.*C*S*A(IP,IQ) + C*C*A(IQ,IQ)
       A(IP,IQ) = C*S*(A(IP,IP) - A(IQ,IQ)) + (C*C - S*S)*A(IP,IQ)
       A(IP,IP) = H
       A(IQ,IQ) = G
       IF  (IP-1)  15,15, 5
    05 IPLSS1 = IP-1
       DO  10  J = 1, IPLSS1
         H = C*A(J,IP) - S*A(J,IQ)
         A(J,IQ) = S*A(J,IP) + C*A(J,IQ)
         A(J,IP) = H
    10 CONTINUE
    15 IF  (IQ-IP-1) 40, 25, 16
    16 IPPLS1 = IP+1
       IQLSS1 = IQ - 1
       DO  20  J = IPPLS1, IQLSS1
         H = C*A(IP,J) - S*A(J,IQ)
         A(J,IQ) = S*A(IP,J)  + C*A(J,IQ)
         A(IP,J) = H
    20 CONTINUE
    25 IF  (N-IQ)  40,40,26
    26 IQPLS1 = IQ + 1
       DO  30  J = IQPLS1, N
         H = C*A(IP,J) - S*A(IQ,J)
         A(IQ,J) = S*A(IP,J) + C*A(IQ,J)
         A(IP,J) = H
    30 CONTINUE
    40 RETURN
       END


C         FORTRAN SUBROUTINE NO. 9. JACOBI 1.
C      THIS SUBROUTINE COMPUTES EIGENVALUES OF MATRIX A BY
C      CLASSICAL JACOBI METHODS...
C         ALWAYS USE SUBROUTINE ROTATN WITH THIS SUBROUTINE.
       SUBROUTINE  JACOB1(N,EPS,A)
       DIMENSION A(N,N)
       NLSS1 = N-1
C      COMMENCE SEARCH FOR LARGEST ELEMENT OF A...
    10 XMAX = 0.
       DO  30  I = 1, NLSS1
         IPLS1 = I+1
         DO  30  J = IPLS1, N
           IF  (ABS(A(I,J)) - XMAX)  30, 30, 20
    20     XMAX = ABS(A(I,J))
           IP = I
           IQ = J
    30 CONTINUE
C      TEST WHETHER SUFFICIENTLY CLOSE TO SOLUTION...
       IF  (N*XMAX - EPS)  100, 40, 40
C      NO, IT IS NOT. CONTINUE COMP≠S...
    40 THETA = 0.5*(A(IQ,IQ) - A(IP,IP))/A(IP,IQ)
       IF  (THETA)  60, 50, 60
    50 T = 1.
       GO TO 70
    60 T = 1./(THETA + SIGN(1.,THETA)*SQRT(1. + THETA**2))
```

```
   70 C = 1./SQRT(1. + T*T)
      S = C*T
      CALL  ROTATN(N,C,S,IP,IQ,A)
      GO TO 10
  100 RETURN
      END


C         FORTRAN SUBROUTINE NO. 10. JACOBI 2.
C     THIS SUBROUTINE COMPUTES EIGENVALUES OF MATRIX A BY
C     CYCLIC JACOBI METHODS....
C         ALWAYS USE SUBROUTINE ROTATN WITH THIS SUBROUTINE.
      SUBROUTINE  JACOB2(N,EPS,A)
      DIMENSION A(N,N)
      NLSS1 = N-1
C     ESTABLISH SS,A CRITERION OF SIZE OF OFF-DIAGONAL ELEMENTS...
   10 SS = 0.
      DO  20   I = 1, NLSS1
        IPLS1 = I+1
        DO 20  J = IPLS1, N
          SS = SS + A(I,J)**2
   20 CONTINUE
C     TEST WHETHER SS SUFFICIENTLY SMALL...
      IF  (2.*SS - EPS**2)  100,30, 30
C     NO, IT IS NOT. KEEP COMPUTING...
   30 DO  80  IP = 1, NLSS1
        IPPLS1 = IP + 1
        DO  80   IQ =IPPLS1, N
          IF   (A(IP,IQ))  40, 80, 40
C     APPLY  IP-IQ ROTATION....
   40     THETA = 0.5*(A(IQ,IQ) - A(IP,IP))/A(IP,IQ)
          IF   (THETA)  60, 50, 60
   50     T = 1.
          GO TO 70
   60     T = 1. /(THETA + SIGN(1.,THETA)*SQRT(1. + THETA**2))
   70     C = 1./SQRT(1. + T*T)
          S = C*T
          CALL ROTATN (N,C,S,IP,IQ,A)
   80 CONTINUE
      GO TO 10
  100 RETURN
      END


C         FORTRAN SUBROUTINE NO. 11. GIVENS.
C     THIS SUBROUTINE USES GIVENS≠ METHOD FOR SIMPLIFYING MATRIX A
C     INTO TRIDIAGONAL FORM, WITH FINAL RESULTS STORED IN MATRIX A...
C              ALWAYS USE SUBROUTINE ROTATN WITH THIS SUBROUTINE.
      SUBROUTINE  GIVENS(N,A)
      DIMENSION A(N,N)
      NLSS2 = N-2
      DO 20   J = 1, NLSS2
        JPLS2 = J + 2
        DO  20   K = JPLS2, N
          IF   (A(J,K))   10, 20, 10
   10     W = SQRT(A(J,J+1)**2 + A(J,K)**2)
          C = A(J,J+1)/W
          S = -A(J,K)/W
          CALL ROTATN(N,C,S,J+1,K,A)
   20 CONTINUE
      RETURN
      END
```

```
C           FORTRAN SUBROUTINE NO. 12. HOUSEHOLDER.
C      THIS SUBROUTINE USES HOUSEHOLDER≠S METHOD FOR SIMPLIFYING MATRIX A
C      INTO TRIDIAGONAL FORM, WITH FINAL RESULTS STORED IN MATRIX A...
       SUBROUTINE  HSHLDR(N,A)
C      NOTE.IF N GREATER THAN 200,ALTER DIMENSION STATEMENT, RECOMPILE.
       DIMENSION A(N,N), W(200), P(200), Q(200)
       NLSS2 = N-2
       DO  100  IT = 1, NLSS2
         SIGMA = 0.
         ITPLS1 = IT + 1
C      ESTABLISH SIGMA,A CRITERION FOR SIZE OF OFF-DIAGONAL ELEMENTS...
         DO  10  K = ITPLS1,N
           SIGMA = SIGMA + A(IT,K)**2
  10     CONTINUE
         IF  (SIGMA)  20, 100, 20
  20     IF  (A(IT,IT+1))  30, 30, 25
  25     S = SQRT(SIGMA)
         GO TO 40
  30     S = -SQRT(SIGMA)
C      FIND NORMALIZED VECTOR W...
  40     W(IT) = 0.
         W(IT+1) = SQRT(0.5*(1. + A(IT,IT+1)/S))
         H = S*W(IT+1)
         ITPLS2 = IT+2
         DO  50  K = ITPLS2,N
           W(K) = A(IT,K)/(2.*H)
  50     CONTINUE
C      FIND VALUES OF VECTOR P...
         P(IT) = H
         DO  71  J = ITPLS1,N
           P(J) = 0.
           DO  60  K = ITPLS1, J
             P(J) = P(J) + W(K)*A(K,J)
  60       CONTINUE
           JPLS1 = J+1
       IF (JPLS1 - N) 65, 65, 71
  65       DO  70  K = JPLS1, N
             P(J) = P(J) + W(K)*A(J,K)
  70     CONTINUE
  71   CONTINUE
         G = 0.
         DO  80  K = ITPLS1, N
           G = G + P(K)*W(K)
  80     CONTINUE
C      FIND VALUES OF VECTOR Q, THEN CORRECT THE MATRIX A...
         DO  90  I = IT, N
           Q(I) = 2.*(P(I) - G*W(I))
  90     CONTINUE
         DO  95  I = IT, N
           DO  95  J = I, N
             A(I,J) = A(I,J) - W(I)*Q(J) - W(J)*Q(I)
  95     CONTINUE
 100   CONTINUE
       RETURN
       END
```

```
C        FORTRAN SUBROUTINE NO. 13. BISECTION.
C     THIS SUBROUTINE FINDS THE K-TH EIGENVALUE OF THE TRIDIAGONAL
C     MATRIX HAVING DIAGONAL ELEMENTS A AND ADJACENT ELEMENTS B,
C     USING METHOD OF BISECTION...
      SUBROUTINE  BISECT(N,A,B,IT, K,EIG)
C     IF N GREATER THAN 200,ALTER DIMENSION STATEMENT,RECOMPILE.
      DIMENSION A(N), B(N), B2(200)
C     SET UP COARSE LIMITS FOR EIGENVALUE...
      EIGMIN = A(1) - ABS(B(1))
      EIGMAX = A(1) + ABS(B(1))
      B2(1) = B(1)**2
      DO 40  I = 2, N
        B2(I) = B(I)**2
C     INVOKE GERSCHGORIN≠S CIRCLE THEOREM...
        R = ABS(B(I-1)) + ABS(B(I))
        IF  (EIGMIN - A(I) + R)  20, 20, 10
   10   EIGMIN= A(I) - R
   20   IF  (A(I) + R - EIGMAX)  40, 40, 30
   30   EIGMAX = A(I) + R
   40 CONTINUE
      DO 80  L = 1, IT
C     BISECT EXISTING LIMITS ON THE EIGENVALUE...
        ALAMDA = (EIGMAX + EIGMIN)/2.
        P = 1.
        Q = ALAMDA - A(1)
        IV = ABS(SIGN(1., P)  - SIGN(1.,Q))
        DO 50  I= 2,N
          R = (ALAMDA - A(I))*Q - B2(I-1)*P
          IV = IV + ABS(SIGN(1.,Q) - SIGN(1.,R))
          P = Q
          Q = R
   50   CONTINUE
        IV = IV/2
        IF (IV - K)  70, 60, 60
   60   EIGMIN = ALAMDA
        GO TO 80
   70   EIGMAX = ALAMDA
   80 CONTINUE
      EIG =(EIGMIN + EIGMAX)/2.
      RETURN
      END
```

```
C          FORTRAN SUBROUTINE NO. 14. LRCHOLBAND
C          ALWAYS USE SUBROUTINE ≠CHLSKB≠ WITH THIS SUBROUTINE.
C       THIS SUBROUTINE USES LR-CHOLESKY METHOD ON A GIVEN BAND MATRIX A
C       AND FINDS SMALLEST EIGENVALUE(S)...
        SUBROUTINE LRCHLB(NDIM,MPLS1,A,R, EPS, NEIG)
        DIMENSION  A(NDIM,MPLS1),  R(NDIM, MPLS1)
        M = MPLS1 - 1
        Z = 0.
        KEIG = 0
        PHI = 0.5
C          BEGIN ITERATION.
     05 Y = PHI*A(N,1)
        DO 10  K=1,N
          A(K,1) = A(K,1) - Y
     10 CONTINUE
C          ATTEMPT AT CHOLESKY DECOMPOSITION.SUMMON NESTED SUBROUTINE...
     15 CALL CHLSKB(A,NDIM, N, MPLS1, R, MISSL)
        IF  (MISSL)  20, 20, 35
C       SUCCESSFUL CHOLESKY DECOMPOSITION...
     20 Z = Z+Y
        IF (Y) 25, 45, 25
     25 PHI = (1. + PHI)*0.5
        GO TO 45
C          FAILURE, ARRAY NOT POSITIVE DEFINITE.
     35 PHI = PHI*0.5
        DO 40  K= 1,N
          A(K,1) = A(K,1)  + Y
     40 CONTINUE
        Y = 0.
        GO TO 15
C          ALTERATION OF ARRAY A.
     45 DO 80  K=1,N
        MIN = MINO(N-K,M)
     60   MINPS1 = MIN +1
          DO  80  J = 1, MINPS1
            A(K,J) = 0.
            DO 70  I = J, MINPS1
              A(K,J) = A(K,J) + R(K,I)*R(K+J-1,I-J+1)
     70     CONTINUE
C       PERTURBATION OF ZERO CELLS....
          IF  (A(K,J))  80, 75, 80
     75   A(K,J) = EPS*0.001
     80 CONTINUE
        IF  (A(N,1) - EPS) 85, 85, 05
C       EIGENVALUE EXTRACTED...
     85 KEIG = KEIG + 1
        A(N,1) = A(N,1) + Z
        DO 95  J = 1, M
          IF  (N-J)  95, 95, 90
     90   A(N-J, J +1) = 0.
     95 CONTINUE
        N = N- 1
        PHI = 0.9
        IF  (NEIG-KEIG)  110, 110, 100
    100 IF  (N-1)  05, 85, 05
    110 DO 115  K = 1, N
          A(K,1) = A(K,1) + Z
    115 CONTINUE
        RETURN
        END
```

```
C         FORTRAN SUBROUTINE NO. 15. UEBERRELAX.
C      THIS SUBROUTINE CONDUCTS AN ITERATION CYCLE FOR OVERRELAXATION
C      APPLIED TO RECTANGULAR GRID...
       SUBROUTINE OVRLAX(N,MPLS2,H, OMEGA, F, U)
       DIMENSION  F(MPLS2,N), U(MPLS2,N)
       M = MPLS2 - 2
       NMLSS1 = N+M-1
       DO 70  IZ = 1, NMLSS1
         MAX = MAXO(IZ-N+1, 1)
         MIN = MINO(IZ,M)
         DO 70  I = MAX,MIN
           K = IZ - I + 1
C         THREE-WAY FORK TO CHOOSE CORRECT DELTA COMPUTATION.
           IF  (K-1)   40, 30, 40
   30      DELTA = (F(I+1,K) - 0.5*(U(I,K) + U(I+2,K))-U(I+1,K+1))/2.
           GO TO 65
   40      IF  (K-N)  60, 50, 60
   50      DELTA = (F(I+1,K) - 0.5*(U(I,K) + U(I+2,K))-U(I+1,K-1))/(2.+H)
           GO TO 65
   60      DELTA = (F(I+1,K) - U(I+1,K+1)-U(I+1,K-1)-U(I,K) -U(I+2,K))/4.
   65      U(I+1,K) = U(I+1,K) - OMEGA*(U(I+1,K) + DELTA)
   70  CONTINUE
       RETURN
       END


C         FORTRAN SUBROUTINE NO. 16. OP.
C      THIS SUBROUTINE EVALUATES Z=AP IN 9 ZONES OF FIG. 27,
C      WITH VERTICAL NUMBERING SCHEME...
       SUBROUTINE  OP(N,P,Z)
       DIMENSION  P(77), Z(77)
C         EQUATION OP-ONE.
       Z(1) = 2.*P(1) - 0.5*P(2) - P(8)
C         EQUATION OP-TWO.
       DO  10  I=2,6
         Z(I) = 2.*P(I) - 0.5*(P(I-1) +P(I+1)) - P(I+7)
   10  CONTINUE
C         EQUATION OP-THREE.
       Z(7) = 2.*P(7) - 0.5*P(6) - P(14)
       DO  30  I = 8,64,7
C         EQUATION OP-FOUR.
         Z(I) = 4.*P(I) - P(I-7) - P(I+7) - P(I+1)
C         EQUATION OP-FIVE.
         IPLS1 = I+1
         IPLS5 = I+5
         DO  20  K = IPLS1, IPLS5
           Z(K) = 4.*P(K) - P(K-1) - P(K-7) - P(K+1) - P(K+7)
   20    CONTINUE
C         EQUATION OP-SIX.
         Z(I+6) = 4.*P(I+6) - P(I+5) - P(I-1) - P(I+13)
   30  CONTINUE
C         EQUATION OP-SEVEN.
       Z(71) = 2.5*P(71) - 0.5*P(72) - P(64)
C         EQUATION OP-EIGHT.
       DO  40  I = 72,76
         Z(I) = 2.5*P(I) - 0.5*(P(I-1) + P(I+1)) - P(I-7)
   40  CONTINUE
C         EQUATION OP-NINE.
       Z (77) = 2.5*P(77) - 0.5*P(76) - P(70)
       RETURN
       END
```

BIBLIOGRAPHY

[1] AITKEN, A. "Studies in practical mathematics II. The evaluation of the latent roots and vectors of a matrix. "*Proc. Roy. Soc. Edinburgh*, A57 (1937) 269–304.

[2] BAUER, F. L.; HEINHOLD, J.; SAMELSON, K.; SAUER, R. *Moderne Rechenanlagen*. Stuttgart, Teubner, 1965.

[3] BAUMANN, R.; FELICIANO, M.; BAUER, F. L.; SAMELSON, K. *Introduction to ALGOL*. Englewood Cliffs, N.J., Prentice-Hall, Inc., 1964.

[4] BAUMANN, R. "ALGOL-Manual der ALCOR-Gruppe." München, Oldenbourg, 1967.

[5] BENOIT. "Note sur une méthode de résolution des équations normales etc. (Procédé du commandant Cholesky)." *Bull. géodésique* **3** (1924) 67–77.

[6] BIRKHOFF, G.; VARGA, R. S. "Implicit alternating direction methods." *Trans. Amer. Math. Soc.* **92** (1959) 13–24.

[7] BIRKHOFF, G.; VARGA, R. S.; YOUNG, D. "Alternating direction implicit methods." *Advances in computers*, **3** (1962) 189–273.

[8] DE BOOR, C. M.; RICE, J. R. "Chebyshev approximation by a $\prod (x - r_j)/(x + s_j)$ and application to ADI iteration." *J. Soc. Industr. Appl. Math.*, **11** (1963) 159–169.

[9] COLLATZ, L. "Über die Konvergenzkriterien bei Iterationsverfahren für lineare Gleichungssysteme." *Math. Z.*, **53** (1950) 149–161.

[10] COLLATZ, L. *The Numerical Treatment of Differential Equations*, 3rd ed. Berlin, Springer, 1960.

[11] COURANT, R., HILBERT, D. *Methods of Mathematical Physics*. New York, Interscience, 1953.

[12] Dijkstra, E. W. *A Primer to ALGOL 60 Programming*. London-New York, Academic Press, Inc., 1962.

[13] Engeli, M.; Ginsburg, Th.; Rutishauser, H.; Stiefel, E. "Refined iterative methods for the computation of the solution and the eigenvalues of selfadjoint boundary value problems." Basel-Stuttgart 1959. Mitt. Inst. f. angew. Math. ETH Zürich, Nr. 8.

[14] Engeli, M. "Automatisierte Behandlung elliptischer Randwertprobleme." Dissertation. Zürich, 1962.

[15] Faddeeva, V. N. *Computational Methods of Linear Algebra*. New York, Dover Publications, Inc., 1959.

[16] Faddeev, D. K., Faddeeva, V. N. *Computational Methods of Linear Algebra*. San Francisco, W. H. Freeman and Co., Publishers, 1963.

[17] Forsythe, G. E.; Wasow W. R. *Finite-Difference Methods for Partial Differential Equations*. New York, John Wiley & Sons, Inc., 1960.

[18] Forsythe, G. E.; Henrici, P. "The cyclic Jacobi method for computing the principal values of a complex matrix." *Trans. Amer. Math. Soc.*, **94** (1960) 1–23.

[19] Fox, L. *Numerical Methods in Linear Algebra*. Oxford, Clarendon Press, 1964.

[20] Francis, J. F. G. "The QR transformation. A unitary analogue to the LR transformation, Parts I and II." *Computer J.*, **4** (1961/62) 265–271; 332–345.

[21] Gerschgorin, S. "Abgrenzung der Eigenwerte einer Matrix." *Bul. Acad. Sci.* USSR, Leningrad, classe math. **7** (1931) 749–754.

[22] Ginsburg, Th. "The conjugate gradient method." *Numer. Math.*, **5** (1963) 191–200.

[23] Givens, W. "Numerical computation of the characteristic values of a real symmetric matrix." Oak Ridge Nat. Lab. Report ORNL-1574 (1954).

[24] Goodwin, E. T. *Modern Computing Methods*. London, Her Majesty's Stationery Office, 1961.

[25] Gröbner, W. *Matrizenrechnung*. München, Oldenbourg, 1956.

[26] Grossmann, W. *Grundzüge der Ausgleichsrechnung*. Berlin, Springer, 1961.

[27] Hansen, E. R. "On cyclic Jacobi methods." *J. Soc. Industr. Appl. Math.*, **11** (1963) 448–459.

[28] Henrici, P. "The quotient-difference algorithm." Nat. Bur. Standards Appl. Math. Ser. 49, 1958, 23–46.

[29] Henrici, P. "On the speed of convergence of cyclic and quasicyclic Jacobi methods for computing eigenvalues of Hermitian matrices." *J. Soc. Industr. Appl. Math.*, **6** (1958) 144–162.

[30] Henrici, P. "Some applications of the quotient-difference algorithm." *Proc. Symposia in Appl. Math.*, **15** (1963) 159–183.

[31] HESTENES, M.; STIEFEL, E. "Methods of conjugate gradients for solving linear systems." *J. Res. Nat. Bur. Standards*, **49** (1952) 409–436.

[32] HOUSEHOLDER, A. S. *Principles of Numerical Analysis*. New York, McGraw-Hill Book Company, 1953.

[33] HOUSEHOLDER, A. S. *The Theory of Matrices in Numerical Analysis*. New York-Toronto-London, Blaisdell, 1964.

[34] JACOBI, C. G. J. "Über ein leichtes Verfahren, die in der Theorie der Säkularstörungen vorkommenden Gleichungen numerisch aufzulösen." *Crelle's J.*, **30** (1846) 51–94.

[35] KAHAN, W. "Gauss-Seidel methods for solving large systems of linear equations." Dissertation. Toronto 1958.

[36] LÄUCHLI, P. "Iterative Lösung und Fehlerabschätzung in der Ausgleichsrechnung." *Z. f. angew. Math. u. Phys.*, **10** (1959) 245–280.

[37] MÜLLER, D. *Programmierung elektronischer Rechenanlagen*. 2. Aufl., Mannheim, Bibliographisches Institut, 1965.

[38] MURDOCH, D. C. *Linear Algebra for Undergraduates*. New York-London, John Wiley & Sons, Inc., 1957.

[39] NAUR, P., ED. "Revised report on the algorithmic language ALGOL 60." *Numer. Math.*, **4** (1963) 420–453; *Commun. Ass. Comp. Mach.*, **6** (1963) 1–17.

[40] OSTROWSKI, A. "Über das Nichtverschwinden einer Klasse von Determinanten und die Lokalisierung der charakteristischen Wurzeln von Matrizen." *Compositio Math.*, **9** (1951) 209–226.

[41] PEACEMAN, D. W.; RACHFORD, H. H. "The numerical solution of parabolic and elliptic differential equations." *J. Soc. Industr. Appl. Math.*, **3** (1955) 28–41.

[42] RUTISHAUSER, H. "Der Quotienten-Differenzen-Algorithmus." *Z. f. angew. Math. u. Phys.*, **5** (1954) 233–251.

[43] RUTISHAUSER, H. *Der Quotienten-Differenzen-Algorithmus*. Basel-Stuttgart 1957. Mitt. Inst. f. angew. Math. ETH Zürich, Nr. 7.

[44] RUTISHAUSER, H. "Solution of eigenvalue problems with the LR-transformation." Nat. Bur. Standards Appl. Math. Ser. 49, 1958, 47–81.

[45] RUTISHAUSER, H. "Über eine kubisch konvergente Variante der LR-Transformation." *Z. f. angew. Math. u. Mech.*, **40** (1960) 49–54.

[46] RUTISHAUSER, H. "Stabile Sonderfälle des Quotienten-Differenzen-Algorithmus." *Numer. Math.*, **5** (1963) 95–112.

[47] RUTISHAUSER, H. "On Jacobi rotation patterns." *Proc. Symposia in Appl. Math.*, **15** (1963) 219–239.

[48] RUTISHAUSER, H.; SCHWARZ, H. R. "The LR-transformation method for symmetric matrices." *Numer. Math.*, **5** (1963) 273–289.

[49] RUTISHAUSER, H. "The Jacobi method for real symmetric matrices." *Numer. Math.*, **9** (1966) 1–10.

[50] SCHLENDER, B. "Grundzüge der algorithmischen Formelsprache ALGOL." *Der math. u. naturwiss. Unterricht*, **13** (1961) 451–458.

[51] SCHMEIDLER, W. *Vorträge über Determinanten und Matrizen*. Berlin, Akadamie-Verlag, 1949.

[52] SCHÖNHAGE, A. "Zur Konvergenz des Jacobi-Verfahrens." *Numer. Math.*, **3** (1961) 374–380.

[53] SCHRÖDER, G. "Über die Konvergenz einiger Jacobi-Verfahren zur Bestimmung der Eigenwerte symmetrischer Matrizen." Schriften d. Rheinisch-Westfälischen Inst. f. instr. Math. Universität Bonn. Ser. A. Nr. **5** (1964).

[54] SCHWARZ, H. A. "Gesammelte mathematische Abhandlungen." Bd. 1. Berlin 1890, 241–265.

[55] SCHWARZ, H. R. "An Introduction to ALGOL." *Comm. Ass. Comp. Mach.*, **5**, 1962, S. 82–95.

[56] SCHWARZ, H. R. "Die Reduktion einer symmetrischen Bandmatrix auf tridiagonale Form." *Z. f. angew. Math. u. Mech.* (Sonderheft), **45** (1965) T76–T77.

[57] SHAW, F. S. *Relaxation Methods*. New York, Dover Publications, Inc., 1953.

[58] SHELDON, J. W. "On the numerical solution of elliptic difference equations." *Math. Tables and Other Aids to Comp.*, **9** (1955) 101–112.

[59] SOUTHWELL, R. V. *Relaxation Methods in Engineering Science*. London, Oxford University Press, 1940.

[60] SOUTHWELL, R. V. *Relaxation Methods in Theoretical Physics*. Oxford, Clarendon Press, 1956.

[61] STIEFEL, E. "Über einige Methoden der Relaxationsrechnung." *Z. f. angew. Math. u. Phys.*, **3** (1952) 1–33.

[62] STIEFEL, E. "Ausgleichung ohne Aufstellung der Gaußschen Normalgleichungen." *Wiss. Z. Technische Hochschule Dresden*, **2** (1952/53) 441–442.

[63] STIEFEL, E. "Relaxationsmethoden bester Strategie zur Lösung linearer Gleichungssysteme." *Comment. Math. Helv.*, **29** (1955) 157–179.

[64] STIEFEL, EDUARD L. *An Introduction to Numerical Mathematics*, transl. W. C. and Cornelie J. Rheinboldt. New York, Academic Press Inc., 1963.

[65] SYNGE, J. L. *The Hypercircle in Mathematical Physics*. Cambridge, Cambridge University Press, 1957.

[66] TODD, J. *A Survey of Numerical Analysis*. New York-Toronto-London, McGraw-Hill Book Company., 1962.

[67] UNGER, H. "Nichtlineare Behandlung von Eigenwertaufgaben." *Z. f. angew. Math. u. Mech.*, **30** (1950) 281–282.

[68] VARGA, R. S. *Matrix Iterative Analysis*. Englewood Cliffs, N.J., Prentice-Hall, Inc., 1962.

[69] WACHSPRESS, E. L. "CURE: A generalized two-space-dimension multigroup coding for the IBM-704." Report KAPL-1724 (Knolls Atomic Power Laboratory) Schenectady, New York, 1957.

[70] WACHSPRESS, E. L. "Optimum alternating-direction implicit iteration parameters for a model problem." *J. Soc. Industr. Appl. Math.*, **10** (1962) 339–350.

[71] WACHSPRESS, E. L. *Iterative Solution of Elliptic Systems and Applications to the Neutron Diffusion Equations of Reactor Physics.* Englewood Cliffs, N.J., Prentice-Hall, Inc., 1966.

[72] WIELANDT, H. "Bestimmung höherer Eigenwerte durch gebrochene Iteration." Bericht der aerodynamischen Versuchsanstalt Göttingen. 44/J/37 (1944).

[73] WILKINSON, J. H. "Note on the quadratic convergence of the cyclic Jacobi process." *Numer. Math.*, **4** (1962) 296–300.

[74] WILKINSON, J. H. *Rounding Errors in Algebraic Processes.* London, Her Majesty's Stationery Office, 1963.

[75] WILKINSON, J. H. *The Algebraic Eigenvalue Problem.* Oxford, Clarendon Press, 1965.

[76] WILKINSON, J. H. "Convergence of the LR, QR, and related algorithms." *Computer J.*, **8** (1965) 77–84.

[77] WILKINSON, J. H. "The QR algorithm for real symmetric matrices with multiple eigenvalues." *Computer J.*, **8** (1965) 85–87.

[78] WYNN, P. "Acceleration techniques for iterated vectors and matrix problems." *Math. Comput.* **16** (1962) 301–322.

[79] YOUNG, D. "Iterative methods for solving partial differential equations of elliptic type." *Trans. Amer. Math. Soc.*, **76** (1954) 92–111.

[80] ZURMÜHL, R. *Matrizen.* 4th edition. Berlin, Springer-Verlag, 1964.

[81] FORSYTHE, G., MOLER, C. B. *Computer Solution of Linear Algebraic Systems*, Englewood Cliffs, N.J., Prentice-Hall, Inc., 1967.

[82] RUTISHAUSER, H. *Description of ALGOL 60, Handbook for Automatic Computation*, Vol. I, Part a. Berlin-Heidelberg-New York, Springer-Verlag, 1967.

[83] WILKINSON, J. H., REINSCH, C. *Handbook for Automatic Computation*, Vol. II, Linear Algebra. Berlin-Heidelberg-New York, Springer-Verlag, 1971.

INDEX

Q

R

S

T